MATERIALS PROCESSING FUNDAMENTALS

MATERIALS PROCESSING FUNDAMENTALS

Proceedings of a symposium sponsored by the
TMS Process Technology and Modeling Committee and
Extraction and Processing Division

Held during the
TMS 2013 Annual Meeting & Exhibition

March 3-7, 2013

Edited by

Lifeng Zhang
Antoine Allanore
Cong Wang
James A. Yurko
Justin Crapps

A John Wiley & Sons, Inc., Publication

Published by John Wiley & Sons, Inc., Hoboken, New Jersey.
Published simultaneously in Canada.

Library of Congress Cataloging-in-Publication Data is available.

ISBN 978-1-11860-598-1

10 9 8 7 6 5 4 3 2 1

A John Wiley & Sons, Inc., Publication

TMS

TABLE OF CONTENTS
Materials Processing Fundamentals

Process Metallurgy of Steel

Physical Metallurgy of Metals

Metallurgy of Non-Ferrous Metals

Process Metallurgy of Non-Ferrous Metals

Recirculation of Materials and Environments

Poster Session

Preface

The key interest areas to be covered in the symposium of Materials Processing Fundamentals are all aspects of the fundamentals, synthesis, analysis, design, monitoring, and control of metals, materials, and metallurgical processes and phenomena.

Topics will include:

- The experimental, analytical, physical and computer modeling of physical chemistry and thermodynamics
- Transport phenomena in materials and metallurgical processes involving iron, steel, non-ferrous metals, and composites
- Second phase particles in metals and processes, such as non-metallic inclusions and bubbles in metals (steel, aluminum, silicon, magnesium, etc.) or gas bubbles in slag or electrolyte (foaming, gas evolution, or injection); the fundamentals (experimental studies or theoretical studies) on the nucleation, growth, motion and removal of these second phase particles from the molten metal or reactors
- Physical chemistry, thermodynamics and kinetics for the production and refining of rare earth metals
- Control of industrial processes in the field of extraction and processing of metals and materials: novel sensors for hostile-environment materials processes, such as online inclusion detection, temperature, and velocity in molten materials, surface condition of hot moving products, etc.; innovative online sampling and analysis techniques; models for real-time process control and quality monitoring systems

This year more than 60 abstracts and 35 papers were accepted for this symposium.

Lifeng Zhang

Antoine Allanore

Cong Wang

James A. Yurko

Justin Crapps

Editors

Lifeng Zhang currently is a professor and the dean of the School of Metallurgical and Ecological Engineering at University of Science and Technology Beijing. Lifeng received his Ph.D. degree from University of Science and Technology Beijing in 1998 and has 15 years teaching and research work at different universities – University of Science and Technology Beijing, Missouri University of Science and Technology, Norwegian University of Science and Technology, University of Illinois at Urbana-Champaign, Technical University of Clausthal and Tohoku University. Lifeng has compound backgrounds in primary production, refining, casting, and recycling of metals; recycling of electronic wastes and solar grade silicon; and process modeling for metallurgical processes. Lifeng has published over 250 papers and gave over 170 presentations at meetings and conferences. He is a Key Reader (Member of Board of Review) for four journals and a reviewer for over 30 journals. Lifeng is a member of TMS, AIST and ISIJ. He has received several best paper awards from TMS and AIST.

Antoine Allanore is Assistant Professor of Metallurgy at the Massachusetts Institute of Technology (Cambridge, Massachusetts), in the Department of Materials Science and Engineering where he currently holds the Thomas B. King Chair. He earned a chemical engineering degree from the Ecole Nationale Superieure des Industries Chimiques de Nancy and a M.Sc. and Ph.D. in chemical engineering from the Institut National Polytechnique de Lorraine. Prior joining MIT, he worked as a research engineer at ArcelorMittal R&D on the development of new electrolytic processes for primary steel production. Dr. Allanore was a TMS Extraction and Processing Division Young Leader Professional Development Award winner in 2011 and co-recipient of the Vittorio de Nora Prize awarded at TMS 2012.

Cong Wang is Senior Research Engineer of Saint-Gobain Innovative Materials R&D. Prior to joining Saint-Gobain, he worked at the Alcoa Technical Center. He obtained his Ph.D. from the Carnegie Mellon University; M.S. from the Institute of Metal Research, Chinese Academy of Sciences; and B.S. from Northeastern University with distinctions, respectively. His specialties are in materials processing, micro-structure characterization, mechanical testing, and electrochemistry.

James A. Yurko is a Principal Technologist with Materion Brush Beryllium and Composites (Elmore, Ohio), the global leader in beryllium and non-beryllium based metal matrix composites. Dr. Yurko's primary technical focus is in the area of bulk metallic glass processing. Prior to joining Materion, he co-founded Electrolytic Research Corporation (ERC) LLC with Prof. Don Sadoway of MIT to commercialize molten oxide electrolysis (MOE) technology. Before working with ERC, Jim was the R&D team leader and staff metallurgist of BuhlerPrince, Inc. where he was responsible for commercializing the Semi-Solid Rheocasting (SSR) process and various die casting development projects of aluminum, magnesium, and bulk-metallic glass alloys.

Dr. Yurko received a Ph.D. in metallurgy from the Massachusetts Institute of Technology and a B.S.E. in materials science and engineering from the University of Michigan. He is currently a member of TMS and ASM and serves on the University of Michigan Materials Science and Engineering External Advisory Board. In 2010, Dr. Yurko was selected as the TMS EPD Young Leader Professional Development Award winner and was a co-recipient of the Vittorio de Nora Prize awarded at TMS 2012.

Justin Crapps attended Mississippi State University, studying Mechanical Engineering. As an undergraduate, Justin was very involved in leadership and extracurricular activities with several organizations including the student chapter of the American Society of Mechanical Engineers (ASME), Engineering Student Council, and Bulldog Toastmasters. Justin also participated in the Cooperative education and professional internship programs, spending three semesters working at Georgia Pacific Paper in Monticello, MS and three summers working at Eaton Aerospace in Jackson, Mississippi. In 2005, Justin was awarded the prestigious ASME Charles T. Main silver medal for leadership and service to his student section over a period of more than one year. After finishing his undergraduate education, Justin enrolled in the Ph.D. program at Mississippi State University. During graduate school, Justin's research focused on using finite element modeling for process simulation and developing a combined roughness and plasticity induced closure modified strip-yield model to simulate fatigue crack growth. After graduate school, Justin accepted a postdoctoral appointment at Los Alamos National Laboratory where he is working on projects focused on manufacturing process development for bonding cladding to nuclear fuel, reprocessing fuel through a casting process, multiphysics models for nuclear fuel performance, and development of testing methodologies and specimens for nuclear fuel-cladding bond strength measurement.

MATERIALS PROCESSING FUNDAMENTALS

Process Metallurgy of Steel

Session Chair

Lifeng Zhang

Materials Processing Fundamentals
Edited by: Lifeng Zhang, Antoine Allanore, Cong Wang, James A. Yurko, and Justin Crapps
TMS (The Minerals, Metals & Materials Society), 2013

EVOLUTION OF INCLUSIONS IN TI-BEARING ULTRA-LOW CARBON STEELS DURING RH REFINING PROCESS

Wen Yang[1,2,3], Shusen Li[2,4], Yubin Li[5], Xinhua Wang[2]
Lifeng Zhang[1,2], Xuefeng Liu[3], Qinglin Shan[6]

[1]State Key Laboratory of Advanced Metallurgy, University of Science and Technology Beijing, Beijing 100083, China
Correspondence author: Lifeng Zhang, zhanglifeng@ustb.edu.cn
[2]School of Ecological and Metallurgical Engineering, University of Science and Technology Beijing, Beijing 100083, China

[3]School of Material Science and Engineering, University of Science and Technology Beijing, Beijing, 100083, China
[4]Shougang Qian'an Steel Co. Ltd.,
Qian'an City, Hebei Province 064400, China

[5]Steelmaking Department, SINOSTEEL MECC, Beijing 100080, China
Email: liyubin@mecc.sinosteel.com

[6]Steelmaking Department, Shougang Jingtang
United Iron & Steel Co. Ltd, Tangshan 063200, China

Keywords: Non-metallic inclusion, Ultra-low carbon steel, RH refining, Titanium treatment.

Abstract

Ultra-low carbon steel, in which titanium is used to fix the interstitial atoms such as C and N, is well used for automobile. During the RH refining process, the evolution of oxide inclusions was analyzed using ASPEX and acid extraction. It was found that after Ti addition a large amount of Al-Ti-O inclusions generated, and the Ti content in the oxides decreased as the refining time increased. It was estimated that most of the oxides could be removed before the time of 10 min after Ti addition. The average size of alumina increased while that of Al-Ti-O inclusions decreased during the refining process. After Ti addition, the Al_2O_3 clusters transformed to coral shape and there existed three types of Al-Ti-O inclusions. Accordingly, a formation mechanism of Al-Ti-O inclusion was proposed.

Introduction

Ultra-low carbon (ULC) steel, in which titanium is used to fix the interstitial atoms such as C and N, is well used for automobile and thus high surface quality of the steel sheet is required. In the RH refining process of ULC steel, after decarburization, aluminum was added to deoxidize. Several minutes later titanium addition was performed for alloying. During this process, the characteristics of inclusions would significantly change. Inclusions in ULC steel after deoxidation are mainly alumina-based [1, 2]. Large alumina-based inclusions have a detrimental effect on castability [3-5] and the surface quality of the steel sheet [6-8]. After ferro-titanium alloy (Ti-Fe) addition, Al-Ti-O inclusions form and some Al_2O_3. TiO_x clusters exist [2]. Currently, large inclusions in ULC steel are either removed by flow transport [9-15], or become defects of the steel product [8]. There have been many reports on the change of inclusions during

ULC steel refining process [2, 16-18]. However, the previous investigations were based on the two-dimensional detection and with limited inclusion amount, which resulted in incomplete and inaccurate results. In the current study, the evolution of inclusions in Ti-bearing ULC steel was analyzed during RH refining process using methods of ASPEX detection and the acid extraction.

Experimental Methodology

Experiments were carried out in the 300t ladle at Shougang Jingtang Iron & Steel works during RH refining process. After decarburization, the dissolved oxygen in steel measured by oxygen probes was approximately 700 ppm. 488 kg of aluminum was added to deoxidation, after which the dissolved oxygen decreased to 5.9 ppm. And then after 4 min 398 kg of Ti-Fe alloy (containing 70% titanium) was added into the melt. Steel samples were taken with pail samplers (inner diameter is 50 mm, height is 120 mm) at different times: before deoxidation, 3 min after Al addition, 3 min and 10 min after Ti-Fe addition, 10 min after vacuum break. The total oxygen (T.O.) of the samples were analyzed using infrared analysis, the dissolved aluminum and titanium were analyzed using ICP-AES, and the nitrogen content was analyzed using the thermal conductance method. The measured compositions of the steel at different times are listed in Table I.

Table I. Compositions of the steel at different times during RH refining process (%)

Sample	Description	Time (min)		T.O.	Al_s	Ti	N
		t_{Al}	t_{Ti}				
0	before deoxidation	--	--	0.094			0.0019
1	3 min after Al addition	3	--	0.0130	0.026	<0.005	0.0019
2	3 min after Ti addition	7	3	0.0070	0.021	0.073	0.0024
3	10 min after Ti addition	14	10	0.0041	0.030	0.072	0.0025
4	10 min after vacuum break	25	21	0.0033	0.028	0.073	0.0025

*t_{Al} and t_{Ti} respectively represent the time after Al addition and Ti addition.

Firstly, inclusions on the cross-section of each sample were detected using the automated SEM/EDS inclusion analysis (ASPEX). The working magnification was set at ×225 and the minimum detectable inclusion was ~1.0 μm in the current study. Two statistical parameters, area fraction and number density, are defined as the following two formulas to characterize the amount of non-metallic inclusions.

$$AF = \frac{A_{inclusion}}{A_{total}} \tag{1}$$

where, AF is the area fraction of inclusions, ppm; $A_{inclusion}$ is the total area of detected inclusions, μm^2; and A_{total} is the sample detection area, mm^2.

$$ND = \frac{n}{A_{total}} \tag{2}$$

where, ND is the number density of inclusions, per mm^2; and n is the number of detected inclusions on the area of A_{total}.

In the current study, in order to reveal the three-dimensional morphology of inclusions in the steel during RH refining process , inclusions were extracted from steel samples[19, 20]. The extracted inclusions were then observed using SEM after depositing carbon powder over the filter paper under vacuum conditions.

Results and Discussions

Compositional Evolution of Inclusions during RH Refining

The composition of inclusions in each sample was analyzed using ASPEX, and the results after deoxidation are shown in Figure 1, where the detected area of the samples and the number of detected inclusions were given. Since the inclusions before deoxidation were all MnO, the result is not listed here. After 3 min of Al addition, all of the inclusions transformed to Al_2O_3 due to the deoxidation reaction and the reduction reaction of MnO inclusions by [Al]. After 3 min of Ti-Fe addition, besides of Al_2O_3, many other type of inclusions such as Al_2O_3-TiO_x existed, some of them were larger than 15 μm. The detected TiN inclusions were precipitated after steel solidification. As the refining time after Ti addition increasing, the Ti content of some large inclusions increased, as shown in Fig. 1(c). At 10 min of holding time after the vacuum break, the type of inclusions changed little, however, the observed large inclusions were less and the proportion of precipitated TiN increased.

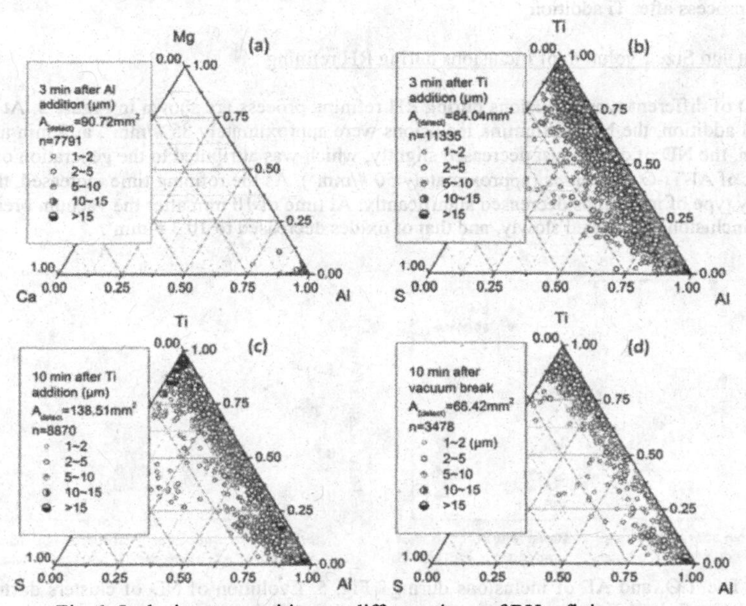

Fig. 1. Inclusion compositions at different times of RH refining process

Figure 2 shows the evolution of average atomic ratio of Ti and Al in oxides during the RH refining process after Ti addition. It is indicated that the atomic ratio of Ti in the oxides was highest in the initial stage after Ti addition, and it was about 27% at time of 3 min after Ti addition. The value decreased gradually in the following refining process. A peak of Ti percentage in oxide was observed at approximately 5 min after the Ti addition by Wang[21], that the peak did not exist in the current study might be resulted from the large interval of sampling time, 7 min, and thus the peak was missed.

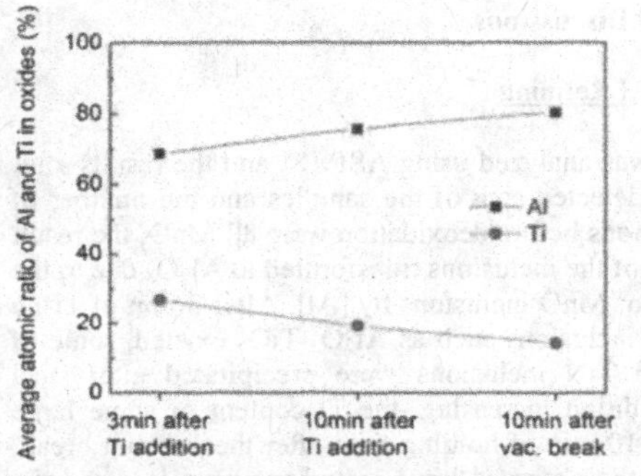

Fig. 2. Evolution of average atomic ratio of Al and Ti in oxides during the refining process after Ti addition

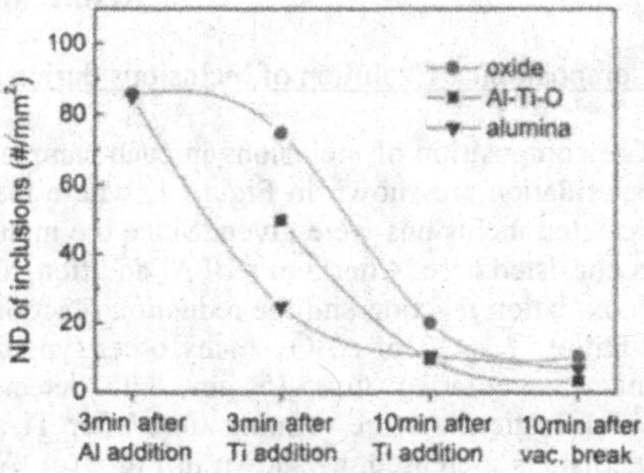

Fig. 3. ND of different type inclusions during RH refining

Amount and Size Evolution of Inclusions during RH refining

The ND of different type inclusions during RH refining process are shown in Figure 3. At 3 min after Al addition, the ND of alumina inclusions were approximately 85 $\#/mm^2$, at 3 min after Ti addition, the ND of oxides just decreased slightly, which was attributed to the generation of large amount of Al-Ti-O inclusions (approximately 50 $\#/mm^2$). As the refining time increased, the ND of every type of inclusions decreased significantly. At time of 10 min after the vacuum break, the ND of inclusions decreased slowly, and that of oxides decreased to 10.3 $\#/mm^2$.

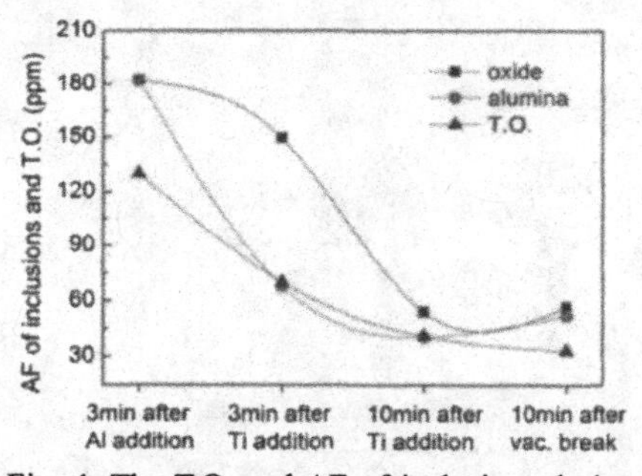

Fig. 4. The T.O. and AF of inclusions during RH refining

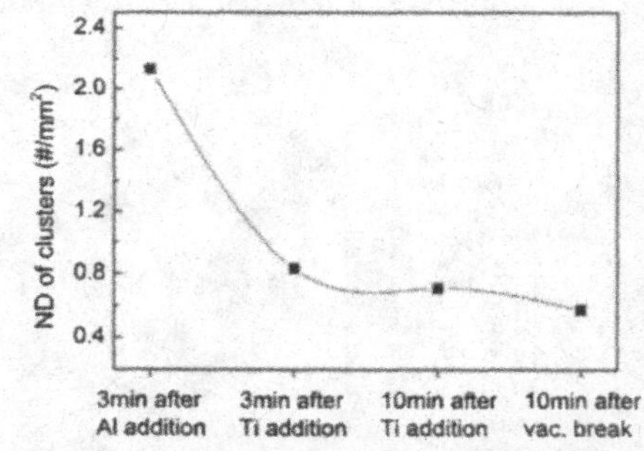

Fig. 5. Evolution of ND of clusters during RH refining process

It was mentioned that the AF of inclusions maybe more accurate to characterize the amount of inclusions, thus the T.O. and the AF of both the oxide and alumina are exhibited together in Figure 4 to make a comparison. It can be seen that they agree well with each other, especially before the time of 10 min after Ti addition, when the contents decreased to a relatively low value, 54 ppm of AF of oxides. The T.O. of the steel was sharply decreased from 940 ppm before deoxidation to 130 ppm at time of 3 min after Al addition, accordingly, it can be estimated that most of the alumina generated by deoxidation reaction were removed in the first 3 min after Al addition.

The removal of clusters is extremely important to ULC steel, thus, the evolution of ND of clusters is studied and shown in Figure 5. At 3 min after Al addition, more than 2.1 $\#/mm^2$ of alumina clusters existed in the melt, the amount decreased rapidly to 0.83 $\#/mm^2$ until time of 3 min after Ti addition, after that it decreased slowly and reached 0.57 $\#/mm^2$ at time of 10 min after the vacuum break. It is indicated that the clusters were mainly removed before Ti addition. The evolution of average size of oxides including alumina and Al-Ti-O is shown in Figure 6. The average size of alumina increased during the whole refining process owing to the aggregation of inclusions especially at the holding stage after vacuum break, however, that of oxides decreased slightly after Ti addition due to the generation of large amount of small Ti-Al-O inclusions. On the contrary, the size of Al-Ti-O inclusions was smaller than that of alumina and decreased with the refining progressing, which may be induced by the floatation of large particles and the resolution of Al-Ti-O inclusions.

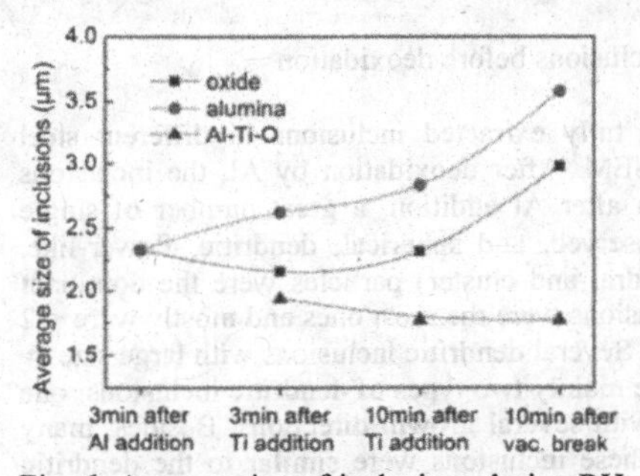

Fig. 6. Average size of inclusions during RH refining process

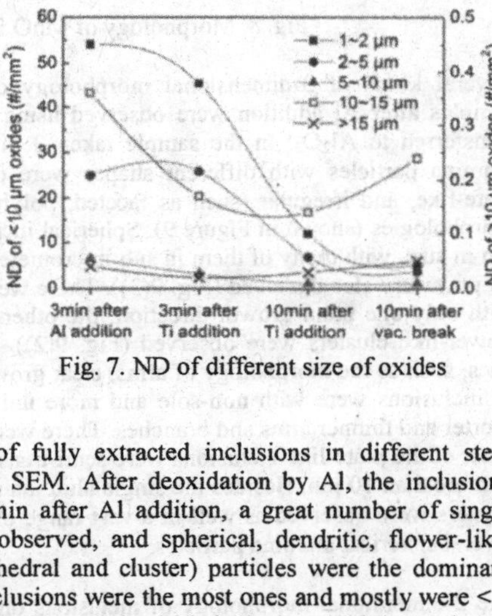

Fig. 7. ND of different size of oxides

Several kinds of 3-dimensional morphology of fully extracted inclusions in different steel samples after Al addition were observed using SEM. After deoxidation by Al, the inclusions transferred to Al2O3. In the sample taken 3 min after Al addition, a great number of single alumina particles with different shapes were observed, and spherical, dendritic, flower-like, plate-like, and irregular (such as faceted, polyhedral and cluster) particles were the dominant morphologies (shown in Figure 9). Spherical inclusions were the most ones and mostly were < 2 μm in size, with many of them in sub-micrometer. Several dendritic inclusions with large size (> 20 μm) were also observed (Fig. 9(3)). There were mainly two types of dendritic inclusions: one
Figure 7 shows the ND of different size of oxides during the refining process, indicating that the ND of most except for the 2~5 μm inclusions decreased during the refining process before vacuum break. As mentioned above, the generation of Ti-Al-O inclusions caused the slow decreasing or even the increasing of ND at 3 min after Ti addition. During the holding time after vacuum break, the ND of >10 μm inclusions increased due to the aggregation of small inclusions.

Morphological Evolution of Inclusions during RH Refining

Before deoxidation, the inclusions were mainly MnO, the morphology of which is shown in Figure 8. Since MnO is soluble in HCl acid, inclusions on cross section of the steel sample taken before deoxidation was directly detected by SEM-EDS. The T.O. of steel before deoxidation was approximately 940 ppm, so the combined oxygen in the melt was about 240 ppm, and it was mainly occupied by MnO inclusions.

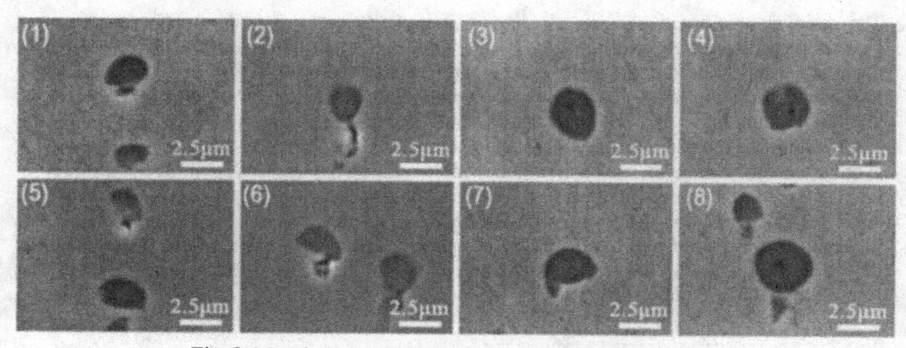

Fig. 8. Morphology of MnO inclusions before deoxidation

Several kinds of 3-dimensional morphology of fully extracted inclusions in different steel samples after Al addition were observed using SEM. After deoxidation by Al, the inclusions transferred to Al_2O_3. In the sample taken 3 min after Al addition, a great number of single alumina particles with different shapes were observed, and spherical, dendritic, flower-like, plate-like, and irregular (such as faceted, polyhedral and cluster) particles were the dominant morphologies (shown in Figure 9). Spherical inclusions were the most ones and mostly were < 2 μm in size, with many of them in sub-micrometer. Several dendritic inclusions with large size (> 20 μm) were also observed (Fig. 9(3)). There were mainly two types of dendritic inclusions: one with only one main growth direction; the other with several growth directions. Besides, many flower-like clusters were observed (Fig. 9(2)). These inclusions were similar to the dendritic ones, such as the morphology of arms, clear growth directions and nucleus particle. But this type of inclusions were with non-sole and more uniform growth directions, non-uniform arm size, shorter and thinner arms and branches. There were also plenty of plate-like inclusions (Fig. 9(3)). Most of the plate-like inclusions were semi-transparent and their size range might be from 1 μm to more than 10 μm. Besides the single alumina particles inclusion mentioned above, aggregated clusters were observed as well in a vast range of sizes, Fig. 9(4) shows a cluster aggregated by several spherical alumina particles.

The 2-dimentional morphology of inclusions on the cross section of sample of 3 min after Al addition were presented as well to compare 2-D and 3-D morphologies, as shown in Figure 10, where the observed clusters consist of several particles close together, and plate-like and dendritic inclusions are revealed. However, because the detected area was extremely limited, inclusions such as dendritic and plate-like ones were not observed, and comparing to the 3-D morphologies, the observed 2-D clusters are discontinued on the appearance and could hardly present the real characteristics of inclusions. The size of a 2-D inclusion is always smaller than its original size, and the statistic amount obtained using the 2-D observation is always larger than its real number. Thus, inclusion extraction is indeed a necessary method to reveal the real morphology of irregular inclusions.

Fig. 9. Three-dimensional morphologies of extracted inclusions observed in the sample taken 3 min after Al addition, white and grey inclusions located on the dark grey filter.

Similar morphologies of inclusions were reported before[22-27], and formation mechanisms of the different shape inclusions in steel after deoxidation were also proposed[23, 25, 26]. From their study, the morphology of single particles after deoxidation greatly depends on the supersaturation of deoxidizing elements such as the activities of dissolved oxygen and aluminum. After Al addition, with the decrease of supersaturation caused by the deoxidation reaction and the diffusion of the dissolved aluminum, the generation order of alumina particles is spherical first, then dendrite, flower-like, plate-like and polyhedron subsequently.

From Fig. 1, it can be known that after Ti-Fe addition there were mainly two types of oxides in the samples, including Al_2O_3 and Al-Ti-O.

Figure 11 shows the large extracted aggregated Al_2O_3 clusters in the sample taken after titanium addition. Most were aggregates of spherical or polyhedral particles before and after sintering. Before sintering, the single particle in the aggregated inclusion touched each other by a limited area, while after sintering the component particles were sintered together with larger connecting area and more curved boundary line. Aggregates in Fig. 11(7~9) indicating that the aggregating boundary line almost disappears, and the surface of inclusions became denser and more smooth, which let inclusions transform to coral shape. The studies by Zhang et al[28] reported that this morphology is induced by the process of Ostwald-Ripening by which small inclusions and the sharp edge of inclusions are dissolved and larger inclusions grow. Besides, some dendritical and flower-like Al2O3 inclusions were observed and are shown in Fig. 11(9). There were also many small individual spherical alumina particles which are not shown in Fig. 11.

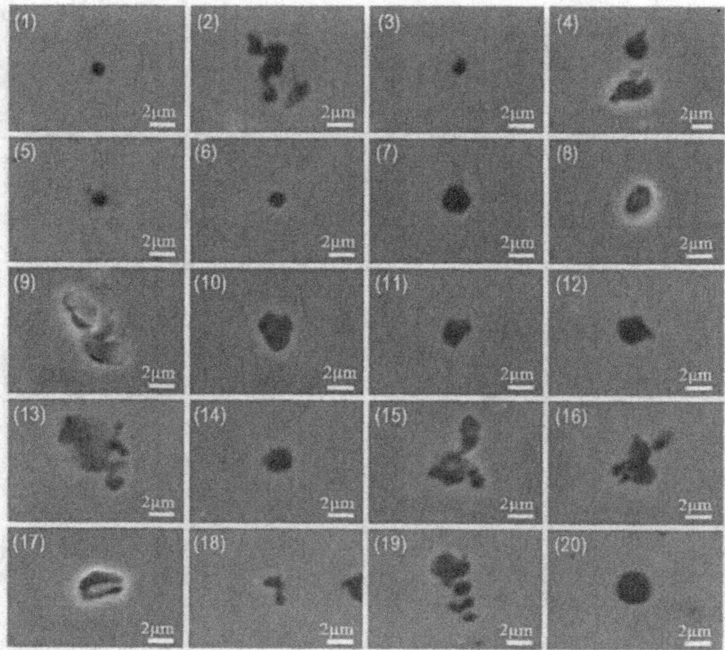

Fig. 10. Two-dimensional morphologies of alumina inclusions 3 min after Al addition

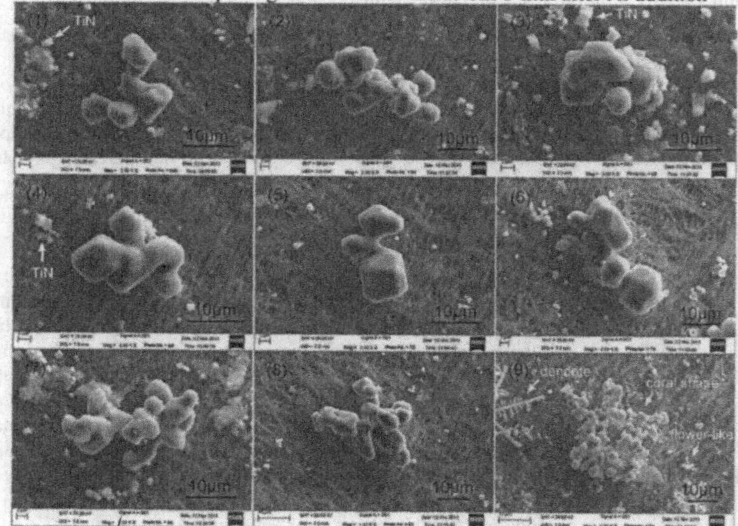

Fig. 11. Three-dimensional morphologies of extracted Al_2O_3 clusters 3 min after Ti-Fe addition

Figure 12 shows the large extracted Al-Ti-O inclusions in the samples taken after titanium addition. In order to reveal the inner structure of the particles, the two-dimensional cross-section morphology of the Al-Ti-O inclusions were analyzed as well and shown in Figure 13, indicating that there were mainly three types of Al-Ti-O inclusions. One type was small ones with homogeneous distribution of Al and Ti elements and the size was usually smaller than 5 µm, as shown in Fig. 12(1) and Fig. 13(1~2).

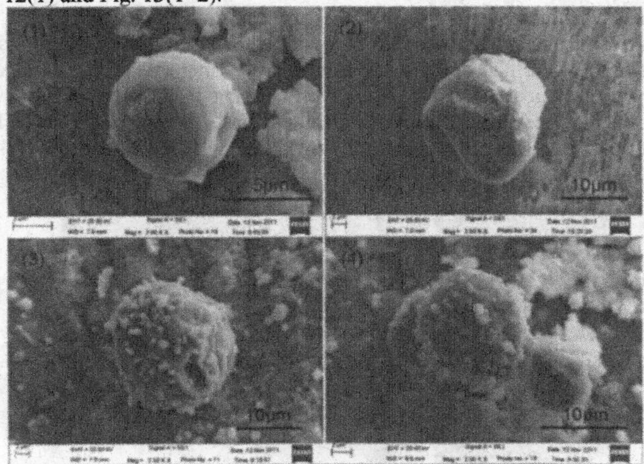

Fig. 12. Three-dimensional morphologies of extracted Al-Ti-O inclusions observed in the samples taken after Ti-Fe addition

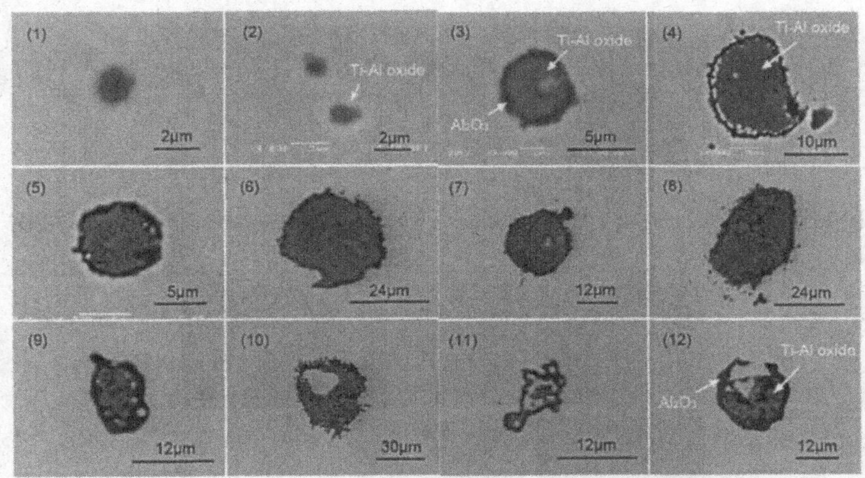

Fig. 13. Two-dimensional morphologies of Al-Ti-O inclusions observed in the samples taken after Ti-Fe addition, the black part of inclusion is alumina, the dark grey part is Ti-Al-O oxide

Figure 14 shows the element mapping of an extracted inclusion of this type, Al and Ti distributed uniformly and some TiN precipitated on the surface of the oxide. The second type was that a Al-Ti-O core with elements homogeneously distributed be surrounded by a Al_2O_3 layer, which is shown in Fig. 13(3~4), and the element mapping is shown in Figure 15. The third type was those with an uneven Ti-Al-O core and a Al_2O_3 surrounding layer, and the size was relatively large, as shown in Fig. 12(3~4) and Fig. 13(5~12). The element distribution of the surface and inner part of this type inclusions are shown in Figure 16 and Figure 17 respectively. The similar morphology of Al-Ti-O inclusions was also observed by Wang[2], however, in his observation the Al-Ti-O was in the outer part and the Al_2O_3 was in the inner part of the inclusions, which is contrary to the current study. Moreover, it is worth noting that there are many small Al_2O_3 protrusions on the surface of some Al-Ti-O inclusions, as shown in Fig. 12(3~4), the corresponding cross-section morphologies are shown in Fig. 13(6~8 and 10). The formation mechanism will be discussed later.

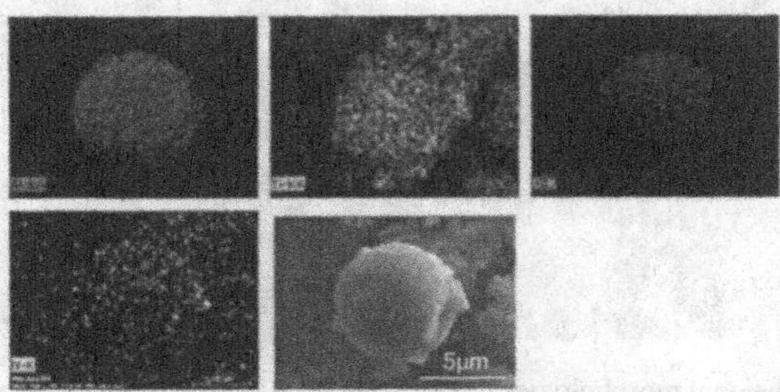

Fig. 14. Element mapping of an extracted Al-Ti-O inclusion, Al and Ti distributed homogeneously and some TiN precipitated on the oxide

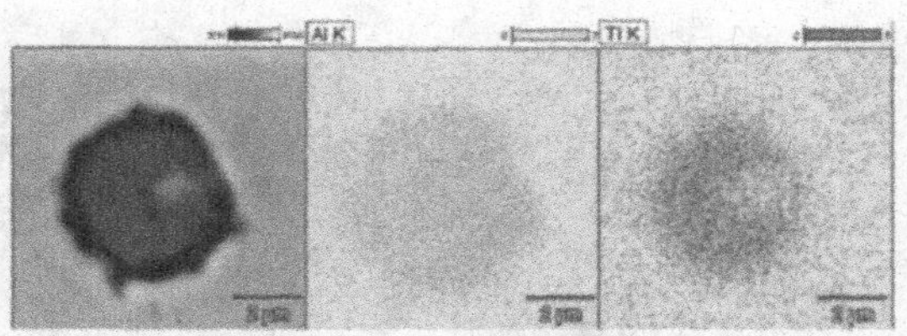

Fig. 15. Element mapping of a Al-Ti-O inclusion with a homogenous Ti-Al-O core and a Al_2O_3 surrounding layer

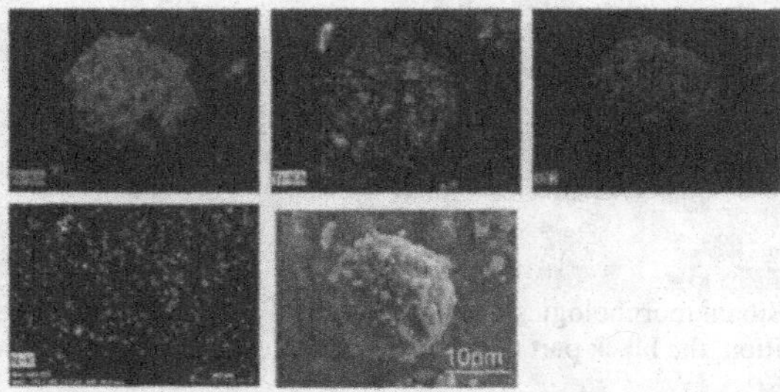

Fig. 16. Element mapping of an extracted Al-Ti oxide with uneven distributing elements

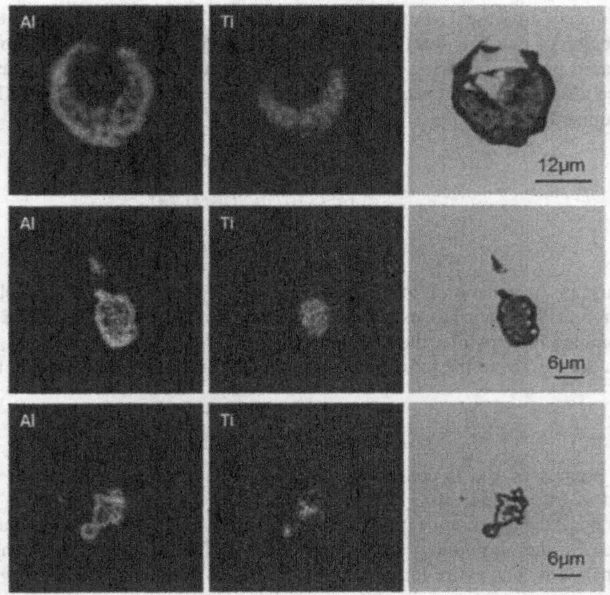

Fig. 17. Element mapping of Al-Ti-O inclusions with a uneven Ti-Al-O core and a Al_2O_3 surrounding layer

Formation Mechanism of Al-Ti-O Inclusion

Figure 18 shows the stable regions for various inclusions and how it changes depending on the soluble contents of Al and Ti[16]. The black dot in the figure represents the steel composition in the current work, indicating that on equilibrium condition Al_2O_3 is the only stable phase under the current steel composition after Ti addition.

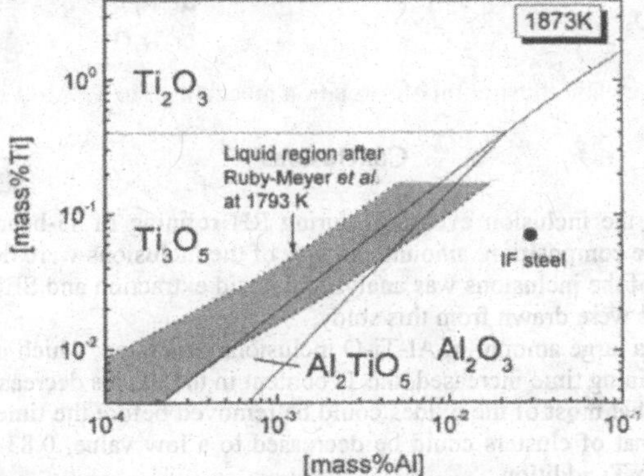

Fig. 18. Oxide phase diagram equilibrated with Fe-Al-Ti melt at 1873K[16]

However, immediately after the Ti-Fe alloy feeding into the melt, the local concentration would be high, and a region of rich [Ti] forms, which provides the possibility of generation of Al_2TiO_5 or even Ti_xO based on the following five reactions, some of them are liquid phase. Since the dissolved oxygen after deoxidation was low to 5.9 ppm in the current study, maybe Reactions (6) and (7) were dominant.

$$x[Ti]+[O]=(Ti_xO) \tag{3}$$

$$2[Al]+[Ti]+5[O]=(Al_2TiO_5) \tag{4}$$

$$(Al_2O_3)+[Ti]+2[O]=(Al_2TiO_5) \tag{5}$$

$$(Al_2O_3)+3x[Ti]=3(Ti_xO)+2[Al] \tag{6}$$

$$5(Al_2O_3)+3[Ti]=3(Al_2TiO_5)+4[Al] \tag{7}$$

After that, because of the generation of Ti-oxide and the diffusion of [Ti] in the melt, a region of low [Ti]/[Al] around the inclusion exists. Thus, the generation of Al_2O_3 around the Ti-Al-O inclusion is promoted, meanwhile, the transformation of Ti-Al-O inclusion to Al_2O_3 would be carried out from the surface to the inner part of the Ti-Al-O inclusion according to the following reactions:

$$4[Al]+3(Al_2TiO_5)=3[Ti]+5(Al_2O_3) \tag{8}$$

$$2[Al]+3(Ti_xO)=3x[Ti]+(Al_2O_3) \tag{9}$$

As the reaction progresses, the Ti content in the oxides decreases, which matches well with the result illustrated in Fig. 2, indicating that the reduction of Ti based on Reaction (8) or (9) is achieved. As a result, the appearance of a Ti-Al-O core surrounding with a Al_2O_3 layer forms. In some unstable cases, the morphology of an uneven Ti-Al-O core and surface containing small Al_2O_3 protrusions exists. The entire formation process is shown schematically in Figure 19.

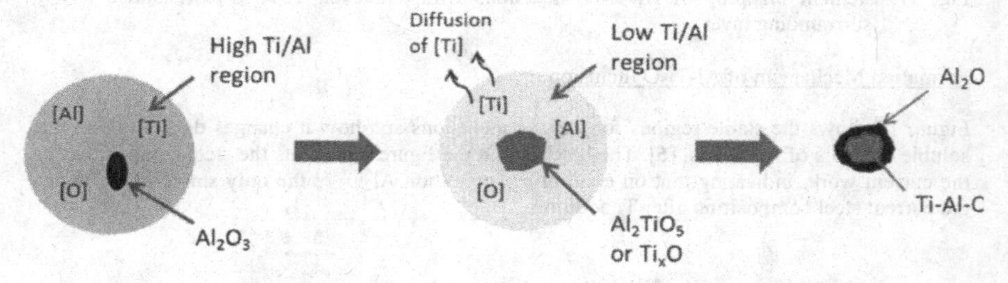

Fig. 19. Schematic illustration of formation mechanism of Ti-Al-O inclusions

Conclusions

In the present study, the inclusion evolution during RH refining of Ti-bearing ULC steel has been investigated. The composition, amount and size of the inclusions were detected by ASPEX, and the morphology of the inclusions was analyzed by acid extraction and SEM observation. The following conclusions were drawn from this study.

1) After Ti addition a large amount of Al-Ti-O inclusions generated, which induced high ND of oxides. As the refining time increased, the Ti content in the oxides decreased.
2) It was estimated that most of the oxides could be removed before the time of 10 min after Ti addition. While that of clusters could be decreased to a low value, 0.83 #/mm2, before the time of 3 min after Ti addition.

3) The average size of alumina increased during the whole refining process owing to the inclusion aggregation. On the contrary, the size of Al-Ti-O inclusions was smaller and decreased to 1.77 μm with the refining progressing.
4) After Al addition, some aggregates as well as a great number of single alumina particles with different shapes were observed. After Ti addition there were mainly two types of oxides: Al_2O_3 and Al-Ti-O. The Al_2O_3 clusters transformed to coral shape owing to the effects of sintering and Ostwald-Ripening. There were mainly three types of Al-Ti-O inclusions. One type was small ones with homogeneous distribution of Al and Ti elements, the second type was that a Al-Ti-O core with elements homogeneously distributed be surrounded by a Al_2O_3 layer, the third type was those with an uneven Ti-Al-O core and a Al_2O_3 surrounding layer, it was noticed that many small Al_2O_3 protrusions grew on some of the surfaces.
5) Al-Ti-O inclusions formed due to the local high [Ti] region after the feeding of Ti-Fe alloy even though the mean composition located in the Al_2O_3 stable region. Ti oxides were reduced again by the diffusion and consumption of [Ti], however, resulting in the change of inclusion composition and morphology.

Acknowledgement

The authors are grateful for the support from the National Science Foundation China (Grant No. 51274034), the Laboratory of Green Process Metallurgy and Modeling (GPM^2), the High Quality Steel Consortium at University of Science and Technology Beijing (China), and Shougang Jingtang United Iron & Steel Co. (China) for industrial trials.

References

1. Zhang, L. and B.G. Thomas, *State of the Art in Evaluation and Control of Steel Cleanliness.* ISIJ International, 2003. **43**(3): p. 271-291.
2. Wang, M., et al., *The Composition and Morphology Evolution of Oxide Inclusions in Ti-bearing Ultra Low-carbon Steel Melt Refined in the RH Process.* ISIJ International, 2010. **50**(11): p. 1606-1611.
3. Dekkers, R., et al., *A Morphological Comparison between Inclusions in Aluminum Killed Steels and Deposits in Submerged Entry Nozzle.* Steel Research, 2003. **74**(6): p. 351-355.
4. Wang, B., et al., *Analysis on Cause of Formation of Clogging Materials around the Tundish Nozzle in the Process of Continuously Casting Al Steel.* Steelmaking (in Chinese), 2008. **24**(6): p. 41-43.
5. Dekkers, R., 2002, Katholieke University Leuven: Leuven.
6. Gao, W., *Formation and Prevention of Sliver Defects on the Surface of Cold-rolled Strip.* Advanced Materials Research, 2012. **402**: p. 221-226.
7. P. Rocabois and J. Pontoire, *Different Slivers Type Observed in Sollac Steel Plants and Improved Practice to Reduce Surface Defects on Cold Roll Sheet.* ISSTech 2003 Conference Proceedings, 2003: p. 995.
8. Zhu, G., et al., *Study of Surface Defects of Cold-rolled IF Steel Sheet.* Iron and Steel (in Chinese), 2004. **39**(4): p. 54-56.
9. Zhu, W., D. Jin, and B. Li, *Removal of Inclusions from Steel by RH Treatment.* Iron and Steel (in Chinese), 1991. **26**(2): p. 22-25.
10. Xue, Z., et al., *Inclusion Removal from Molten Steel by Attachment Small Bubbles.* Acta Metallurgica Sinica (in Chinese), 2003. **39**(4): p. 431-434.

11. Wang, Y. and L. Zhang, *Fluid Flow Related Transport Phenomena in Steel Slab Continuous Casting Strands under Electromanetic Brake.* Metallurgical and Materials Transactions B, 2011. **42B**(6): p. 1319-1351.
12. Zhang, L., Y. Wang, and X. Zuo, *Flow Transport and Inclusion Motion in Steel Continuous-Casting Mold under Submerged Entry Nozzle Clogging Condition.* Metal. & Material Trans. B., 2008. **39B**(4): p. 534-550.
13. Zhang, L., et al., *Investigation on the Fluid Flow and Steel Cleanliness in the Continuous Casting Strand.* Metallurgical and Materials Transactions B, 2007. **38B**(1): p. 63-83.
14. Zhang, L., et al. *Physical, Numerical and Industrial Investigation of Fluid Flow and Steel Cleanliness in the Continuous Casting Mold at Panzhihua Steel.* in *AISTech 2004 - Iron and Steel Technology Conference Proceedings, September 15, 2004 - September 17, 2004.* 2004. Nashville, TN, United states: Association for Iron and Steel Technology, AISTECH.
15. Zhang, L., S. Taniguchi, and K. Cai, *Fluid Flow and Inclusion Removal in Continuous Casting Tundish.* Metallurgical and Materials Transactions B: Process Metallurgy and Materials Processing Science, 2000. **31**(2): p. 253-266.
16. Matsuura, H., et al., *The Transient Stages of Inclusion Evolution During Al and/or Ti Additions to Molten Iron.* ISIJ International, 2007. **47**(9): p. 1265-1274.
17. Van Ende, M.-A., et al., *Formation and Evolution of Al-Ti Oxide Inclusions during Secondary Steel Refining.* ISIJ International, 2009. **49**(8): p. 1133-1140.
18. Wang, C., et al., *A Study on the Transient Inclusion Evolution during Reoxidation of a Fe-Al-Ti-O Melt.* ISIJ International, 2011. **51**(3): p. 375-381.
19. Fernandes, M., N. Cheung, and A. Garcia, *Investigation of Nonmetallic Inclusions in Continuously Cast Carbon Steel by Dissolution of the Ferritic Matrix.* Materials Characterization, 2002. **48**(4): p. 255-261.
20. Li, S., L. Zhang, and X. Zuo, *Observation of Inclusions in the Steel Using Partial Acid Extraction and SEM*, in *Proccedings of Materials Science and Technology (MS&T)* 2008, AIST, Warrandale, PA: Pittsburgh. p. 1259-1269.
21. Wang, C., N.T. Nuhfer, and S. Sridhar, *Transient Behavior of Inclusion Chemistry, Shape, and Structure in Fe-Al-Ti-O Melts: Effect of Titanium Source and Laboratory Deoxidation Simulation.* Metallurgical and Materials Transactions B: Process Metallurgy and Materials Processing Science, 2009. **40**(6): p. 1005-1021.
22. Dekkers, R., B. Blanpain, and P. Wollants, *Steel Cleanliness at Sidmar.* ISSTech 2003 Conference Proceedings, 2003: p. 197-209.
23. Dekkers, R., B. Blanpain, and P. Wollants, *Crystal Growth in Liquid Steel during Secondary Metallurgy.* Metallurgical and Materials Transactions B, 2003. **34**(2): p. 161-171.
24. Dekkers, R., et al., *Non-metallic Inclusions in Aluminium Killed Steels.* Ironmaking and Steelmaking, 2003. **29**(6): p. 437-446.
25. Van Ende, M.A., et al., *Morphology and Growth of Alumina Inclusions in Fe-Al Alloys at Low Oxygen Partial Pressure.* Ironmaking and Steelmaking, 2009. **36**(3): p. 201-8.
26. Wasai, K., K. Mukai, and A. Miyanaga, *Observation of Inclusion in Aluminum Deoxidized Iron.* ISIJ International, 2002. **42**(5): p. 459-466.
27. Jin, Y., Z. Liu, and R. Takata, *Nucleation and Growth of Alumina Inclusion in Early Stages of Deoxidation: Numerical Modeling.* ISIJ International, 2010. **50**(3): p. 371-379.
28. Zhang, L. and W. Pluschkell, *Nucleation and Growth Kinetics of Inclusions during Liquid Steel Deoxidation.* Ironmaking and Steelmaking, 2003. **30**(2): p. 106-110.

Materials Processing Fundamentals
Edited by: Lifeng Zhang, Antoine Allanore, Cong Wang, James A. Yurko, and Justin Crapps
TMS (The Minerals, Metals & Materials Society), 2013

MASS ACTION CONCENTRATION MODEL OF CaO-MgO-FeO-Al_2O_3-SiO_2 SLAG SYSTEMS AND ITS APPLICATION TO THE FORMATION MECHANISM OF $MgO{\cdot}Al_2O_3$ SPINEL TYPE INCLUSION IN CASING STEEL

Haiyan Tang[1,2], Jingshe Li[1,2], Chuanbo Ji[2]

1State Key Laboratory of Advanced Metallurgy, University of Science and Technology Beijing, No.30 Xueyuan Road, Haidian District, Beijing, 100083, P.R.China
2School of metallurgical and ecological engineering, University of Science and Technology Beijing, No.30 Xueyuan Road, Haidian District, Beijing, 100083, P.R.China

Keywords: $MgO{\cdot}Al_2O_3$ spinel type inclusion, Slag basicity, Mass action concentration model, Formation mechanism

Abstract

Large size $MgO{\cdot}Al_2O_3$ spinel type inclusions in casing steel are harmful to the surface quality and performance of casing. In order to effectively control it in production, its existence in casing steel was analysed by scanning electron microscope and energy dispersive spectrometer, and its formation mechanisms were studied using coexistence theory of slag structure. The mass action concentration model of CaO-MgO-FeO-Al_2O_3-SiO_2 slag systems was established. The effect of slag basicity, aluminium content in molten steel, degree of vacuum during vacuum treatment of molten steel, and furnace lining on the magnesium content in molten steel was discussed. The results show that magnesium content increases with increasing slag basicity and aluminium content, and decreases with increasing CO partial pressure.

Introduction

Casing is an important material for oil fields. It can reinforce the wall of an oil well and protect the hole, and its destruction can lead to the failure of a complete well[1]. Therefore, not only are high strength, uniform and stable qualities, and strong corrosion and wear resistance required for casings, but also the capacity to support all kinds of loads such as pulling, pressing, twisting, and bending.[2]
Published test-results show that $MgO{\cdot}Al_2O_3$ spinel type inclusions in the tube are one of the important factors that affect the overall quality of the casing[3]. Therefore, it is necessary to study its formation mechanism and the effect factors in order to control it in production.
There have been several investigations[4-6] into the mechanism of the $MgO{\cdot}Al_2O_3$ spinel in stainless steel melts, but systematic theoretical analyses are few. Because it is difficult to measure and calculate the activity of every component of a multi-component slag, the activities of some components are often assumed to be 1.0 in most investigations. In fact, it is impossible for their activities to be 1.0 for these components under high temperatures. This will cause large errors in the analysis and application process.
The coexistence theory of slag structure[7] was used to calculate the activity of the slag component in the paper. The mass action concentration of the component was equivalent to the activity of the component in this theory. According to the actual production, the mass action concentration model of CaO-MgO-FeO-Al_2O_3-SiO_2 slag systems was chosen to calculate the activities of components relative to the

formation of $MgO{\cdot}Al_2O_3$ spinel, and to further study the effect of the slag component, vacuum treatment, and furnace lining on the magnesium content in molten steel.

Assumption of Coexistence Theory of Slag Structure

The ionic and molecular coexistence theory of slag structure was originally put forward by ЧУЙКО Н. М. in the former Soviet Union, and it has been developed and modified by Zhang J. in China in recent decades. Zhang J. and his group have dealt with many problems, such as the oxidation ability of slag, manganese distribution between slag and molten steel, the desulphurization and dephosphorization ability of slag, and calculation of the sulphur distribution ratio between slag and hot metal with his theory, and achieved uniform results which match reality.[8-12]

It is assumed in the ionic and molecular coexistence theory as follows.

(1) Molten slags are composed of simple ions such as Na^+, Ca^{2+}, Mg^{2+}, Mn^{2+}, Fe^{2+}, O^{2-}, S^{2-}, F^-, etc. and simple molecules such as Al_2O_3, SiO_2 and complicated molecules such as silicates, aluminates, ferrites, etc.

(2) There exist such dynamic equilibriums when simple ions and simple molecules forming complicated molecules such as:

$$2(Me^{2+}+O^{2-}) + SiO_2 = 2MeO{\cdot}SiO_2$$
$$3(Me^{2+}+O^{2-}) + SiO_2 = 3MeO{\cdot}SiO_2$$
$$\vdots$$
$$2(Me^{2+}+O^{2-}) + SiO_2 + Al_2O_3 = 2MeO{\cdot}SiO_2{\cdot}Al_2O_3$$

Where $(Me^{2+}+O^{2-})$ denote that the MeO molecule can be divided into two structural units Me^{2+} and O^{2-}, but they participate in the chemical reaction in the form of an ionic couple $(Me^{2+}+O^{2-})$.

(3) The activity of component i in slag melts is equivalent to the mass action concentration of structure unit i defined by Zhang J, herein denoted by N_i whose physical meaning is the equilibrium mole fraction of structure unit i.

(4) The chemical reactions as shown in the equations obey the mass conservation law.

The coexistence theory of slag structure provides another method to calculate the activity of multi-component slag systems.

Model for Calculating Mass Action Concentrations of Structural Units or Ionic Couples in CaO-MgO-FeO-Al2O3-SiO2 Slag Melts

CaO-MgO-FeO-Al_2O_3-SiO_2 slag systems are widely used in steelmaking production, so has a long track record as a subject for investigation.

According to the ionic and molecular coexistence theory, there are four simple ions as Ca^{2+}、Mg^{2+}、Fe^{2+} and O^{2-}, two simple molecules as SiO_2 and Al_2O_3 in CaO-MgO-FeO-Al_2O_3-SiO_2 slags at metallurgical temperature. According to the reported phase diagrams[13] of CaO-MgO-SiO_2, CaO-Al_2O_3-SiO_2, CaO-Al_2O_3, MgO-Al_2O_3, Al_2O_3-SiO_2, CaO-FeO-SiO_2, CaO-MgO-Al_2O_3 and MgO-Al_2O_3-SiO_2 slags at steelmaking temperature, 20 kinds of complex molecules listed in **Table 1**, such as $CaO{\cdot}SiO_2$, $2CaO{\cdot}SiO_2$, $3CaO{\cdot}SiO_2$, $MgO{\cdot}SiO_2$, $2MgO{\cdot}SiO_2$, $MgO{\cdot}Al_2O_3$, $CaO{\cdot}Al_2O_3$, $CaO{\cdot}2Al_2O_3$, $CaO{\cdot}6Al_2O_3$, $3CaO{\cdot}Al_2O_3$, $12CaO{\cdot}7Al_2O_3$, $3Al_2O_3{\cdot}2SiO_2$, $CaO{\cdot}MgO{\cdot}SiO_2$, $CaO{\cdot}MgO{\cdot}2SiO_2$, $2CaO{\cdot}MgO{\cdot}2SiO_2$, $3CaO{\cdot}MgO{\cdot}2SiO_2$, $CaO{\cdot}Al_2O_3{\cdot}2SiO_2$, $2CaO{\cdot}Al_2O_3{\cdot}SiO_2$, $FeO{\cdot}Al_2O_3$ and $2FeO{\cdot}SiO_2$, can be formed in CaO-MgO-FeO-Al_2O_3-SiO_2 slags in the steelmaking temperature range. Briefly, the chosen 20 kinds of complex molecules can be considered to be stable at 1873 K as equilibrium temperature between CaO-MgO-FeO-Al_2O_3-SiO_2 slags and molten steel. All simple ions, simple and complex molecules in CaO-

MgO-FeO-Al_2O_3-SiO_2 slags equilibrated with molten steel at metallurgical temperature are summarized and assigned exclusive numbers in **Table 1**.Every ion or molecule is regarded as one structural unit.

Table 1 Expression of structure units as ion couples, simple or complex molecules, their mole number and mass action concentrations in 100g of CaO-MgO-FeO-Al_2O_3-SiO_2 slags at metallurgical temperature based on the ion and molecule coexistence theory

Item	Structural units as ion couples or molecules	Mole number of structural units	Mass action concentration of structural units or ion couple
Simple cation and anion	$Ca^{2+}+O^{2-}$	$n_1 = n_{Ca^{2+},CaO} = n_{O^{2-},CaO} = n_{CaO}$	$N_1 = 2n_1/\sum n_i = N_{CaO}$
	$Mg^{2+}+O^{2-}$	$n_2 = n_{Mg^{2+},MgO} = n_{O^{2-},MgO} = n_{MgO}$	$N_2 = 2n_2/\sum n_i = N_{MgO}$
	$Fe^{2+}+O^{2-}$	$n_3 = n_{Fe^{2+},FeO} = n_{O^{2-},FeO} = n_{FeO}$	$N_3 = 2n_3/\sum n_i = N_{FeO}$
Simple molecules	Al_2O_3	$n_4 = n_{Al_2O_3}$	$N_4 = n_4/\sum n_i = N_{Al_2O_3}$
	SiO_2	$n_5 = n_{SiO_2}$	$N_5 = n_5/\sum n_i = N_{SiO_2}$
Complex molecules	$CaO\cdot SiO_2$	$n_6 = n_{CaO\cdot SiO_2}$	$N_6 = n_6/\sum n_i = N_{CaO\cdot SiO_2}$
	$2CaO\cdot SiO_2$	$n_7 = n_{2CaO\cdot SiO_2}$	$N_7 = n_7/\sum n_i = N_{2CaO\cdot SiO_2}$
	$3CaO\cdot SiO_2$	$n_8 = n_{3CaO\cdot SiO_2}$	$N_8 = n_8/\sum n_i = N_{3CaO\cdot SiO_2}$
	$MgO\cdot SiO_2$	$n_9 = n_{MgO\cdot SiO_2}$	$N_9 = n_9/\sum n_i = N_{MgO\cdot SiO_2}$
	$2MgO\cdot SiO_2$	$n_{10} = n_{2MgO\cdot SiO_2}$	$N_{10} = n_{10}/\sum n_i = N_{2MgO\cdot SiO_2}$
	$MgO\cdot Al_2O_3$	$n_{11} = n_{MgO\cdot Al_2O_3}$	$N_{11} = n_{11}/\sum n_i = N_{MgO\cdot Al_2O_3}$
	$CaO\cdot Al_2O_3$	$n_{12} = n_{CaO\cdot Al_2O_3}$	$N_{12} = n_{12}/\sum n_i = N_{CaO\cdot Al_2O_3}$
	$CaO\cdot 2Al_2O_3$	$n_{13} = n_{CaO\cdot 2Al_2O_3}$	$N_{13} = n_{13}/\sum n_i = N_{CaO\cdot 2Al_2O_3}$
	$CaO\cdot 6Al_2O_3$	$n_{14} = n_{CaO\cdot 6Al_2O_3}$	$N_{14} = n_{14}/\sum n_i = N_{CaO\cdot 6Al_2O_3}$
	$3CaO\cdot Al_2O_3$	$n_{15} = n_{3CaO\cdot Al_2O_3}$	$N_{15} = n_{15}/\sum n_i = N_{3CaO\cdot Al_2O_3}$
	$12CaO\cdot 7Al_2O_3$	$n_{16} = n_{12CaO\cdot 7Al_2O_3}$	$N_{16} = n_{16}/\sum n_i = N_{12CaO\cdot 7Al_2O_3}$
	$3Al_2O_3\cdot 2SiO_2$	$n_{17} = n_{3Al_2O_3\cdot 2SiO_2}$	$N_{17} = n_{17}/\sum n_i = N_{3Al_2O_3\cdot 2SiO_2}$
	$CaO\cdot MgO\cdot SiO_2$	$n_{18} = n_{CaO\cdot MgO\cdot SiO_2}$	$N_{18} = n_{18}/\sum n_i = N_{CaO\cdot MgO\cdot SiO_2}$
	$CaO\cdot MgO\cdot 2SiO_2$	$n_{19} = n_{CaO\cdot MgO\cdot 2SiO_2}$	$N_{19} = n_{19}/\sum n_i = N_{CaO\cdot MgO\cdot 2SiO_2}$
	$2CaO\cdot MgO\cdot 2SiO_2$	$n_{20} = n_{2CaO\cdot MgO\cdot 2SiO_2}$	$N_{20} = n_{20}/\sum n_i = N_{2CaO\cdot MgO\cdot 2SiO_2}$
	$3CaO\cdot MgO\cdot 2SiO_2$	$n_{21} = n_{3CaO\cdot MgO\cdot 2SiO_2}$	$N_{21} = n_{21}/\sum n_i = N_{3CaO\cdot MgO\cdot 2SiO_2}$
	$CaO\cdot Al_2O_3\cdot 2SiO_2$	$n_{22} = n_{CaO\cdot Al_2O_3\cdot 2SiO_2}$	$N_{22} = n_{22}/\sum n_i = N_{CaO\cdot Al_2O_3\cdot 2SiO_2}$
	$2CaO\cdot Al_2O_3\cdot SiO_2$	$n_{23} = n_{2CaO\cdot Al_2O_3\cdot SiO_2}$	$N_{23} = n_{23}/\sum n_i = N_{2CaO\cdot Al_2O_3\cdot SiO_2}$
	$FeO\cdot Al_2O_3$	$n_{24} = n_{FeO\cdot Al_2O_3}$	$N_{24} = n_{24}/\sum n_i = N_{FeO\cdot Al_2O_3}$
	$2FeO\cdot SiO_2$	$n_{25} = n_{2FeO\cdot SiO_2}$	$N_{25} = n_{25}/\sum n_i = N_{2FeO\cdot SiO_2}$

Let $a_1 = n^0_{Al_2O_3}$, $a_2 = n^0_{SiO_2}$, $b_1 = n^0_{CaO}$, $b_2 = n^0_{MgO}$ and $b_3 = n^0_{FeO}$ present the initial mole number of every composition of the slags.

The defined equilibrium mole numbers n_i of all above mentioned structural units in CaO-MgO-FeO-Al_2O_3-SiO_2 slags equilibrated with molten steel at a metallurgical temperature are in **Table 1**. According to the ionic and molecular coexistence theory, each ion couple is electroneutral and can be electrolyzed into a cation and anion based on the electrovalence balance principle. Hence, the equilibrium mole number of each ion couple is defined as the sum of the equilibrium mole number of the separated cation and anion. Choosing an ion couple (Ca^{2+}+O^{2-}) as an example, (Ca^{2+}+O^{2-}) can be separated into $Ca^{2+}$and O^{2-}, the equilibrium mole number of $Ca^{2+}$and O^{2-} can be expressed as $n_1 = n_{Ca^{2+},CaO} = n_{O^{2-},CaO} = n_{CaO}$. Therefore, ion couple (Ca^{2+}+O^{2-}) with a fixed amount under equilibrium conditions can produce two times the amount of structural units, *i.e.*, $n_{Ca^{2+},CaO} + n_{O^{2-},CaO} = 2n_{CaO} = 2n_1$.

In ionic and molecular coexistence theory, mass action concentration of a structural unit is defined as a ratio of equilibrium mole number of structural unit i to the total equilibrium mole number of all structural units in a close system with a fixed amount, the mass action concentration of structural units i in molten slags, N_i, can be calculated as follows.[8~12]

$$N_i = n_i / \sum n_i \tag{1}$$

Hence, the physical meaning of N_i is the equilibrium mole fraction of structural unit i.

It should be emphasized that mass action concentrations of all structural units in the form of ions, simple and complex molecules can be directly calculated from Eq.(1); however, mass action concentrations of ion couples, such as (Me^{2+}+O^{2-}), should be calculated as follows[9~12].

$$N_{MeO} = N_{Me^{2+},MeO} + N_{O^{2-},MeO} = \frac{2n_{MeO}}{\sum n_i} \tag{2}$$

Each simple and complex molecule can provide only one structural unit under equilibrium conditions, and their definitions[9~13] for mass action concentration in CaO-MgO-FeO-Al_2O_3-SiO_2 slags are listed in **Table 1**. Meanwhile, the mass action concentrations of ion couples, such as (Ca^{2+}+O^{2-}), (Mg^{2+}+O^{2-}) and (Fe^{2+}+O^{2-}) defined by Eq.(2) are also represented in **Table 1**.

The chemical reaction formulas of 20 kinds of potentially formed complex molecules, their standard molar Gibbs free energy changes $\Delta_r G^{\ominus}_{m,ci}$ as a function of absolute temperature T, and representation of mass action concentration of all complex molecules expressed by using $K_i^{\ominus}$, N_{CaO} (N_1), N_{MgO} (N_2), N_{FeO} (N_3), N_{Al2O3}(N_4) and N_{SiO2}(N_5) are summarized in **Table 2**.

The mass conservation equations for five components in 100 g of CaO-MgO-FeO-Al_2O_3-SiO_2 slags equilibrated with molten steel can be established from equilibrium mole numbers n_i and mass action concentrations N_i of all structural units listed in **Table 1 and Table 2,** as follows:

$$\begin{aligned} b_1 &= (0.5N_1 + N_6 + 2N_7 + 3N_8 + N_{12} \\ &+ N_{13} + N_{14} + 3N_{15} + 12N_{16} + N_{18} + N_{19} \\ &+ 2N_{20} + 3N_{21} + N_{22} + 2N_{23}) \sum n_i \end{aligned} \tag{3}$$

$$\begin{aligned} b_2 &= (0.5N_2 + N_9 + 2N_{10} + N_{11} + N_{18} \\ &+ N_{19} + N_{20} + N_{21}) \sum n_i \end{aligned} \tag{4}$$

$$b_3 = (0.5N_3 + N_{24} + 2N_{25}) \sum n_i \tag{5}$$

$$\begin{aligned} a_1 &= (N_4 + N_{11} + N_{12} + 2N_{13} + 6N_{14} \\ &+ N_{15} + 7N_{16} + 3N_{17} + N_{22} + N_{23} + N_{24}) \sum n_i \end{aligned} \tag{6}$$

$$a_2 = (N_5 + N_6 + N_7 + N_8 + N_9 + N_{10} + 2N_{17} + N_{18} + 2N_{19} + 2N_{20} + 2N_{21} + 2N_{22} + N_{23} + N_{25})\sum n_i \quad (7)$$

According to the principle that the sum of the mole fraction for all structural units in CaO-MgO-FeO-Al_2O_3-SiO_2 slags with a fixed amount under equilibrium condition is equal to 1.0, the following expression can be obtained

$$\sum N_i = 1 \quad (8)$$

The expressions of N_i in Table 3 and the equations (3)~(8) are the calculating model of mass action concentrations of the structural units in CaO-MgO-FeO-Al_2O_3-SiO_2 slag melts. Programming in basic language was used to solve the above equations, the mass action concentration of every structural unit under different temperatures and compositions of the slag was obtained.

In order to verify the accuracy of the model assumptions and calculations, mass action concentrations model of the structural units of CaO-SiO_2-Al_2O_3-MnO-FeO slags were also established in a similar way, in which the calculated mass action concentrations of MnO were compared with the experimental activity from literature[17], as in **Figure 1**, which shows they are in agreement. So, the model assumptions are considered reasonable. It provides another method to calculate the activity of multi-component slags.

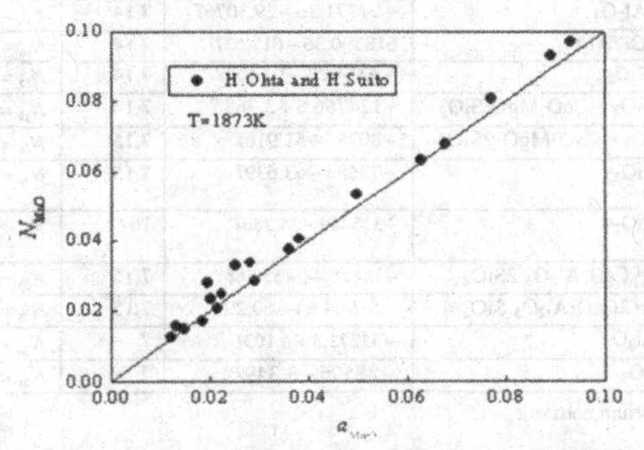

Figure 1 The comparison of calculated mass action concentrations of MnO and the measured activities

The above slag systems were chosen for verification because there have not had the experimental activity values of CaO-SiO_2-Al_2O_3-MgO-FeO slags determined to the authors' knowledge, up to now, and MnO and MgO are both basic oxides and have a similar structure.

Application of Mass Action Concentration Model to Thermodynamic Analysis on Formation Mechanism of MgO·Al_2O_3 Spinel Type Inclusion in Casing Steel

The forms of MgO·Al_2O_3 spinel type inclusion observed in casing steel

The composition of the casing steel used in the research (wt%) is C 0.39, Si 0.22, Mn 1.52, P 0.015, S 0.007, Mo 0.10, Al 0.020, V 0.12, Cu 0.25, Cr 0.04. It was produced by the 150t EAF-LF-VD-CC process route. Si-Fe and Mn-Fe alloys were added into the ladle at tapping and Al wire during LF refining. Ca-Si wire was fed into the molten steel to modify Al_2O_3 and MnS inclusions after VD treatment. The average composition of the final refined slag is 61.27%CaO-22.54%SiO_2-10.58%Al_2O_3-5.10%MgO-0.51%FeO in mass percent. The total oxygen of the billet is 9ppm.

Table 2 Chemical reaction formulas of possibly formed complex molecules, their standard mole Gibbs free energy, and mass action concentrations in CaO-MgO-FeO-Al_2O_3-SiO_2 slag systems at metallurgical temperature

Reactions	$\Delta_r G^{\ominus}_{m,i}/(J \bullet mol^{-1})$	Ref.	N_i
$(Ca^{2+}+O^{2-})+SiO_2= CaO \cdot SiO_2$	$-81416-10.498T$	7	$N_6 = K_1^{\ominus} N_1 N_5$
$2(Ca^{2+}+O^{2-})+SiO_2 = 2CaO \cdot SiO_2$	$-160431+4.106T$	7	$N_7 = K_2^{\ominus} N_1^2 N_5$
$3(Ca^{2+}+O^{2-})+SiO_2 = 3CaO \cdot SiO_2$	$-93366-23.03T$	14	$N_8 = K_3^{\ominus} N_1^3 N_5$
$(Mg^{2+}+O^{2-})+SiO_2 = MgO \cdot SiO_2$	$-36425+1.675T$	7	$N_9 = K_4^{\ominus} N_2 N_5$
$2(Mg^{2+}+O^{2-})+SiO_2 = 2MgO \cdot SiO_2$	$-77403+11.0T$	7	$N_{10} = K_5^{\ominus} N_2^2 N_5$
$(Mg^{2+}+O^{2-})+Al_2O_3 = MgO \cdot Al_2O_3$	$-35530-2.09T$	8	$N_{11} = K_6^{\ominus} N_2 N_4$
$(Ca^{2+}+O^{2-})+Al_2O_3 = CaO \cdot Al_2O_3$	$-18120-18.62T$	7,14	$N_{12} = K_7^{\ominus} N_1 N_4$
$(Ca^{2+}+O^{2-})+2Al_2O_3 = CaO \cdot 2Al_2O_3$	$-16400-26.80T$	7,14	$N_{13} = K_8^{\ominus} N_1 N_4^2$
$(Ca^{2+}+O^{2-})+6Al_2O_3 = CaO \cdot 6Al_2O_3$	$-17430-37.2T$	7,14	$N_{14} = K_9^{\ominus} N_1 N_4^6$
$3(Ca^{2+}+O^{2-})+Al_2O_3 = 3CaO \cdot Al_2O_3$	$-21771.36-29.3076T$	7,14	$N_{15} = K_{10}^{\ominus} N_1^3 N_4$
$12(Ca^{2+}+O^{2-})+7Al_2O_3=12CaO \cdot 7Al_2O_3$	$618390.36-612.53T$	7,14	$N_{16} = K_{11}^{\ominus} N_1^{12} N_4^7$
$3Al_2O_3 + 2SiO_2 = 3 Al_2O_3 \cdot 2SiO_2$	$-4354.27-10.467T$	7,14	$N_{17} = K_{12}^{\ominus} N_4^3 N_5^2$
$(Ca^{2+}+O^{2-})+(Mg^{2+}+O^{2-})+SiO_2 = CaO \cdot MgO \cdot SiO_2$	$-124766.6+3.768T$	7,14	$N_{18} = K_{13}^{\ominus} N_1 N_2 N_5$
$(Ca^{2+}+O^{2-})+(Mg^{2+}+O^{2-})+2SiO_2 = CaO \cdot MgO \cdot 2SiO_2$	$-80387-51.916T$	7,15	$N_{19} = K_{14}^{\ominus} N_1 N_2 N_5^2$
$2(Ca^{2+}+O^{2-})+(Mg^{2+}+O^{2-})+2SiO_2= 2CaO \cdot MgO \cdot 2SiO_2$	$-73688-63.639T$	7,15	$N_{20} = K_{15}^{\ominus} N_1^2 N_2 N_5^2$
$3(Ca^{2+}+O^{2-})+(Mg^{2+}+O^{2-})+2SiO_2= 3CaO \cdot MgO \cdot 2SiO_2$	$-315469+24.786T$	16	$N_{21} = K_{16}^{\ominus} N_1^3 N_2 N_5^2$
$(Ca^{2+}+O^{2-})+Al_2O_3+2SiO_2 = CaO \cdot Al_2O_3 \cdot 2SiO_2$	$-13816.44-55.266T$	7,15	$N_{22} = K_{17}^{\ominus} N_1 N_4 N_5^2$
$2(Ca^{2+}+O^{2-})+Al_2O_3+SiO_2 = 2CaO \cdot Al_2O_3 \cdot SiO_2$	$-61964.64-60.29T$	7,15	$N_{23} = K_{18}^{\ominus} N_1^2 N_4 N_5$
$(Fe^{2+}+O^{2-})+Al_2O_3 = FeO \cdot Al_2O_3$	$-33272.8+6.103T$	7	$N_{24} = K_{19}^{\ominus} N_3 N_4$
$2(Fe^{2+}+O^{2-})+SiO_2 = 2FeO \cdot SiO_2$	$-28596-3.349T$	7	$N_{25} = K_{20}^{\ominus} N_3^2 N_5$

$K_i^{\ominus}$ presents reaction equilibrium constant

There are three kinds of forms of $MgO \cdot Al_2O_3$ spinel in billets of casing steel. One is the dependent spherical $MgO \cdot Al_2O_3$, mainly containing Mg, Al, and O elements as in **Figure 2** with in small quantities. Another is the composite oxide inclusion with Mg, Al, Si, Ca, and O as shown in **Figure 3,** it is the main existence form for of spinel inclusion. The third is the $MgO \cdot Al_2O_3$ spinel as core wrapped with sulphide as shown in **Figure 4**. The energy spectrums given in **Figures 2-4** are all for point 1.

The process samples before and after LF and VD were also taken and analysed. It was found that there was very little $MgO \cdot Al_2O_3$ spinel type inclusions observed before LF, but they began to occur after LF, mainly in the third form. The quantity of $MgO \cdot Al_2O_3$ spinel type inclusion increased after VD treatment and occurred mostly in the composite oxide form.

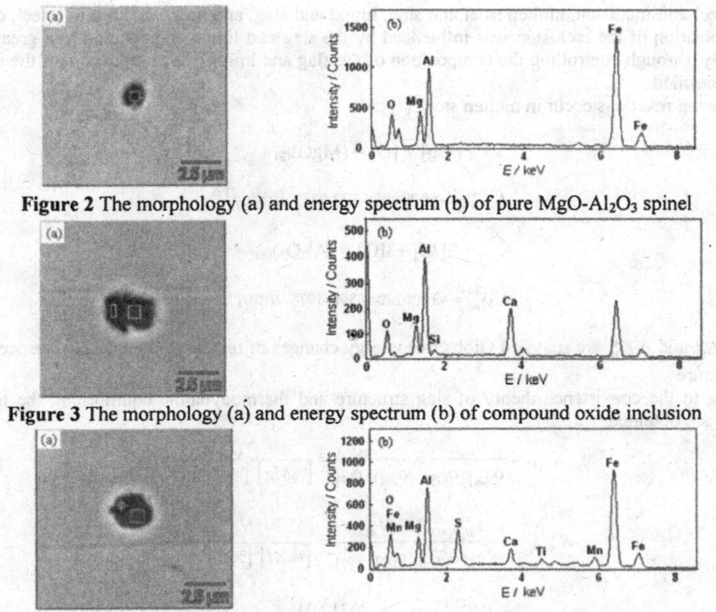

Figure 2 The morphology (a) and energy spectrum (b) of pure MgO-Al_2O_3 spinel

Figure 3 The morphology (a) and energy spectrum (b) of compound oxide inclusion

Figure 4 The morphology (a) and energy spectrum (b) of MgO·Al_2O_3 covered by sulphid

The Formation Mechanism of MgO·Al_2O_3 Spinel

There are four main sources for magnesium in molten steel: (1) MgO in slag is reduced by aluminium in molten steel; (2) MgO in slag is reduced by carbon in vacuum treatment; (3) MgO in furnace lining is reduced by carbon in vacuum treatment; (4) MgO in furnace lining is reduced by aluminium in molten steel. Then, magnesium in molten steel is oxidized into MgO, which combines with Al_2O_3 into MgO·Al_2O_3 spinel. The reactions are expressed as [4]:

$$3(\mathrm{MgO}) + 2[\mathrm{Al}] = (\mathrm{Al_2O_3}) + 3[\mathrm{Mg}]$$

$$(\mathrm{MgO}) + [\mathrm{C}] = [\mathrm{Mg}] + \mathrm{CO(g)}$$

$$[\mathrm{Mg}] + [\mathrm{O}] = (\mathrm{MgO})$$

$$(\mathrm{MgO}) + (\mathrm{Al_2O_3}) = (\mathrm{MgO \cdot Al_2O_3})_{(S)}$$

Theoretical Model of the Influence of Slag Compositions on Magnesium Content. Reference [18] showed that the composition of the inclusion was close to that of slag at the steel-slag equilibrium. In actual production, absolute equilibrium hardly existed, but a local quasi-equilibrium state between molten steel and inclusion, molten steel and slag, lining and slag, and lining and molten steel, did exist. The composition of the inclusion was influenced by the slag and lining composition to a great degree. Conversely, through controlling the composition of the slag and lining, the composition of the inclusion can be controlled.

The following reactions occur in molten steel:

$$[Mg] + [O] = (MgO)_{(S)} \tag{9}$$

$$\Delta_r G_9^{\ominus} = -5009908 + 122.97T,\ \text{J/mol}\ ^{[19]}$$

$$2[Al] + 3[O] = (Al_2O_3)_{(S)} \tag{10}$$

$$\Delta_r G_{10}^{\ominus} = -1206220 + 390.39T,\ \text{J/mol}\ ^{[19]}$$

where $\Delta_r G_9^{\ominus}$ and $\Delta_r G_{10}^{\ominus}$ are standard Gibbs free energy changes of reactions (9) and (10), respectively; T is temperature.

According to the coexistence theory of slag structure and thermodynamic equilibrium, the following equations are obtained:

$$K_{MgO} = \frac{a_{MgO}}{a_{[Mg]} \cdot a_{[O]}} = \frac{N_{MgO}}{a_{[Mg]} \cdot a_{[O]}} = \frac{N_{MgO}}{[\%Mg] \cdot [\%O] f_{Mg} f_o} \tag{11}$$

$$K_{Al_2O_3} = \frac{a_{Al_2O_3}}{a_{[Al]}^2 \cdot a_{[O]}^3} = \frac{N_{Al_2O_3}}{a_{[Al]}^2 \cdot a_{[O]}^3} = \frac{N_{Al_2O_3}}{[\%Al]^2 [\%O]^3 f_{Al}^2 f_O^3} \tag{12}$$

$$\lg\{[\%Mg] \times [\%O]\} = \lg N_{MgO} - \frac{26212.04}{T} - \lg f_{Mg} - \lg f_O + 6.43 \tag{13}$$

$$\lg\{[\%Al]^2 \times [\%O]^3\} = \lg N_{Al_2O_3} - \frac{63109}{T} - 2\lg f_{Al} - 3\lg f_O + 20.42 \tag{14}$$

$$\lg f_i = \sum_j e_i^j [\%j] \tag{15}$$

where K_{MgO} and $K_{Al_2O_3}$ are equilibrium constants of reactions (9) and (10), respectively; a_{MgO} and $a_{Al_2O_3}$ are activities of MgO and Al_2O_3 in slag, respectively; N_{MgO} and $N_{Al_2O_3}$ are mass action concentrations of MgO and Al_2O_3 in slag, respectively; a_{Mg}, a_O and a_{Al} are activities of Mg, O and Al in molten steel, respectively; [%Mg], [%O] and [%Al] are mass percent concentrations of Mg, O and Al in molten steel, respectively; f_i is activity coefficient of component i.

Substituting the chemical components of casing steel into Eq.(15), it is derived as: at 1873 K, $f_{Al} = 1.062$, $f_{Mg} = 0.767$, $f_O = 0.492$.

Theoretical models of the effect of slag compositions on magnesium content at 1873 K are derived as Eqs. (16) and (17) by substituting activity coefficients into Eqs. (13) and (14):

$$\lg\{[\%Mg] \times [\%O]\} = \lg N_{MgO} - 7.1419 \tag{16}$$

$$\lg\{[\%Al]^2\times[\%O]^3\}=\lg N_{Al_2O_3}-12.4034 \quad (17)$$

The Effect of Slag Compositions on [Al]–[O] Equilibrium. It can be seen from Eq.(14), if $N_{Al_2O_3}=1$, then

$$\lg\{[\%Al]^2\times[\%O]^3\}=-\frac{63109}{T}-2\lg f_{Al}-3\lg f_O+20.42;$$

if $N_{Al_2O_3}\neq 1$, then

$$\lg\{[\%Al]^2\times[\%O]^3\}=\lg N_{Al_2O_3}-\frac{63109}{T}-2\lg f_{Al}-3\lg f_O+20.42.$$

In the paper, thirteen groups of slags were designed with their binary basicity (B = (CaOmass%)/(SiO$_2$mass%) from 1.0 to 7.0., N_{MgO} and $N_{Al_2O_3}$ were calculated by the mass action concentration model mentioned in Section 3. The programmes and results are shown in **Table 3**. It is seen that basicity has a large effect on the activity of Al_2O_3. The latter decreases as the former increases, which agrees with reality. When basicity increased from 1.0 to 7.0, the mass action concentration of Al_2O_3 decreases from 0.0811 to 0.00063. As a result, it was not appropriate to assume that the activity of Al_2O_3 in multi-component slag was 1.0 as in some references.**Table 4** lists the calculated [Al]–[O] equilibrium values at 1873 K with and without considering the slag effect. From **Table 4**, the aluminium content in molten steel and slag basicity has a large effect on oxygen content. The higher the aluminium content, the lower the oxygen content. When not considering the effect of slag (*i.e.* $N_{Al_2O_3}=1$), if [%Al]=0.02, [%O]=9.96×10^{-4}; if [%Al]=0.05, [%O]=5.41×10^{-4}. Therefore, the control of the aluminium content is necessary to make high purity clean steel. It can also be seen that the oxygen contents have large differences when between considering and not considering slag effects at the same aluminium content. When not considering slag effect, [%Al]=0.02, [%O]=9.96×10^{-4}; whereas when considering slag effect, [%Al]=0.02, [%O] varies from 4.31×10^{-4} to 0.90×10^{-4} with slag basicity increasing from 1.0 to 6.0. As a result, slag composition has a large effect on the oxygen content in molten steel and high basicity slag is favourable for decreasing the oxygen content of molten steel. In actual production, the binary basicity of refining slag was 2.72 and the acid soluble aluminium content was 0.02%, so the calculated soluble oxygen was 0.000151%, which is in good agreement with 0.0002% of measured value.

Table 3. Designed slag composition (%) and calculated N_{MgO} and $N_{Al_2O_3}$

B	CaO	SiO_2	MgO	Al_2O_3	FeO	N_{MgO}	$N_{Al_2O_3}$
1.0	42.25	42.25	5.00	10.00	0.50	0.0103	0.0811
1.5	50.70	33.80	5.00	10.00	0.50	0.0063	0.0580
2.0	56.33	28.17	5.00	10.00	0.50	0.0250	0.0132
2.5	60.36	24.14	5.00	10.00	0.50	0.0765	0.0044
3.0	63.38	21.12	5.00	10.00	0.50	0.1200	0.0025
3.5	65.72	18.78	5.00	10.00	0.50	0.1463	0.0018
4.0	67.60	16.90	5.00	10.00	0.50	0.1501	0.0014
4.5	69.14	15.36	5.00	10.00	0.50	0.1440	0.0011
5.0	70.42	14.08	5.00	10.00	0.50	0.1382	0.00095
5.5	71.50	13.00	5.00	10.00	0.50	0.1332	0.00083
6.0	72.43	12.07	5.00	10.00	0.50	0.1290	0.00074
6.5	73.23	11.27	5.00	10.00	0.50	0.1254	0.00068
7.0	73.94	10.56	5.00	10.00	0.50	0.1222	0.00063

Table 4. [Al]–[O] equillibrium values at 1873K with and without considering slag action

[%Al]	[%O] / 10^{-4}								
	Without considering slag action	With considering slag action							
		B=1.0	B=1.5	B=2.0	B=2.5	B=3.0	B=4.0	B=5.0	B=6.0
0.001	73.37	31.26	28.40	17.34	12.02	9.96	8.21	7.20	6.65
0.005	25.09	10.86	9.71	5.93	4.11	3.41	2.81	2.46	2.27
0.010	15.81	6.84	6.12	3.74	2.59	2.15	1.77	1.55	1.43
0.014	12.63	5.47	4.89	2.99	2.07	1.71	1.41	1.24	1.14
0.020	9.96	4.31	3.85	2.35	1.63	1.35	1.11	0.98	0.90
0.025	8.58	3.72	3.32	2.03	1.41	1.16	0.96	0.84	0.78
0.030	7.60	3.29	2.94	1.80	1.25	1.03	0.85	0.75	0.69
0.040	6.27	2.72	2.43	1.48	1.03	0.85	0.70	0.62	0.57
0.050	5.41	2.34	2.09	1.28	0.89	0.73	0.60	0.53	0.49
0.060	4.79	2.07	1.85	1.13	0.78	0.65	0.54	0.47	0.43

The Effect of Aluminium Content on Magnesium Content in Molten Casing Steel. Eq.(18) of [Al]–[Mg] relationship at 1873K is derived by the solution of Eqs. (16) and (17):

$$[\%Mg] = 10^{-3.0074} \cdot N_{MgO} \cdot \left([\%Al]^2 / N_{Al_2O_3}\right)^{\frac{1}{3}} \quad (18)$$

For eight groups of slags chosen from **Table 3**, the calculated [Al]–[Mg] relationship is plotted in **Figure 5**. From **Figure 5**, magnesium content in molten steel increases with increasing aluminium content. When B is in the range of 1.0-1.5, the trend of increase is not obvious. Whereas when B is in the range of 2.5-6.0, the trend of increase is very obvious, which might lead to increasing magnesium-aluminium spinel. The process samples test shows that there was very little $MgO{\cdot}Al_2O_3$ spinel type inclusion observed before LF, they began to occur after LF mainly in the third form (as in **Figure 4**), which was the results of MgO in slag and furnace lining reduced by the soluble Al deoxidation agent to form $MgO{\cdot}Al_2O_3$, then wrapped with sulphide.

The Effect of Slag Basicity and Al_2O_3 Content on Magnesium Content in Molten Casing Steel. **Figure 6** shows the relationship between slag basicity and [O] in the condition of (MgO) = 5%, (FeO) = 0.5% and [Al] = 0.02%; **Figure 7** shows the relationship between slag basicity and [Mg] in the same condition. It can be seen that [O] decreases and [Mg] increases with increasing basicity. And the higher the Al_2O_3 in slag, the higher [O] is and the lower [Mg] is. That is because of mass action concentration of Al_2O_3, *i.e.*, activity decreases when basicity increases. According to [Al]-[O] equilibrium, the [O] decreases, further when [Mg] increases. When (Al_2O_3) = 10 % and B = 2.5, it can be seen [O]=0.00018% and [Mg]=0.000038 % in the **Figures 6** and **7**, which are closer to the measured values in production.

The Effect of Refractory on Magnesium Content in Casing Steel. At present, most refractories contain MgO. When aluminium in molten steel is higher, MgO in the refractory might be reduced, and further magnesium-aluminium spinel might be formed. The equations can be expressed by

$$2[Al] + 3(MgO)_{(S)} = 3[Mg] + (Al_2O_3)_{(S)} \quad (19)$$

$$\Delta G^{\ominus} = 296752.4 + 21.69T, \text{ J/mol}$$

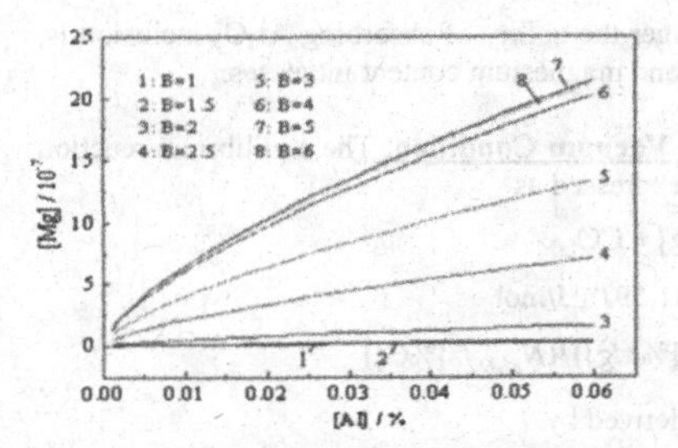

Figure 5 [Al]–[Mg] equilibrium at different basicities

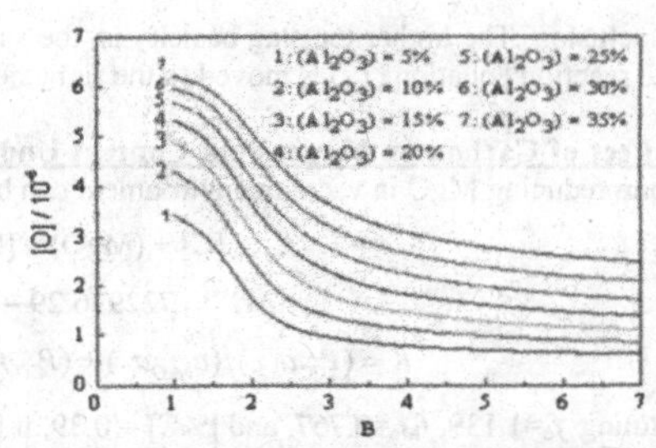

Figure 6 Relationship between slag basicity and [O] in the condition of (MgO) = 5%, (FeO) = 0.5% and [Al] = 0.02%

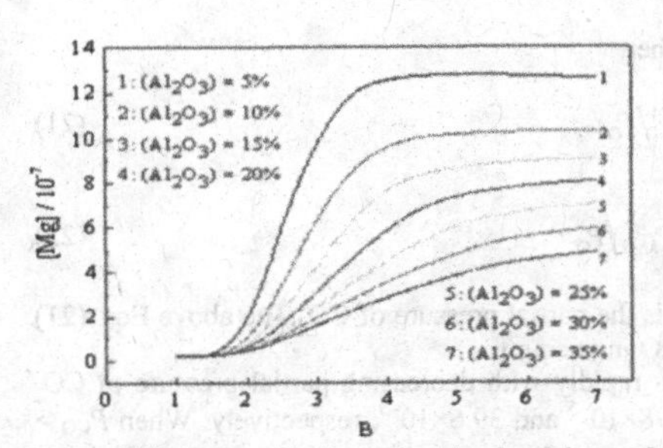

Figure 7 Relationship between slag basicity and [Mg] in the condition of (MgO) = 5%, (FeO) = 0.5% and [Al] = 0.02%

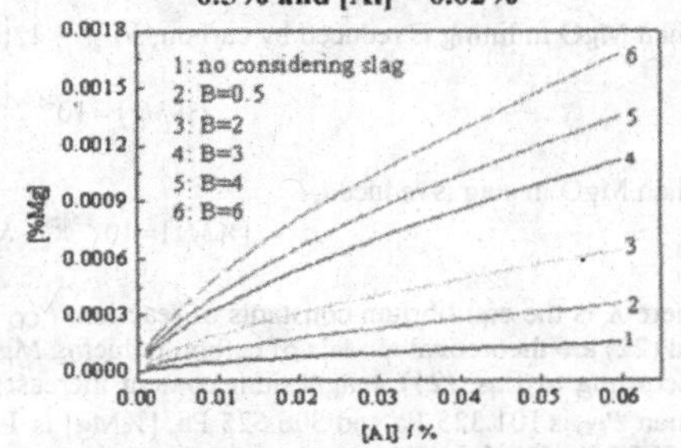

Figure 8 Relationship of [Mg]-[Al] for Al reducing MgO in lining

For refractory, $N_{MgO} = 1$

$$K = a_{[Mg]}^3 \cdot N_{Al_2O_3} / a_{[Al]}^2 = \left(f_{Mg}^3 [\%Mg]^3 \cdot N_{Al_2O_3} / f_{Al}^2 [\%Al]^2 \right)$$

At $T = 1873$ K,

$$[\%Mg] = 10^{-3.1358} [\%Al]^{2/3} / N_{Al_2O_3}^{1/3} \tag{20}$$

This is the theoretical model of aluminium reducing MgO in lining. According to the actual compositions of slag, assuming (MgO) = 5%, (Al_2O_3) = 10%, (FeO) = 0.5%, B = 0.5, 2, 3, 4, 6, and T = 1873 K, calculate $N_{Al_2O_3}$ under different basicities using mass action concentration model of CaO-MgO-FeO-Al_2O_3-SiO_2 slag systems, then substitute into Eq. (20) to obtain the relationship of magnesium content and aluminium content when Al reducing MgO in furnace lining, as shown in **Figure 8.** It can be seen that the action of molten steel on the lining increases with increasing aluminium content, which makes magnesium content increase. At the same aluminium content, magnesium content in molten steel is higher when taking into consideration slag than when not considering it. With the increase of slag basicity, magnesium content increases. This is because slag has the ability to absorb

Al_2O_3 inclusion. The higher the slag basicity is, the stronger the ability of absorbing Al_2O_3 inclusion is, thus the reaction equation (19) is moved to the right side and magnesium content increases.

The Effect of Carbon on Magnesium Content Under Vacuum Condition. The equilibrium reaction of carbon reducing MgO in vacuum environment can be expressed as

$$[C] + (MgO) = [Mg] + CO_{(g)}$$

$$\Delta G^{\ominus} = 722976.29 - 281.59T \text{, J/mol}$$

$$K = (P_{CO}a_{Mg})/(a_{MgO}a_C) = (P_{CO}f_{Mg}[\%Mg])/(N_{MgO}f_C[\%C])$$

Substituting f_c=1.139, f_{Mg}=0.767, and [%C] = 0.39, it is derived by

1873 K, $$\lg P_{CO}[\%Mg] = \lg N_{MgO} - 5.70284$$

When MgO in lining is reduced by carbon, N_{MgO} = 1, then

$$[\%Mg] = 10^{-5.70284}/P_{CO} \quad (21)$$

When MgO in slag is reduced,

$$[\%Mg] = 10^{-5.70284} \cdot N_{MgO}/P_{CO} \quad (22)$$

where K is the equilibrium constants of reaction, P_{CO} is the partial pressure of CO. The above Eqs. (21) and (22) are theoretical models of carbon reducing MgO in vacuum.
According to Eqs. (21), magnesium content increases rapidly with decreasing partial pressure of CO. When P_{CO} is 101.325 Pa and 506.625 Pa, [%Mg] is 198×10^{-5} and 39.6×10^{-5}, respectively. When P_{CO} > 5066.25 Pa, [%Mg] ≈ 0.

Table 5. Relationship of slag basicity and magnesium content at different P_{CO}

B	[%Mg] / 10^{-5}				
	101.325 Pa	506.625 Pa	1013.25 Pa	10132.5 Pa	101325 Pa
0.5	1.070	0.214	0.107	0.0107	0.00107
1.0	2.042	0.408	0.204	0.0204	0.00204
2.0	4.956	0.991	0.496	0.0496	0.00496
3.0	23.787	4.757	2.379	0.2379	0.02379
4.0	29.750	5.951	2.975	0.2975	0.02975
5.0	27.395	5.479	2.740	0.2740	0.02740
6.0	25.571	5.114	2.557	0.2557	0.02557

Table 5 shows the relationship of slag basicity and CO partial pressure and magnesium content when carbon reduces MgO in slag at (MgO)=5%, (FeO)=0.5%, and (Al_2O_3)=10%. It can be seen that magnesium content at first increases with increasing basicity at the same CO partial pressure. Magnesium content reaches the largest value when B=4. Later on, magnesium content decreases with increasing slag basicity. At the same basicity, magnesium content decreases with increasing CO partial pressure. When CO partial pressure is 101.325 Pa and 506.625 Pa, [%Mg] is $(0.214\sim29.75)\times10^{-5}$, far less than that carbon reduced MgO in lining. As a result, there is a larger effect on magnesium content

when carbon reduces MgO in the lining than in slag under vacuum conditions. The inclusion test of samples shows that the quantity of $MgO \cdot Al_2O_3$ spinel type inclusion increased after VD treatment and most occurred in the composite oxide form, which indicates that MgO in furnace lining was partly reduced by carbon during vacuum treatment, to form $MgO \cdot Al_2O_3$, then modified by Ca-Si into the composite oxide, as in Figure 3, some of them had not been modified, existing in dependent spherical $MgO \cdot Al_2O_3$.

Conclusions

(1) There are three forms for $MgO \cdot Al_2O_3$ spinel type inclusion in casing steel. One is pure $MgO \cdot Al_2O_3$ spinel, another is composite oxide of Mg-Al-Ca-Si-O system, and the third is complex with an oxide core, covered by sulphide.
(2) The formation of $MgO \cdot Al_2O_3$ spinel type inclusion in casing steel can be interpreted as, MgO in slag or lining is reduced by aluminium in molten steel or by carbon under vacuum to form magnesium. Then magnesium combines with oxygen to form MgO. Finally, MgO reacts with Al_2O_3 to form magnesium-aluminium spinel.
(3) Mass action concentration model of CaO-MgO-FeO-Al_2O_3-SiO_2 slag systems is established based on the coexistence theory of ion and molecule of slag structure. The effect of the slag component and degree of vacuum and refractory on magnesium content in molten casing steel are studied. The results show that magnesium content increases with increasing slag basicity and aluminium content, and decreases with increasing CO partial pressure.

Acknowledgements

The authors are very grateful for the financial support of the National Natural Science Foundation of China(No.51074021) and Educational Ministry Fund（No.20100006120008）

References:

[1] X.M. Dong, Q.C. Tian,.Q.A. Zhang, Corrosion behaviour of oil well casing steel in H_2S saturated NACE solution. *Corrosion Engineering Science and Technology*, 45(2010), n 2, 181-184.
[2] M. Huang, Y. Wang; X.Y. Zhang. Aluminizing oil casing steel N80 by a low-temperature pack processing modified with zinc addition. *Surface Review and Letters*, 18(2011), No.3-4, 141-146
[3] H.Y. Tang, Ph. D. Thesis, University of Science and Technology Beijing, 2008, p155.
[4] S.K. Jo, B. Song, and S.H. Kim, Thermodynamics on the formation of spinel ($MgO \cdot Al_2O_3$) inclusion in liquid iron containing chromium, *Metallurgical and Materials Transactions B*, 33B(2002), 709-713.
[5] H.P. Joo, Formation mechanism of spinel-type inclusions in high-alloyed stainless steel melts, *Metallurgical and Materials Transactions B*, 38B (2007), 657-663.
[6] H.P. Joo and S.K. Dong, Effect of CaO-Al_2O_3-MgO slags on the formation of $MgO \cdot Al_2O_3$ inclusions in ferritic stainless steel, *Metallurgical and Materials Transactions B*, 36B(2005), 495-502.
[7] J. Zhang, *Computational Thermodynamics of Metallurgical Melts and Solutions*, in: Metallurgy Industry Press, Beijing, 2007.
[8] J. Zhang, Coexistence theory of slag structure and its application to calculation of oxidizing capability of slag melts, *Journal of Iron and Steel Research International*, 10 (2003), No.1, 1-10.
[9] J. Zhang, The equilibrium of manganese between FeO-MnO-MgO-SiO_2 slag system and liquid iron, *Journal of University of Science and Technology Beijing* (in Chinese), 14(1992), No.5, 496-501.
[10] J. Zhang, Application of the law of mass action in combination with the coexistence theory of slag

structure to multicomponent slag systems, *Acta Metallurgica Sinica* (English Letters), 14(2001) , No.3, 177-190.
[11] J. Zhang, Applicability of law of mass action to distribution of manganese between slag melts and liquid iron, *Transaction of Nonferrous Metals Society of China*, 11 (2001), No.5, 778-783.
[12] J. Zhang, Applicability of mass action law to sulphur distribution between slag melts and liquid iron, *J. Univ. Sci. Technol. Beijing*, 9(2002) , No.2, 90-98.
[13] J. X. Chen. Handbook of common figures, tables and data for steelmaking, *Metallurgy Industry Press*, Beijing, 1984
[14] Allibert M. and Chatilion C.. Mass-spectrometric and electrical studies of thermodynamic properties of liquid and solid phases in system CaO-Al_2O_3. *J Am Ceram Soc*, 64(1981), No.5, 307-314.
[15] Rein R. H. and Chipman J., Activities in liquid solution SiO_2-CaO-MgO-Al_2O_3 at 1600°C. *Trans. Metall Soc. AIME*, (1965), 233, 415-425
[16] Barin I., Knacke O. and Kubaschewski O., Thermochemical properties of inorganic substances, *Springer-Verlag*, Berlin-Heidelberg-New York, 1977, pp128.
[17] H. Ohta and H. Suito. Activities of MnO in CaO-SiO2-Al2O3-MnO (<10Pct)-FetO(<3Pct) slags saturated with liquid iron. *Metallurgical and Materials Transactions B*, 26B (1995), 295-303.
[18] H. Suito, Thermodynamics on control of inclusions compositions in ultra-clean steel, *ISIJ Int.*, 36(1996), No.5, 528-532.
[19] C.M. Yu, X.D. Miu, C.M. Shi, *et al.*, Behaviour of spinel $MgO \cdot Al_2O_3$ inclusions in ball bearing steel, *Journal of University of Science and Technology Beijing* (in Chinese), 27(2005), Suppl.2, 37-40.

Materials Processing Fundamentals
Edited by: Lifeng Zhang, Antoine Allanore, Cong Wang, James A. Yurko, and Justin Crapps
TMS (The Minerals, Metals & Materials Society), 2013

THEORY ANALYSIS OF STEEL CLEANLINESS CONTROL DURING ELECTROSLAG REMELTING

Cheng-bin Shi[1], Xi-chun Chen[2], Yi-wa Luo[1], Han-jie Guo[1]
1) School of Metallurgical and Ecological Engineering, University of Science and Technology Beijing; No.30 Xueyuan Road, Haidian District, Beijing, P. R. China.
2) Research Institute of High Temperature Materials, Central Iron and Steel Research Institute; No. 76 Xueyuannan Road, Haidian District, Beijing, P. R. China.
Corresponding author; e-mail: guohanjie@ustb.edu.cn; shicb09@hotmail.com
Keywords: electroslag remelting, oxygen control, sulfide inclusions, oxide inclusions, carbides

1. Introduction

Although the steel cleanliness can be significantly improved through electroslag remelting (ESR) process, with the increasing demand for more excellent comprehensive performance of steel, the higher cleanliness of ESR steel is required. Impurity elements content and inclusions control is two important aspects of steel cleanliness control [1]. The inclusions control includes reducing their content and controlling their composition, size distribution and morphology. The investigation of steel cleanliness control during ESR process not only has significant scientific meaning, but also important guiding significance for the clean steel and superalloy production.

Based on the previous experimental studies [2-6], oxygen behavior and control during P-ESR process were firstly discussed. Next, the unstable inclusions evolution during ESR process was reviewed. The calcium modification of $MgO\cdot Al_2O_3$ spinel inclusions in ESR process was investigated. On the base of calcium modification of $MgO\cdot Al_2O_3$ spinels in steel, the calcium modification of $MgO\cdot Al_2O_3$ spinel inclusions in Inconel 718 superalloy and its effect on primary carbonitrides were analyzed. This work aims to provide useful information for steel cleanliness control during ESR process.

2. Oxygen Control during ESR Process

The iron oxide formed on the electrode surface before and during ESR process as well as the atmospheric oxygen in remelting atmosphere is considered to be responsible for affecting the oxygen content in liquid steel during ESR process, besides the initial oxygen content in consumable electrode. In the case of ESR operation that was performed under atmosphere, the oxygen in atmosphere would affect the oxygen content in liquid steel in two different ways: (i) atmospheric oxygen permeates directly through liquid slag to liquid metal pool by diffusion of physically dissolved oxygen; (ii) atmospheric oxygen reacts with electrode steel to form iron oxide on the electrode surface at elevated temperatures.

The formed iron oxide is indirectly transferred to the slag-metal pool interface, thereafter FeO transports oxygen into the liquid metal film and liquid metal pool as expressed as follows

$$(FeO)=[O]+[Fe] \tag{1}$$

Meanwhile, in the case where CaO-containing slag is employed in ESR process, the dissolved oxygen [O] as the product of desulfurization reaction in ESR process, as expressed in Eq. (2), will induce the increase of oxygen content in liquid steel

$$[S]+(CaO)=(CaS)+[O] \quad (2)$$

The mechanism of oxygen transfer in ESR process is schematically shown in Fig. 1. In the case where Ar gas atmosphere is employed in ESR process, the formation of FeO caused by the chemical reaction between atmospheric oxygen and the electrode taking place on the electrode surface, which can indirectly lead to oxygen pick-up in liquid steel, would be prevented (at least substantially reduced).

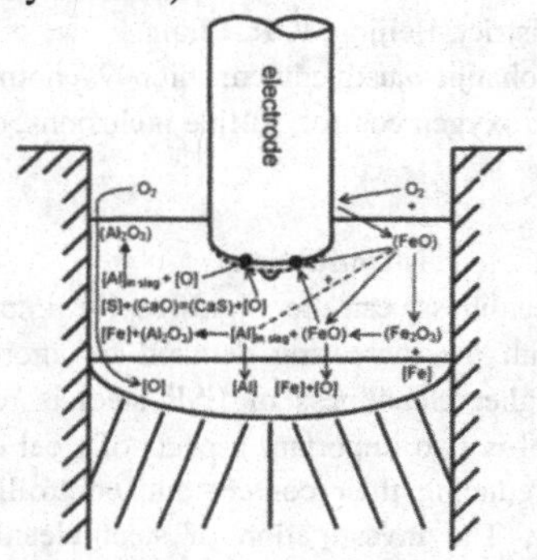

Figure 1. Schematic illustration of the mechanisms of oxygen transfers and deoxidation in ESR process.

The kinetic model of oxygen transfer between molten slag and liquid steel phases during protective gas electroslag remelting process established based on the penetration theory indicates that there is a critical value of FeO content in slag during ESR process. The model developed in previous study [7] is expressed as

$$J = \frac{1}{t_e}\int_0^{t_e}\sqrt{\frac{D}{\pi t}}(c_s - c_b)dt = 2\sqrt{\frac{D_{FeO}}{\pi t_e}}\left(c_{s,FeO} - c_{b,FeO}\right) \quad (3)$$

where J is diffusion flux, ($mol/m^2 \cdot s$); $c_{s,FeO}$ is concentration of FeO at slag-liquid steel interface, (mol/m^3); $c_{b,FeO}$ is concentration of FeO in slag bulk, (mol/m^3); D_{FeO} is diffusion coefficient (m^2/s); t_e is residence time, (s). The detailed descriptions of the kinetic model establishment have been given elsewhere [7].

The present kinetic model was applied to the laboratory electroslag remelting of S136 experiments [4]. Figure 2 shows the relation between the transfer rate of oxygen between slag and liquid steel phases, and FeO concentration in slag. It can be seen that, in the case where FeO content is lower than this critical value, the oxygen transfer is from liquid steel at electrode tip to molten slag. Otherwise, the oxygen transfer is from molten slag to liquid steel at the electrode tip. The critical value of FeO concentration in slag calculated using the kinetic model is 0.26%, 0.45%, 0.42% and 0.70%, respectively, as shown in Fig. 2.

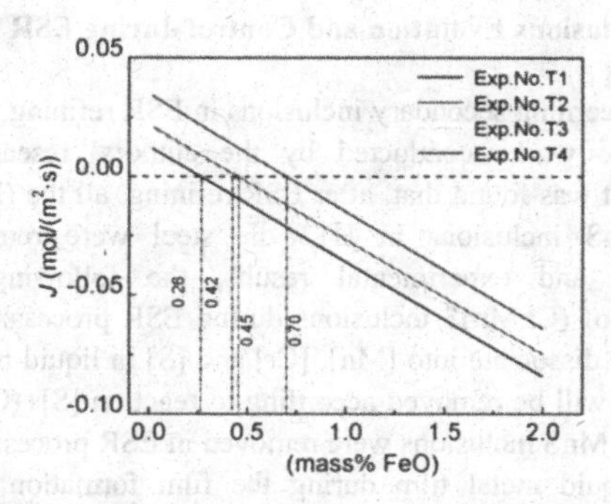

Figure 2. Relation between the transfer rate of oxygen between slag and liquid steel phases and FeO concentration in slag [7].

The diffusion flux of oxygen transfer from liquid steel to molten slag decreases with increasing the FeO concentration in slag. Whereas the diffusion flux of oxygen transfer from molten slag to liquid steel increases with increasing the FeO concentration in slag, provided that FeO concentration in slag is greater than this critical value. Therefore, ferrous oxide plays an important role in transporting oxygen to the metal phase. The control of FeO concentration in slag is the key in controlling oxygen content in liquid steel in ESR process. It is crucial for restraining the oxygen content in liquid steel at extra-low level through reducing FeO concentration in slag.

In the case where Al-containing deoxidant was added into slag pool during ESR process, with the melt of deoxidant, the aluminum dissolved in slag pool $[Al]_{in\ slag}$ would react with FeO in slag to lower the oxygen potential as follows

$$2[Al]_{in\ slag}+3(FeO)=(Al_2O_3)+3[Fe] \quad (4)$$

As shown in Eq. (4), the concentration of FeO in slag would be reduced substantially, which would suppress the transportation of oxygen by FeO according to the reaction represented by Eq. (1) to liquid steel. Meanwhile, the dissolved Al in slag pool $[Al]_{in\ slag}$ brought by deoxidant addition can also directly react with the oxygen in liquid metal film, as represented in the following reaction, to reduce the oxygen content in liquid steel

$$2[Al]_{in\ slag}+3[O]=(Al_2O_3) \quad (4)$$

The mechanism of deoxidation by Al-containing deoxidant in ESR process is also schematically presented in Fig. 1.

Some researchers reported that, in the case of electroslag remelting of the electrode with low oxygen content, oxygen pick-up would occur in this process [8-10]. In the authors' previous studies, the total oxygen content can be reduced from 34 ppm in NAK80 die steel to 10 ppm [7,8], from 89 ppm in S136 die steel to 12 ppm [4], and from 45 ppm in H13 die steel to 8 ppm after ESR process by simultaneously employing protective Ar gas remelting operation combined with specially designed slag deoxidation treatment [11]. It is considered that aluminum added during ESR process reduces not only FeO content in slag pool but also the dissolved oxygen in liquid metal film [4].

3. Non-Metallic Inclusions Evolution and Control during ESR Process

3.1 Secondary Inclusions

No other studies concerning secondary inclusions in ESR refining process were reported in literatures, except the works conducted by the authors' research team. Based on experimental results [4], it was found that, after ESR refining, all the (Mn,Cr)S inclusions in S136 die steel and MnS inclusions in H13 die steel were removed. According to thermodynamic analysis and experimental results, the following reaction sequence summarizes the removal of (Cr,Mn)S inclusions during ESR process: (i) sulfide inclusions (Mn,Cr)S in the electrode dissociate into [Mn], [Cr] and [S] in liquid metal film, and (ii) the sulfur in liquid phase [S] will be removed according to reaction [S]+(CaO)=[O]+(CaS). It is in the consistent way that MnS inclusions were removed in ESR process. It is the dissociation of AlN inclusions in liquid metal film during the film formation at electrode tip that contributes to AlN inclusions removal during ESR of NAK80 die steel [3].

After ESR refining process, no sulfide inclusions were found in ESR ingot. This could be due to the low segregation degree of sulfide inclusions-forming elements at solidifying front during liquid steel solidification, which is due to the fact that small quantity of liquid steel solidifies at the bottom of shallow liquid metal pool in a short time during ESR process. Meanwhile, the very low sulfur content in liquid metal pool resulting from excellent desulfurization in ESR process is also adverse to sulfide inclusions precipitation. However, AlN inclusions precipitate in residual liquid phase at solidifying front during the cooling of liquid steel until the product of $[\%Al]\times[\%N]$ exceeds the calculated critical value for AlN precipitation, due to the enrichment of Al and N in residual liquid steel between solid steel dendritic arms at solidifying front resulting from the difference in solubility of solute between liquid and solid phases. It has been confirmed during laboratory scale ESR of NAK80 die steel that pure AlN inclusions with small size form in ESR ingot [2,3].

3.2 Oxide Inclusions

Extensive studies concerning oxide inclusions removal during ESR process have been made by various researchers [12-15]. It is quite necessary, but hardly achieved, to completely remove oxide inclusions in ESR process. A complementary approach is to modify oxide inclusions to low melting point inclusions in order to reduce their detriment to steel product, such as $MgO\cdot Al_2O_3$ spinel inclusions which have a high melting point and hardness. Some researchers [16-19] have confirmed that $MgO\cdot Al_2O_3$ spinel inclusions can be effectively modified to liquid or partially liquid calcium aluminates by calcium treatment in ordinary steelmaking plant trails or laboratory scale experiments, rather than ESR process.

Figure 3 presents the typical $MgO\cdot Al_2O_3$ spinel inclusions observed in H13 die steel used as consumable electrodes in ESR experiments. After P-ESR refining process combined with calcium treatment, all $MgO\cdot Al_2O_3$ spinel inclusions in the electrode were modified to low-melting-point CaO–MgO–Al_2O_3 or CaO–Al_2O_3 inclusions with homogeneous chemical compositions. It is the incomplete or complete reduction of MgO from the spinels by calcium that contribute to the modification of $MgO\cdot Al_2O_3$ spinels to $CaO-MgO-Al_2O_3$ and $CaO-Al_2O_3$ inclusions, respectively.

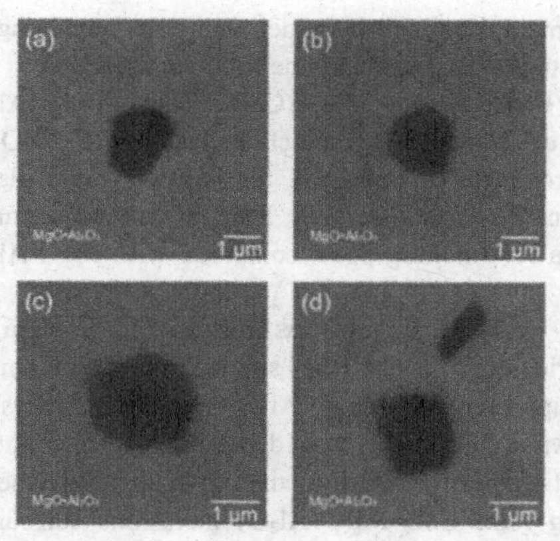

Figure 3. The typical original inclusions in the consumable electrode.

Figure 4 shows the composition distribution of inclusions after calcium treatment on ternary phase diagram of CaO – MgO – Al_2O_3. It can be seen from Fig. 4(a) that, in the case of small amount of calcium addition in P-ESR process (calcium addition rate: 5 kg/t), most of the oxide inclusions are in the liquid region of CaO – MgO – Al_2O_3. After proper calcium treatment (calcium addition rate: 10 kg/t), all the oxide inclusions are in the low melting temperature region, as shown in Fig. 4(b). The inclusions are mainly low melting temperature CaO – MgO – Al_2O_3 and some CaO – Al_2O_3.

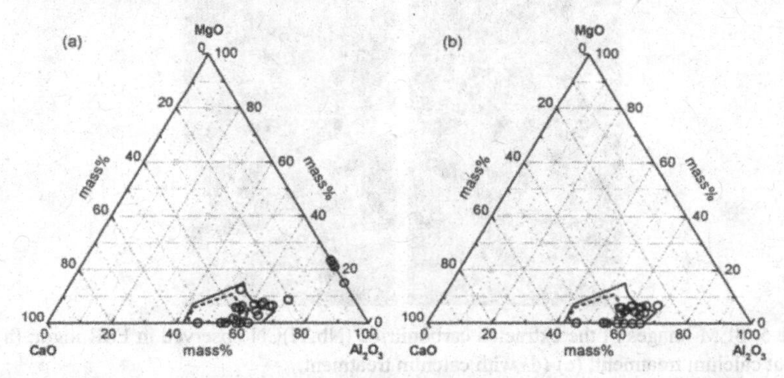

Figure 4. Composition distributions of oxide inclusions in ESR ingot on CaO–MgO–Al_2O_3 ternary diagram, (dashed line < 1773K, solid line < 1873K).

3.3 Carbonitrides in Superalloy Refined through ESR Process

It was found in the authors' previous study [5] that all the observed oxide inclusions in Inconel 718 superalloy ESR ingots were nucleation sites of outer (Nb,Ti)CN layers. Oxide inclusions in the consumable electrode are mainly $MgO\cdot Al_2O_3$ spinels. In the case of proper amount of calcium addition in ESR process, the original $MgO\cdot Al_2O_3$ spinels can be effectively modified to the complex oxide inclusions in the form of $MgO\cdot Al_2O_3$ core surrounded by CaO – MgO – Al_2O_3 layer in superalloy ingot. This is because, in the case of calcium addition treatment, $MgO \cdot Al_2O_3$ spinel inclusions remaining in liquid metal phase react with dissolved calcium, as expressed by the following reaction

$$[Ca]+(x+1/y)(yMgO \cdot Al_2O_3)=(CaO \cdot xyMgO \cdot (x+1/y)Al_2O_3)+[Mg] \quad (5)$$

The modification of $MgO \cdot Al_2O_3$ spinel inclusions to $CaO–MgO–Al_2O_3$ system inclusions by calcium treatment is beneficial for improving the plasticity of the oxygen inclusions remaining in nickel-base superalloy ESR ingot because many $CaO–MgO–Al_2O_3$ system inclusions have a low melting point compared with $MgO \cdot Al_2O_3$ spinels and MgO inclusions.

Calcium treatment in ESR process has resulted in a significant change in the morphology of carbonitrides (Nb,Ti)CN from single octahedral form or clustered block to skeleton-like shape through modifying $MgO \cdot Al_2O_3$ spinel which acts as the nucleation site for carbonitrides, as shown in Fig. 5. This could be due to the fact that, with proper amount of calcium addition during ESR of Inconel 718 superalloy, the oxide inclusions were modified to low-melting-point $CaO–MgO–Al_2O_3$ inclusions surrounding on $MgO \cdot Al_2O_3$ spinel core. The carbide formed on the nitride which precipitates on $CaO–MgO–Al_2O_3$ inclusion has no sufficient time to reach or near the equilibrium morphology (*i.e.*, octahedral form) [5]. The precipitation of primary carbonitrides (Nb,Ti)CN during solidification can be suppressed by reducing the oxide inclusion content through reducing oxygen content in the superalloy, which would indirectly decrease the carbonitrides (Nb,Ti)CN nucleation sites.

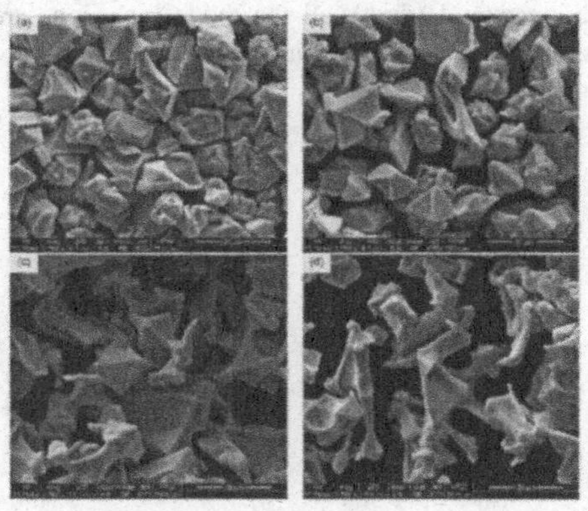

Figure 5. SEM images of the extracted carbonitride (Nb,Ti)CN observed in ESR ingot: (a)-(b) without calcium treatment, (c)-(d) with calcium treatment.

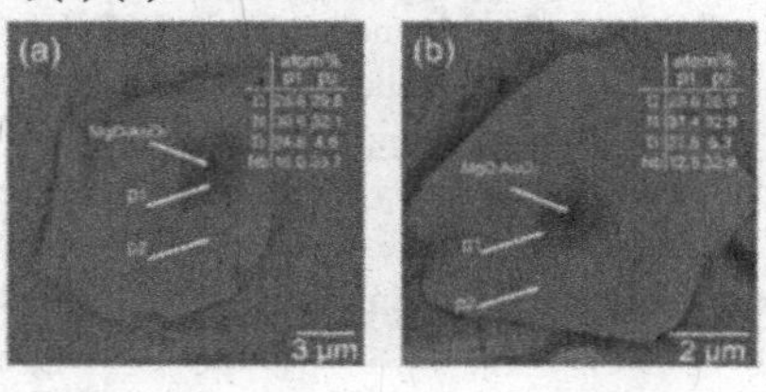

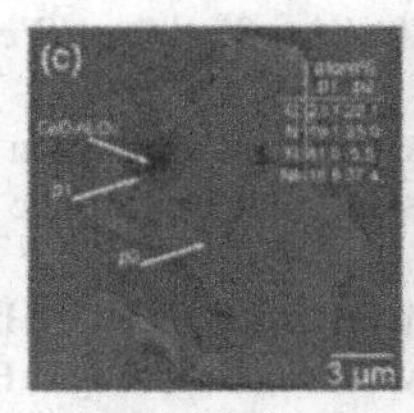
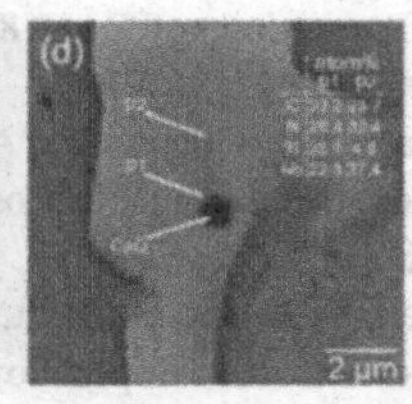

Figure 6. The typical extracted inclusions in ESR ingots: (a)-(b) without calcium treatment, (c)-(d) with calcium treatment.

All carbonitrides (Nb,Ti)CN precipitating around oxide inclusion core exhibit two-layer structure as shown in Fig. 6, except for those with single (Nb,Ti)CN layer containing a small amount of Ti and N in the ingot produced with proper amount of calcium addition in ESR process. For carbonitrides (Nb,Ti)CN with two-layer structure, the outer (Nb,Ti)CN layer has a higher atomic percentage of Nb than that of Ti as well as a small amount of Ti and N. The inner (Nb,Ti)CN layer has higher atomic percentage of Ti than that of Nb, and contains a small amount of Nb and C. The atomic ratio of Nb/Ti in different carbonitrides is relatively constant.

4. Conclusions

(1) There is a critical value of FeO concentration in slag pool during ESR process. In the case where FeO content is lower than this critical value, the oxygen transfer is from liquid steel at electrode tip to molten slag. Otherwise, the oxygen transfer is from molten slag to liquid steel.

(2) Protective Ar gas remelting operation combined with specially designed slag deoxidation treatment in ESR process can effectively reduce oxygen content to extremely low level.

(3) It is the dissociation of sulfide inclusions and AlN inclusions in liquid metal film during the film formation at electrode tip that contributes to these inclusions removal during ESR process.

(4) Calcium treatment is effective in near fully or completely modifying $MgO\cdot Al_2O_3$ spinels in steel to low-melting-point $CaO-MgO-Al_2O_3$ or $CaO-Al_2O_3$ inclusions in P-ESR process. It is the incomplete or complete reduction of MgO from $MgO\cdot Al_2O_3$ spinel by calcium that contributes to the modification of $MgO\cdot Al_2O_3$ spinels.

(5) Calcium treatment in ESR process can refine carbonitrides in Inconel 718 superalloy through modifying $MgO\cdot Al_2O_3$ spinel which acts as the nucleation site for carbonitrides.

Acknowledgments

The financial support provided by the International Science and Technology Cooperation and Exchange of Special Projects (Grant No. 2010DFR50590) is greatly acknowledged.

References

[1] L.F. Zhang, "State of the Art in the Control of Inclusions in Tire Cord Steels–a Review," *Steel Res. Int.*, 77 (3) (2006), 158–169.

[2] C.B. Shi, X.C. Chen, and H.J. Guo, "Characteristics of Inclusions in High-Al Steel during Electroslag Remelting Process," *Int. J. Miner. Metall. Mater.*, 19 (4) (2012), 295–302.

[3] C.B. Shi, X.C. Chen, and H.J. Guo, "Oxygen Control and Its Effect on Steel Cleanliness

during Electroslag Remelting of NAK80 Die Steel," *AISTech 2012 Conference Proceeding*, Atlanta, GA. 2012, 947–957.
[4] C.B. Shi, X.C. Chen, and H.J. Guo et al., "Assessment of Oxygen Control and Its Effect on Inclusion Characteristics during Electroslag Remelting of Die Steel," *Steel Res. Int.*, 83 (5) (2012), 472–486.
[5] X.C. Chen, C.B. Shi, and H.J. Guo et al., "Investigation of Oxide Inclusions and Primary Carbonitrides in Inconel 718 Superalloy Refined through Electroslag Remelting Process," *Metall. Mater. Trans. B*, 43 (6) (2012), 1596-1607.
[6] C.B. Shi, X.C. Chen, and H.J. Guo, "Control of $MgO\cdot Al_2O_3$ Spinel Inclusions during Protective Gas Electroslag Remelting of Die Steel," *Metall. Mater. Trans. B*, 2012, (under review).
[7] C.B. Shi, H.J. Guo, and X.C. Chen et al, "Kinetic Study on Oxygen Transfer during Protective Gas Electroslag Remelting Process," *Special Steel*, 2012, (in press).
[8] S.F. Medina and A. Cores, "Thermodynamic Aspects in the Manufacturing of Microalloyed Steels by the Electroslag Remelting Process," *ISIJ Int.*, 33 (12) (1993), 1244–1251.
[9] C.S. Wang, S.G. Liu, and M.D. Xu et al., "Reducing Oxygen Content in Electro-Slag Remelted Bearing Steel GCr15," *Special Steel*, 18 (3) (1997), 31–35.
[10] L.Z. Chang, H.S. Yang, and Z.B. Li, "Study on Oxygen Behavior during Electroslag Remelting," *Steelmaking*, 26 (5) (2010), 46–50.
[11] F. Wang, X.C. Chen, and H.J. Guo, "Aluminum Deoxidization of H13 Hot Die Steel through Inert Gas Protection Electroslag Remelting," *AISTech 2012 Conference Proceeding*, Atlanta, GA. 2012, 1005–1015.
[12] D.A.R. Kay and R.J. Pomfret, "Removal of Oxide Inclusions during AC Electroslag Remelting," *J. Iron Steel Inst.*, 209 (12) (1971), 962–965.
[13] J. Fu and J. Zhu, "Change of Oxide Inclusions during Electroslag Remelting Process," *Acta Metall. Sin.*, 7 (3) (1964), 250–262.
[14] A. Mitchell, "Oxide Inclusion Behavior during Consumable Electrode Remelting," *Ironmaking Steelmaking*, 1 (3) (1974), 172–179.
[15] Z.B. Li, W.H. Zhou, and Y.D. Li, "Mechanism of Removal of Non-metallic Inclusions in the ESR Process," *Iron Steel*, 15 (1) (1980), 20–26.
[16] E.B. Pretorius, H.G. Oltmann, and T. Cash, "The Effective Modification of Spinel Inclusions by Ca Treatment in LCAK Steel," *Iron Steel Technol.*, 7 (7) (2010), 31–44.
[17] N. Verma, M. Lind, P.C. Pistorius et al., "Modification of Spinel Inclusions by Calcium in Liquid Steel," *Iron Steel Technol.*, 7 (1) (2010), 189–197.
[18] S.F. Yang, Q.Q. Wang, L.F. Zhang et al., "Formation and Modification of $MgO\cdot Al_2O_3$-Based Inclusions in Alloy Steels," *Metall. Mater. Trans. B*, 43 (4) (2012), 731–750.
[19] N. Verma, P.C. Pistorius, and R.J. Fruehan et al., "Calcium Modification of Spinel Inclusions in Aluminum-Killed Steel Reaction Steps," *Metall. Mater. Trans. B*, 43 (4) (2012), 830–840.

Materials Processing Fundamentals
Edited by: Lifeng Zhang, Antoine Allanore, Cong Wang, James A. Yurko, and Justin Crapps
TMS (The Minerals, Metals & Materials Society), 2013

Formation and Drop of Metal Droplets in Slag Bath of Electroslag Remelting Processes

Baokuan Li, Ruinan Li and Bo Wang
School of Materials and Metallurgy, Northeastern University, Shenyang, Liaoning, 110819, P.R.China

Keywords: electroslag remelting, two-phase flow, metal droplets, slag bath

Abstract

A mathematical model of two-phase flows of molten metal/slag in electroslag remelting (ESR) processes has been developed to analyze the formation and drop of liquid metal droplet in slag bath. The finite volume method with the VOF technique is used to solve the flow equations. Numerical results show that the molten metal film at consumable electrode tip is converged into droplets, and fall from the electrode tip via the slag bath into liquid metal pool. The fall droplets cause the wave of interface of slag/metal pool, the frequency of interface wave of slag/metal depends on the quantity of droplets, and the amplitude of wave depends on the size of droplets.

Introduction

The electroslag remelting (ESR) system has been widely applied in the special steel industry. Knowledge of Two phase flow of slag/metal in liquid pool, which controls the remelting processes and quality of the product, is important for operation of the system. Mathematical modeling and numerical simulation are of great interest because experimental study in such a system is extremely difficult.

The principal components of an ESR system in Figure 1 include a consumable electrode, a molten slag pool, a liquid metal pool, a solidified ingot, and a water-cooled mold. The electric current (AC) is passed from the electrode through the molten slag and the liquid pool to the ingot. The resultant Joule heating of the slag melts the electrode while the droplets formed, fall through the slag and accumulate in the metal pool. Because of the water cooling provided in the mold, solidification occurs continuously at the pool-ingot interface. The solidified ingot is the product of the ESR process.

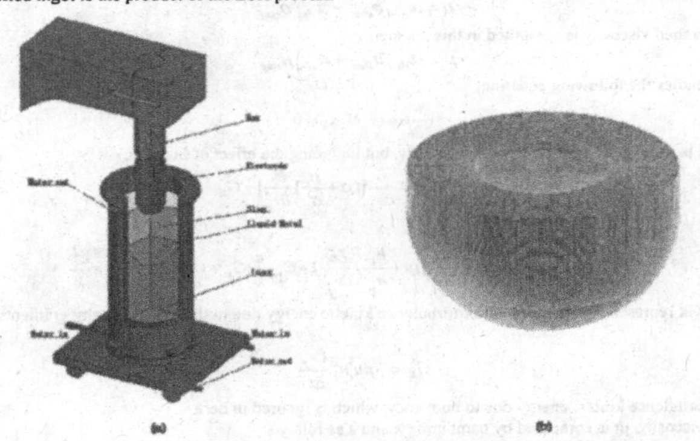

Fig 1 Schematic of the electroslag remelting system, (a) equipment, and (b) mesh.

Dilwari and Szekely [1] developed the general framework for analyzing the electromagnetic fields in ESR. The electromagnetic field was formulated using the Maxwell equations, and a complete set of boundary conditions was also prescribed. They solved the electromagnetic field problem numerically, using the finite difference technique. Jardy et al [2] calculated the flow field in an ESR slag pool using control volume method, and some effects of the fill ratio and the thermophysical properties of slag were studied. Ruckert et al [3] solved the transport phenomena including the fluid flow field, the temperature distribution and the movement of different phases. Kharicha et al [4] developed a numerical model to predict the influence of the frequency of the applied AC current on the electroslag remelting processes. Weber et al [5] developed a two-dimensional axisymmetric transient model of the ESR process, which accounts for coupled electromagnetic, fluid flow, heat transfer, and phase change phenomena. Hernandez-Morales and Mitchell [6] had reviewed the mathematical models of transport in ESR processes and pointed that the more effort is required before the models can be applied to define actual operating conditions.

The purpose of the present work is to understand the formation and Drop of Metal Droplets in Slag Bath of Electroslag Remelting Processes based on the three-dimensional model with the primary variables of two phase flow field. At the same time, parametric studies are also conducted with the effect of melt rates, electrode tip shape, and electrode number in the ESR system.

Mathematical Model

1. Governing equations

In order to investigate the dynamic behavior of two phase flow, VOF (Volume of Fluid) function [7] is used to track the interface between molten steel and slag phase. Therefore, the following governing transport equations including VOF function and turbulence model equations need to be solved.

Continuity equation

$$\frac{\partial \rho}{\partial t}+\nabla\cdot(\rho\vec{V})=0 \tag{1}$$

A single momentum equation is solved throughout the domain, and the resulting velocity field is shared among the phases. The momentum equation, shown below, is dependent on the volume fractions of all phases through the properties ρ and μ.

$$\frac{\partial}{\partial t}(\rho\vec{v})+\nabla\cdot(\rho\vec{v}\vec{v})=-\nabla p+\nabla\cdot[\mu_e(\nabla\vec{v}+\nabla\vec{v}^T)]+\rho\vec{g} \tag{2}$$

One limitation of the shared-fields approximation is that in cases where large velocity differences exist between the phases, the accuracy of the velocities computed near the interface can be adversely affected.

The VOF formulation relies on the fact that two or more fluids (or phases) are not interpenetrating. For each additional phase that you add to your model, a variable is introduced: the volume fraction of the phase in the computational cell. In each control volume, the volume fractions of all phases sum to unity. The fields for all variables and properties are shared by the phases and represent volume-averaged values, as long as the volume fraction of each of the phases is known at each location. Thus the variables and properties in any given cell are either purely representative of one of the phases, or representative of a mixture of the phases, depending upon the volume fraction values.

$$\alpha_{steel}+\alpha_{slag}=1 \tag{3}$$

In other words, if the q th fluid's volume fraction in the cell is denoted as αq, then the following three conditions are possible:

- αq =0: the cell is empty (of the q th fluid).
- αq =1: the cell is full (of the q th fluid)
- 0< αq <1: the cell contains the interface between the q the fluid and one or more other fluids.

The volume-fraction-averaged density takes on the following form:

$$\rho=\alpha_{steel}\rho_{steel}+\alpha_{slag}\rho_{slag} \tag{4}$$

The volume-fraction-averaged viscosity is computed in this manner.

$$\mu=\alpha_{steel}\mu_{steel}+\alpha_{slag}\mu_{slag} \tag{5}$$

The VOF function αq satisfies the following equation:

$$\frac{\partial \alpha_q}{\partial t}+(\vec{v}\cdot\nabla)\alpha_q=0 \tag{6}$$

The standard k - ε model is used to calculated effective viscosity, but including the effect of buoyancy:

$$\frac{\partial}{\partial t}(\rho k)+\frac{\partial}{\partial x_i}(\rho k u_i)=\frac{\partial}{\partial x_j}[(\mu+\frac{\mu_t}{\sigma_k})\frac{\partial k}{\partial x_j}]+G_k+G_b-\rho\varepsilon \tag{7}$$

and

$$\frac{\partial}{\partial t}(\rho\rho\varepsilon+\frac{\partial}{\partial x_i}(\rho\rho\varepsilon_i)=\frac{\partial}{\partial x_j}[(\mu+\frac{\mu_t}{\sigma_\varepsilon})\frac{\partial\varepsilon}{\partial x_j}]+C_{1\varepsilon}\frac{\varepsilon}{k}(G_k+C_{3\varepsilon}G_b)-C_{2\varepsilon}\rho\frac{\varepsilon^2}{k} \tag{8}$$

In these equations, Gk represents the generation of turbulence kinetic energy due to the mean velocity gradients; this term may be defined as

$$G_K=-\rho\overline{u_i'u_j'}\frac{\partial u_j}{\partial x_i} \tag{9}$$

Gb is the generation of turbulence kinetic energy due to buoyancy, which is ignored in here.

The turbulent (or eddy) viscosity, μt is computed by combining k and ε as follows:

$$\mu_t=\rho C_\mu\frac{k^2}{\varepsilon} \tag{10}$$

where C_μ is a constant, $C_{1\varepsilon}=1.44$, $C_{2\varepsilon}=1.92$, $C_\mu=0.09$, $\sigma_k=1.0$, $\sigma_\varepsilon=1.3$.

2. Geometrical and operational conditions

The ESR studied in this work is a 5 ton and 30 ton, where most of the advanced special steels are produced. Input conditions such as geometrical, thermo-physical properties and operating parameters are given in Table 1. Figure 1 (a) displays the actual ESR geometries with one electrode.

Table 1 Geometrical, physical properties and operating parameters

Parameters	5 ton ESR	30 ton ESR
Diameter of mold, m	0.508	1.5
Diameter of electrode, m	0.406	0.406
Thickness of slag cap, m	0.127	0.3
Density of slag, kg/m3	2524	2524
Viscosity of slag, m2/s2	0.015	0.015

Density of molten steel, kg/m3	7013	7013
Viscosity of molten steel, m2/s2	0.006	0.006
Melting rate, kg/s	0.15	0.6

3. Initial and boundary conditions

Initially, the steel bath starts at rest with no melting rate of electrodes. In addition to this, the slag cap rests on top of the ingot. For the liquid drop inlet velocity is calculated by the melting rates by the following equation,

$$V_{in} = \frac{m_L}{\rho_{steel} A} \qquad (11)$$

Where, mL is melting rate at standard condition as shown in Table 1. The free surface of the slag/steel interface is frictionless. An allowance is made for the outflow of liquid metal pool at the bottom.

4. Numerical method

The solutions to the governing equations, boundary conditions, and source terms are obtained using the commercial fluid dynamics package Fluent. The calculation domain is divided by the 347427 nodes for one eletrode and 759040 nodes for three electrodes. The calculations are conducted in the unsteady solution mode using the SIMPLE algorithm to solve the two phase flow problem. A criterion for convergence in all cases simulated here is established when the sum of all residents for the dependent variables is less than 10^{-4}.

Results and Discussion

Figure 1 (b) shows the mesh system for the slag pool and liquid pool of one electrode ESR. Figure 2 displays a comparison of the falls of liquid droplets both plate tip and arc tip of electrode. We can observe that molten steel, after the film melting of the electrode tip, sink to the center position of electrode tip. The recirculation flow in slag pool occurs with the fall droplets as shown in Figure 3. Figure 4 shows the curves of interface wave at different time for two kind of electrode tip, the amplitude of wave for the arc tip of electrode is larger than that of plate tip. Figure 5 illustrate the effect of melt rate on the formation and drop of droplets in slag pool of ESR system. If the melt rate is small and electrode tip is protrusion, the molten steel can concentrate the center position of electrode tip and form a large droplet, which falls to the interface of slag/metal and cause the wave of interface. If the melt rate is medium, the part of molten steel forms the droplets in the midway of sink to the center position and fall to the off- center position of electrode tip. If the melt rate is large, the many droplets are incorporated to form the continuous fall streams. It is not benefit to refining of molten steel. The interface wave of slag/metal is reinforced while the melt rate increases.

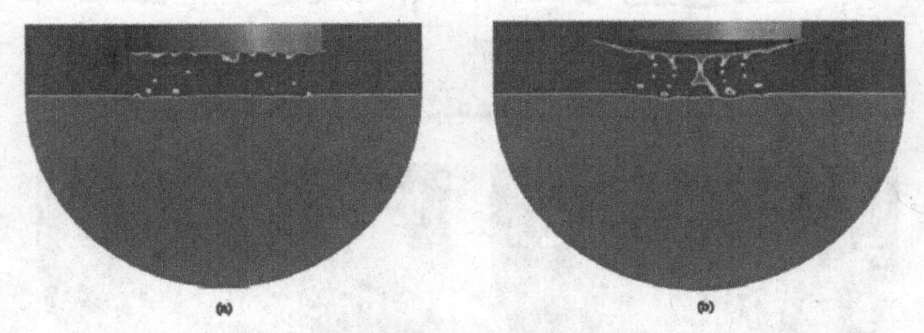

Fig.2 Simulated liquid droplets for different electrode tip shape.

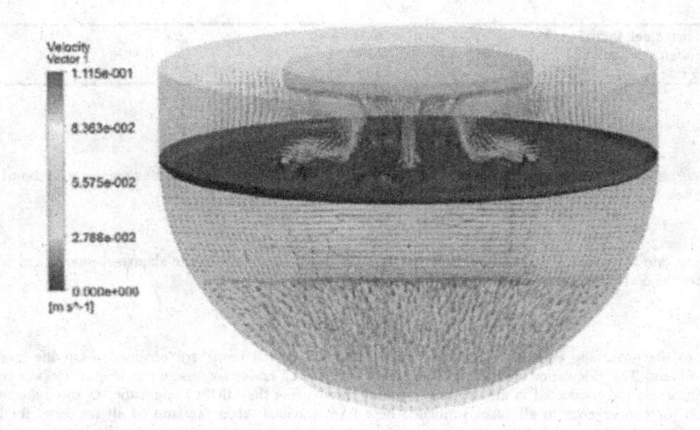

Fig.3 Simulated flow field in slag pool and liquid metal pool of ESR system

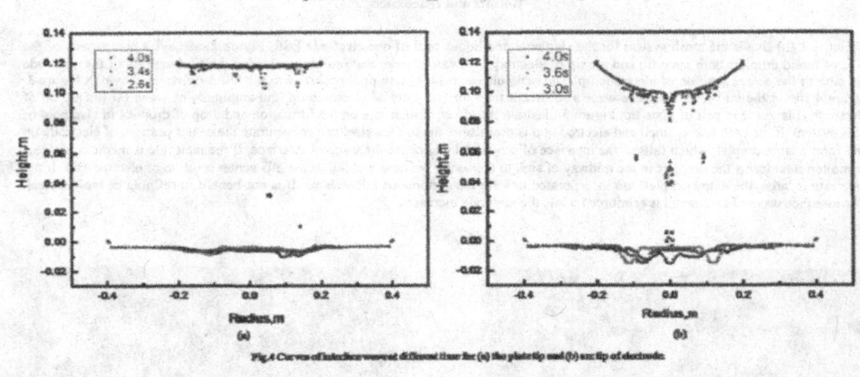

Fig.4 Curves of interface wave at different time for (a) the plate tip and (b) arc tip of electrode

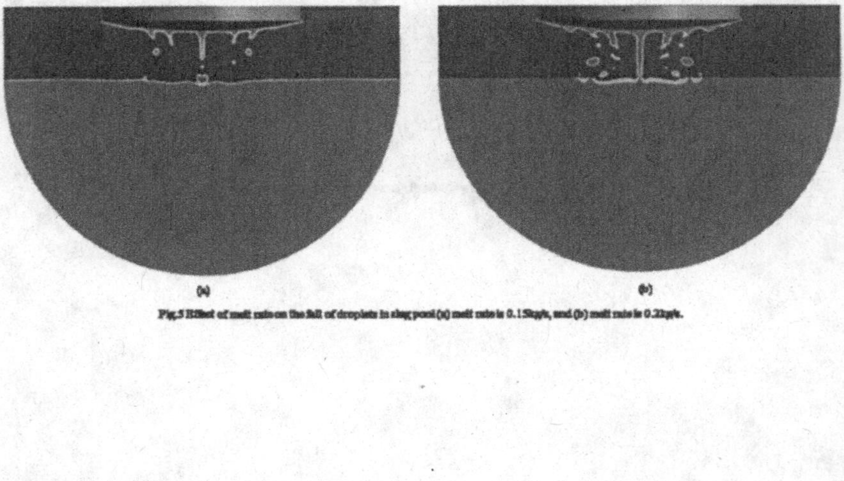

(a) (b)

Fig.5 Effect of melt rate on the fall of droplets in slag pool (a) melt rate is 0.15kg/s, and (b) melt rate is 0.22kg/s.

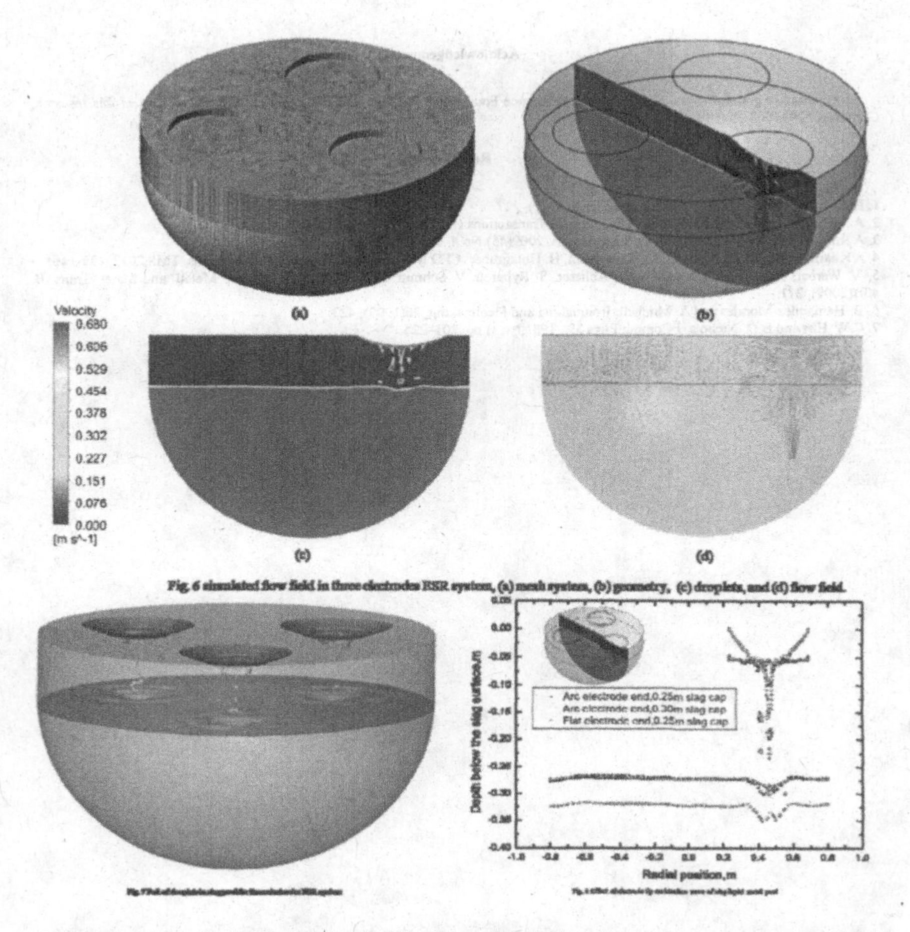

Fig. 6 simulated flow field in three electrodes ESR system, (a) mesh system, (b) geometry, (c) droplets, and (d) flow field.

Formation and fall of droplets for three electrode ESR system are shown in Figure 6, it can observed the complex flow field is produced in the slag pool. The wave of interface of slag/metal is tanglesome, since the droplets of fall comes from three electrode tips in disorder as shown in Figure 7. Figure 8 exhibits the effect of electrode tip shape on the wave of interface of slag/metal in three electrode ESR system. We can find the amplitude of wave is large in the case of protrusion tip and small for plate tip. By comparing with one electrode ESR system, the three electrodes ESR system has the small amplitude of wave and uniform flow field in slag pool.

Conclusions

(1) The molten steel in electrode tip sinks to the center position of electrode tip and the recirculation flow in slag pool occurs with the fall droplets. The amplitude of interface wave of slag/metal for the protrusion tip of electrode is larger than that of plate tip.
(2) If the melt rate is medium, the part of molten steel forms the droplets in the midway of sink to the center position and fall to the off- center position of electrode tip. If the melt rate is large, the many droplets are incorporated to form the continuous fall streams.
(3) The three electrodes ESR system has the small amplitude of wave and uniform flow field in slag pool.

Acknowledgements

Authors are grateful to the National Natural Science Foundation of China and Baosteel Co., Ltd. for support of this research, Grant No. 50934008.

References

1. H. Dilawari and J. Szekely: Metall. Trans. B, 8(1977), 227.
2. A Jardy, D. Ablitzer and J.F.Wadier: Metallurgical Transactions B, 22B, 1991, 111-120.
3. A Ruckert and H.Pfeifer: Magnetohydrodynamics, 2009(45),No.4, 527-533.
4. A Kharicha, M. Wu, A. Ludwig, M. Ramprecht, H. Holzgruber: CFD Modeling and Simulation in Materials, TMS,2012, 139-146.
5. V. Werber, A. Jardy, B. Dussoubs, D. Ablitzer, S. Ryberon, V. Schmitt, S. Hans, and H. Poisson: Metall. and Mater. Trans. B, 40B(2009), 271.
6. B. Hernandez-Morales and A. Mitchell: Ironmaking and Steelmaking, 26(1999), 423.
7. C.W. Hirt and B.D. Nichols, J Comput Phys 39 , 1981,No.1, pp. 201–225.

Materials Processing Fundamentals
Edited by: Lifeng Zhang, Antoine Allanore, Cong Wang, James A. Yurko, and Justin Crapps
TMS (The Minerals, Metals & Materials Society), 2013

STUDY ON FLOW-REACTION DESULFURIZATION OF RH BY PHYSICAL EXPERIMENT

Hongbo Yang[1,2]; Jingshe Li[1,2]; Zengfu Gao[1,2]; Fangfang Song[1,2]; Wanliang Yang[1,2]

[1] School of Metallurgical and Ecological Engineering, University of Science and Technology Beijing, Beijing, 100083, China.

[2] State Key Lab of Advanced Metallurgy, University of Science and Technology Beijing, Beijing, 100083, China.

Keywords: flow-reaction desulfurization, physical simulation experiment, similarity principle.

Abstract

Powder injection desulfurization (PID) in RH was always applied to produce ultra-low sulfur steel, which was used for aerospace industry, marine and petroleum industry. The process of PID was divided into two processes: one was flow-reaction desulfurization and the other one was interface-reaction desulfurization, especially the flow-reaction desulfurization had a huge impact on desulfurization efficiency of RH. In this paper, based on similarity principle, including geometric similarity, dynamic similarity, vacuum similarity and slag-metal interface similarity, a new physical simulation experiment in which water, nitrogen and one special kind of oil were chosen as the substitutes of liquid steel, argon and desulfurizer was carried out to study on the influence rules of vacuum chamber pressure, driving gas flow and flow-reaction desulfurization efficiency, which would have an important guiding significance to the practical production.

Introduction

With the rapid development of science and technology and the increasing steel quality requirements, ultra-low sulfur steel was attracted more and more attentions in marine industry, aviation industry and petroleum transportation [1-7]. For example, the content of sulfur in pipeline steel must be less than 0.003% [7-8]. Nowadays the most common desulfurization equipment during the refining process is LF which has good desulfurization thermodynamics and kinetics conditions [9-11]. While LF could not always keep the content of sulfur meet the requirements of customers, especially for the steel working in the high pressure and corrosive environment. So further desulfurizing by RH was proposed to produce ultra-low sulfur steel. However, desulfurizing process was proceeding in a high temperature and closed environment, which resulted in that it was hardly to observe and analyze the process of desulfurizing.

Physical simulation experiment, especially water simulation experiment, was widely applied to metallurgy industry for studying on molten steel flow, bubble motion behavior and the interface reaction between molten steel and the second phase, which was not only direct for the researchers to observe and analyze the working process of RH but also low cost and short period [12]. In this paper, a new physical simulation experiment, based on the similarity principle, was carried out to study on the desulfurizing process, in which water , nitrogen and one special kind of oil were chosen as the substitutes of liquid steel, argon and desulfurizer.

RH Desulfurizing Principle

When RH begins to work, up leg and down leg should be put into ladle firstly. Then the pressure of vacuum chamber is reduced to 30~50 Pa by vacuumizing, and the molten steel will be pulled into vacuum chamber by the differential pressure between vacuum chamber and outside. After that, the driving gas (argon gas) is blown into molten steel through some tiny holes at one-third of the up leg. Because of the high temperature of molten steel, the driving gas is expanded quickly and makes the molten steel in the up leg move upward into vacuum chamber with a fountain shape and outflow from the down leg under the action of gravity [13], and a circle flow shown in figure 1 appears. The powdery desulfurizer is injected into molten steel through the top lance, flows along with the circle flow and reacts with molten steel quickly to remove the sulfur in molten steel.

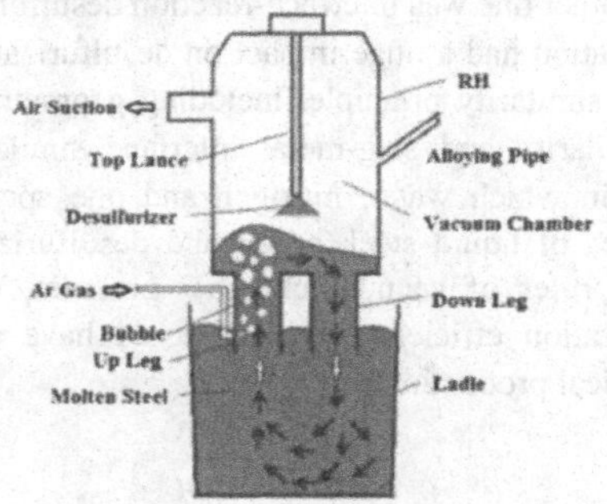

Figure 1. RH working principle schematic picture

The process of desulfurization includes flow-reaction desulfurization and interface-reaction desulfurization [14]. The flow-reaction desulfurization is desulfurization reaction between molten steel and desulfurizer when the desulfurizer is in the flow of molten steel, while the interface-reaction desulfurization is desulfurization reaction between the molten steel and slag of ladle when the desulfurizer floats up into the slag. Because the desulfurizer reacts directly with molten steel in the flow-reaction desulfurization, the flow-reaction desulfurization has a huge impact on desulfurization efficiency of RH. It is very significant and valuable to study on flow-reaction desulfurization.

Physical Simulation Experiment

In this paper, based on similarity principle, including geometric similarity, dynamic similarity, vacuum similarity and slag-metal interface similarity, a new physical simulation experiment that water, nitrogen gas and one special kind of oil were chosen as the substitutes of liquid steel, argon gas and desulfurizer respectively was carried out to study on the flow-reaction desulfurization of RH.

Geometrical Similarity

Geometrical similarity is to ensure that the model and prototype are similar in geometric space, which is the most basic condition of water simulation experiment. It is known that the larger the geometrical similarity ratio (λ) is, the more accurate the simulation results are. Because of the limitation of laboratory space, the biggest geometrical similarity ratio between model and prototype is 1/2.8125 for the experiment. The geometrical parameters of model and prototype were shown in table I.

Table I. The geometry parameters of model and prototype

Items	Prototype	Model
Inner diameter of Ladle/mm	2700	960
Height of ladle/mm	4000	1422.22
Inner diameter of vacuum chamber/mm	1680	597.33
Inner diameter of snorkel/mm	450	160
Outer diameter of snorkel/mm	1100	391.11
Length of snorkel/mm	1700	604.44
Inner diameter of nozzles/mm	6	2.13

Kinematic Similarity

Reynolds number (Re) and Froude number (Fr) are the key dimensionless numbers indicating the movement characteristics of fluid, so the requirements of kinematic similarity would be met if Re and Fr between model and prototype are equal.

Reynolds Number: Only when the geometrical similarity ratio (λ) is 1/1 can the Reynolds numbers and Froude numbers between model and prototype be the same simultaneously. However, the fluid flow is almost the same when the flow belongs to the self-modeling region in which the Reynolds number is more than 10^4 [15].

$$Re(model) = \frac{vD}{\gamma} = \frac{0.37\times0.45}{0.87\times10^{-6}} = 1.91\times10^5 \quad (1)$$

$$Re(model) = \frac{vD}{\gamma} = \frac{0.21\times0.16}{0.84\times10^{-6}} = 4.00\times10^4 \quad (2)$$

v: velocity of down leg (m/s); D: inner diameter of down leg (m); γ: coefficient of dynamic viscosity (m^2/s), and $\gamma_{(water,25\ ^\circ C)}=0.84\times10^{-6}$ (m^2/s) [16], $\gamma_{(molten\ steel,\ 1600^\circ C)}=0.87\times10^{-6}$ (m^2/s) [17]

From the results of formula (1) and (2), it was clear that the fluid flow of model and prototype were both in the self-modeling region, and they were similar.

Froude Number: Because of the bubble flow in the molten steel of RH, the correction Froude number (Fr) was chosen as dimensionless equation shown by formula (3).[18]

$$Fr = \frac{\rho_g v^2}{gL(\rho_l - \rho_g)} \quad (3)$$

Let the Froude numbers of model and prototype are equal,

$$\text{Fr(model)} = \text{Fr(prototype)}$$

$$\frac{\rho_g(model)v(model)^2}{gL(model)\left(\rho_l(model)-\rho_g(model)\right)} = \frac{\rho_g(prototype)v(prototype)^2}{gL(prototype)\left(\rho_l(prototype)-\rho_g(prototype)\right)}$$

$$\frac{v(\text{model})^2}{v(prototype)^2} = \frac{\rho_g(prototype)\text{L(model)}\left(\rho_l(\text{model})-\rho_g(model)\right)}{\rho_g(model)\text{L}(prototype)\left(\rho_l(prototype)-\rho_g(prototype)\right)}$$

And, $\frac{d(model)}{d(prototype)} = \frac{L(model)}{L(prototype)} = \lambda$

So, the relationship between model gas flow and prototype gas flow could be got by the formula (4),

$$\frac{Q(model)}{Q(prototype)} = \frac{n\times\frac{1}{4}\times d(model)^2\times v(model)}{n\times\frac{1}{4}\times d(prototype)^2\times v(prototype)}$$

$$= \sqrt{\lambda^5 \frac{\rho_g(prototype)\left(\rho_l(\text{model})-\rho_g(model)\right)}{\rho_g(model)\left(\rho_l(prototype)-\rho_g(prototype)\right)}} \quad (4)$$

The interpretations and values of the characters in the formula above were shown in table Ⅱ.

Table Ⅱ. The parameters of model and prototype

Items	Interpretation	Parameters of model	Parameters of prototype
ρ_g (kg/m^3)	Density of gas	1.145	1.782
ρ_l (kg/m^3)	Density of liquid	1000	7040
v(m/s)	Velocity of down leg	0.37	0.21
g (m^2/s)	gravity acceleration	9.8	9.8
T_g (K)	Temperature of gas	298	298
T_l (K)	Temperature of liquid	298	1873

Vacuum Similarity

Vacuum similarity mainly ensured that the effect of vacuum to molten steel in prototype and vacuum to water in model were similar.

The pressure in vacuum chamber (P) could be calculated by the formula (5)

$$\text{P} = \text{P}^{\text{o}} - \rho_l\text{gh} \quad (5)$$

P^{o}: normal atmosphere pressure (Pa), h: the liquid level between vacuum chamber and ladle (m).

So

$$\frac{\text{P}^{\text{o}}-P(model)}{\text{P}^{\text{o}}-P(prototype)} = \frac{\rho_l(model)\text{gh}(model)}{\rho_l(prototype)\text{gh}(prototype)} = \frac{\rho_l(model)\text{h}(model)}{\rho_l(prototype)\text{h}(prototype)} =$$

$$\lambda\frac{\rho_l(model)}{\rho_l(prototype)} = \frac{1}{2.8125}\times\frac{1000}{7040} = \frac{1}{19.8}$$

And, the pressure in vacuum chamber of model could be calculated by the following formula (6),

$$P(model) = P^o - \frac{1}{19.8} \times (P^o - P(prototype)) \quad (6)$$

Slag-Steel Interface Similarity

To insure the similarity of slag-steel interface between model and prototype, the Weber numbers (We) of model and prototype must be same [19], and the density of the oil chosen to instead of desulfurizer in water simulation experiment could be calculated by the formula (7).

$$We(model) = We(prototype)$$

$$\frac{v_w{}^2 \rho_w}{g\sigma_{w-o}(\rho_w - \rho_o)^{1/2}} = \frac{v_{st}{}^2 \rho_{st}}{g\sigma_{st-sl}(\rho_{st} - \rho_{sl})^{1/2}}$$

So,

$$\rho_o = \rho_w - \frac{\rho_{Ar}^2 \rho_w^4 \sigma_{st-sl}^2 (\rho_{st} - \rho_{sl}) \lambda^2}{\rho_{N_2}^2 \rho_{st}^4 \sigma_{w-o}^2} \quad (7)$$

In the formula above, w, o, st, and sl are the abbreviation of water, oil, steel and slag receptively and σ means interfacial tension.

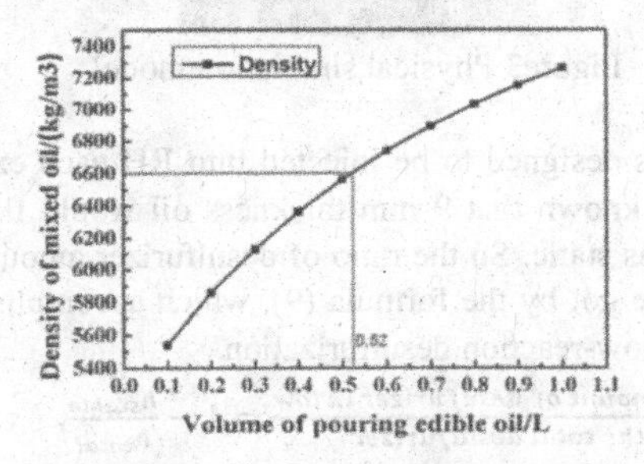

Figure 2 Relation between the volume of edible oil and mixed oil density

Through the calculation by the formula (7), the needed oil density is 661 kg·m^{-3}. However, it could not be found in the market. So the edible oil and clean oil, the densities of which are 901 kg·m^{-3}k and 500 kg·m^{-3} respectively, were chosen to be mixed to get the need oil. First, 1L clean oil was poured into a cup with volume scale, then 0.1L edible oil was poured into the cup at a time. The relation between edible oil volume and mixed oil density was shown in figure 2. It is shown that the density of mixed oil with 1L clean oil and 0.52 L edible oil was 661 kg/m^3, which met the requirement of physical simulation experiment.

Experiment method

In this paper, a physical simulation model of RH and ladle produced with organic glass shown in figure 3 was used to study on flow-reaction desulfurization process. The mixed oil was injected into vacuum chamber and flowed into ladle along with the circle flow of RH. Because oil has a lower density than water, it floated on the water in the ladle. A HTDV camera was used to record the thickness variation of the oil on the water when RH worked shown in figure 4. When the thickness of the oil did not change, it corresponded to that all the desulfurizer moved into the slag of the ladle and the flow-reaction desulfurization was ended in the real production process. The time from the moment that the oil was inject into RH to when the thickness of oil in the ladle was no longer changed was recorded, which indicated how long the flow-reaction desulfurization lasted, and the time was defined as the flow-reaction desulfurization time in this paper.

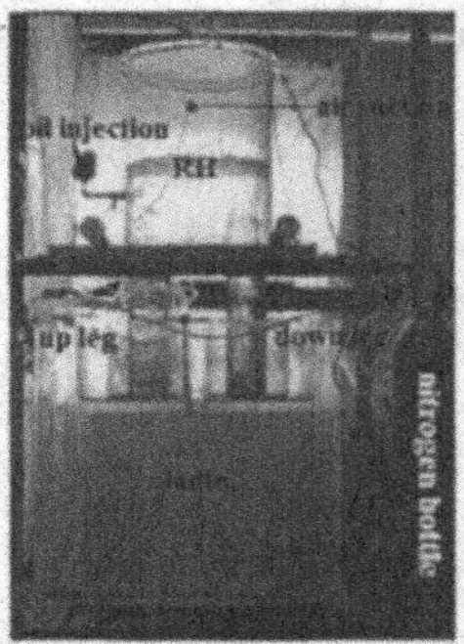

Figure3 Physical simulation model

3.5L mixed oil was designed to be injected into RH each experiment. Through the experiment, as was known that 9 mm thickness oil would float on the water of ladle when the water was static. So the ratio of desulfurizer amount in flow to all the desulfurizer (k) could be got by the formula (9), which accomplished the purpose of qualitative analysis on flow-reaction desulfurization.

$$k = \frac{\textit{the amount of desufurizer in low}}{\textit{the total desulfurizer}} = 1 - \frac{h_{stable}}{h_{total}} \tag{9}$$

h_{stable}—the thickness of oil layer when it was stable, $h_{total} = 9$ mm.

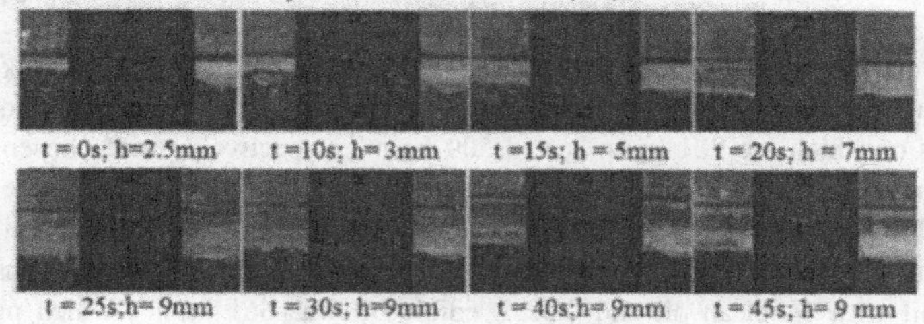

Figure 4. The recording process of oil thickness

t: time; h: thickness of oil. And 2.5mm thick oil was stayed on the water in the previous experiment.

Experiment Results and Analysis

The percentage of desulfurizer in flow and the flow-reaction desulfurization time are two important parameters to demonstrate the desulfurization efficiency of RH. Higher percentage of desulfurizer and longer flow-reaction desulfurization time signify the better desulfurization.

Considering pressure in RH plays an important role in the desuflurization process, the pressure of vacuum chamber in Experiment was successively increased from 92000 Pa to 96000 Pa, and the experiment results were shown in figure 5 a), which indicated that the desulfurizer in flow became less and the flow-reaction desulfurization time was shortened with the pressure increasing. The circle flow would become slow when the pressure in vacuum chamber increased, which led to more desulfurizer moved upward into slag by the buoyancy and extended the time desulfurizer stayed in molten steel. The best pressure of experiment is 96000 Pa by which the RH had the longest flow-reaction desulfurization time and larger percentage of desulfurizer in flow.

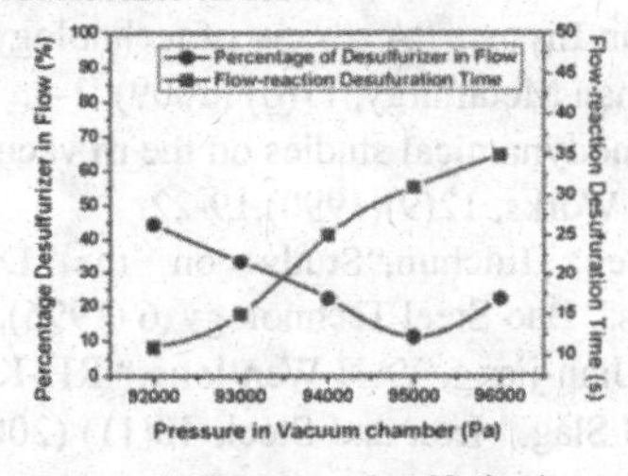

a) Pressure influence on desulfurization

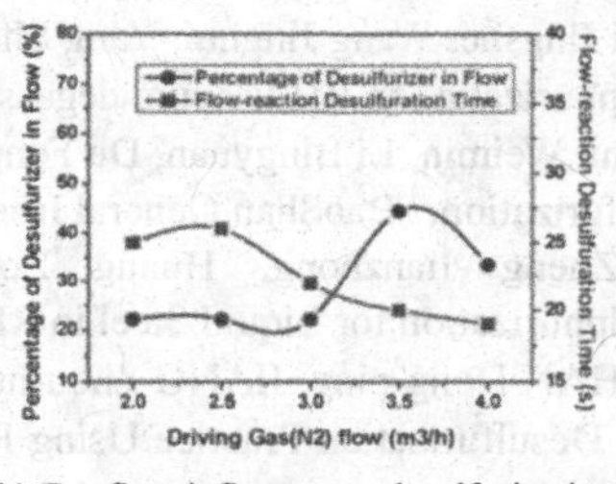

b) Gas flow influence on desulfurization

Figure 5 Pressure and Driving gas flow influences on desulfurizaiton

The diving gas flow is another significant parameter which has great influence on the desulfurization efficiency. The driving gas flow was changed from 2.0 m^3/h to 4.0 m^3/h increasing 0.5 m3/h every time. The experiment results of different diving gas flow was displayed in figure 5 b), which showed that the percentage of desulfurizer in flow had the increasing trend and the flow-reaction desulfurization time was decreasing with increased the driving gas flow. Increasing driving gas flow led to raising the circle flow speed and made more molten steel be flowing, which made more desulfurizer stay in the molten steel in circle flow and shortened the flow-reaction desulfurization time. Considering the both parameters of percentage of desulfurizer in flow and the flow-reaction desulfurization time, 3.0 m^3/h was chosen as the best driving gas flow, which had the larger percentage of desulfurizer in flow and longer flow-reaction desulfurization time.

Conclusions

1) A reasonable physical simulation experiment model to study on flow-reaction desulfurization of RH was built in this paper, which based on similarity principle, including geometric similarity, dynamic similarity, vacuum similarity and slag-metal interface similarity.

2) One kind of special oil mixed with edible oil and clean oil was successfully used to simulate desulfurizer of RH.

3) Percentage of desulfurizer in flow and flow-reaction desulfurization time were defined and quantified through the physical experiment in this paper.

4) When increasing the pressure in vacuum chamber or decreasing driving gas

flow, the percentage of desulfurize in flow decreased and flow-reaction desulfurization time became shorter.

References

[1]Wei Chiho, Zhu Shoujun, Yu Nengwen, "Kinetics of desulphurization by powder injection and blowing in RH refining of molten steel," Acta Metallurgy Sinica,34 (5)(1998),498-505.

[2] Wen Lijuan, Wei Jihe, Jiang Xingyuan, Jiang Qingyuan, Yu Nengwen, "Desulphurization by powder injection in RH refining process of molten steel," Shanghai Metals, 27(4)(2005),54-57.

[3] Li Jingshe, Wang Jinghui, Yang Shufeng, Sun Liyuan, "A review of technology of desulphurization in RH vacuum degassing," Henan Metallurgy, 17(6) (2009), 1-4.

[4] Zhu Weimin, Li Bingyuan, Du Feng, "Thermodynamical studies on the rh vacuum desulfurization, "BaoShan General Iron & Steel Works, 12(9)(1990),19-22.

[5] Zheng Jianzhong, Huang Zongze, Fei Huichun,"Study on the Deep Desulphurization for Liquid Steel in RH Process," Bao Steel Technology, 6 (1996),33.

[6] ZHAN Dong-ping, JIANG Zhou-hua, LUO Jian-jiang, YAN Wen-long, "RH-KTB Deep Desulfurization Practice Using Premelted Slag," Iron and Steel, 40(11) (2005), 27.

[7] Liu Liangtian, "Deep desulphurization by RH vacuum process," Misco Technology, 1 (1990),16-20.

[8]Yang Shengzhou, "Discuss on Ladle Desulfurization Process," GanSu Metallury, 33(5)(2011),7-9.

[9]Guo Hanjie , Physical and chemical conditions of metallurgical process(Beijing:Metallurgical Industry Press,2008),283-286,141-144.

[10]Huang Xiku, Iron and steel metallurgy principle (Beijing:Metallurgical Industry Press ,2005),380-385.

[11] Geng Dianqiao, "Mathematical and Physical Simulation on RH Vacuum Refining Process"(PhD.thesis,Northeastern University,2009).

[12] Feng Juhe,Ai Liqun,and Liu Jianhua, Technology of Molten Iron Pretreatment and Liquid Steel Refining outside the furnace(Beijing:Metallurgical Industry Press ,2008),115-118.

[13]Xiao Xingguo, Xie Yunguo, Metallurgical reaction engineering foundation (Beijing,:Metallurgical Industry Press , 1997), 358-359.

[14] H.Zhou,K.F.Cen,J.R.Fan, "Two-phase flow measurements of a gas-solid jet downstream of fuel rich/lean burner," Energy and Fuels, 19 (2005), 65-74.

[15] Yu Ping, Engineering fluid mechanics (Beijing,Science Press,2008),11.

[16]Chen JiaXiang, Handbook of chart data of steelmaking(Beijing, Metallurgical Industry Press ,1984),397.

[17] Mazmamdar D,Guthire R I L, "The physical and mathematical modeling of gas stirred ladle system,"Application Math Modelling,10(2) (1986),25-32.

[18] Li Xiaohong, Han Lihui, Jia Hongguang, "Water model experiment study on 70t ladle with bottom argon blowing in critical slag entrapment condition, " Journal of Qinghai Normal University(Natural Science), 3(2011),37-41.

Materials Processing Fundamentals
Edited by: Lifeng Zhang, Antoine Allanore, Cong Wang, James A. Yurko, and Justin Crapps
TMS (The Minerals, Metals & Materials Society), 2013

MODELING OF TRANSIENT FLUID FLOW, SOLIDIFICATION PROCESSES AND BUBBLE TRANSPORT IN CONTINUOUS CASTING MOLD

Zhong-qiu Liu, Bao-kuan Li, Mao-fa Jiang

School of Materials and Metallurgy, Northeastern University, Shenyang, Liaoning, 110819, China

Keywords: continuous casting mold, large eddy simulation, transient flow, bubble transport

Abstract

A coupled three-dimensional finite-volume computational model has been developed to simulate fluid flow, solidification processes and bubble transport in a slab continuous casting mold. Transient flow of molten steel in the continuous casting mold is computed using large eddy simulation. A general enthalpy method is presented for the analysis of solidification processes. The transport of bubbles in the liquid pool of the solidified shell is considered according to the dispersed phase model. The model calculations have been compared with the observations of physical water model and ultrasonic testing, and have successfully reproduced many known phenomena and other new predictions. The influence of bubble size and casting speed on the bubble distribution, removal rate and entrapment distribution is considered with this mathematical model. The results show that the ratio of bubble floating to the top surface decreases with casting speed rising, and increases with bubble diameter decreasing. Bubbles with smaller diameter are possible to be entrapped by the solidified shell near the center region.

Introduction

In the current continuous casting processes, argon gas is injected into the submerged entry nozzle (SEN) to prevent clogging with solid inclusions [1]. The argon gas disintegrates into small bubbles of varying diameters as it issues out of the SEN. Large bubbles rise toward the meniscus due to buoyancy and subsequently removed from the mold, while smaller bubbles are carried deep into the mold. The small bubbles can be entrapped from the flowing liquid into the solidifying steel shell to form defects in the final product, such as "pencil pipe" blister, slivers and other costly defects. Therefore, it is important to study the motion of argon bubbles in the mold as well as how to control it. The complex transport phenomenon in continuous casting mold has been shown in **Figure 1**[2].

Heat transfer in the mold is one of the most important phenomena that take place during the continuous casting of steel. Several studies have been performed to analyze the influence of casting parameters on the heat transfer between the strand and the mold wall [3-5]. A standard approach for the numerical modeling of metallurgical solidification processes are so-called "fixed" grid methods [6-7]. The essential feature of these methods is that the evolution of latent heat is accounted for by the definition of enthalpy. The major advantage of this method is that they permit modeling of solidification phase change through simple modifications of existing heat transfer methods.

The asymmetrically distribution of defects in the slab has been found in recent flaw detection of slabs [8]. However, most of the mathematical simulations reported in the literature provide time-averaged values of the turbulent flows; do not provide information about the transient flows and velocity fluctuations that are characteristic of turbulent flow phenomena. The large eddy simulation (LES) was successfully applied to obtain the asymmetrically flow of molten steel in the single phase in the mold [9-10].

The purpose of this study is to develop a mathematical model for the prediction of the position of bubbles trapped in the strand, originated from inters dendritic liquid film, through the fully coupled analysis of solidification-fluid flow-bubble transport in continuous casting process. An enthalpy-porosity formulation is used to simulate the solidification of steel. The turbulence of the molten steel phase is incorporated through the LES model. The Eulerian and Lagrange approach is used respectively for the description of molten steel and argon gas phase. The entrapment distributions, removal rates, and the trajectory of different sized bubbles will be studied with the mathematical model.

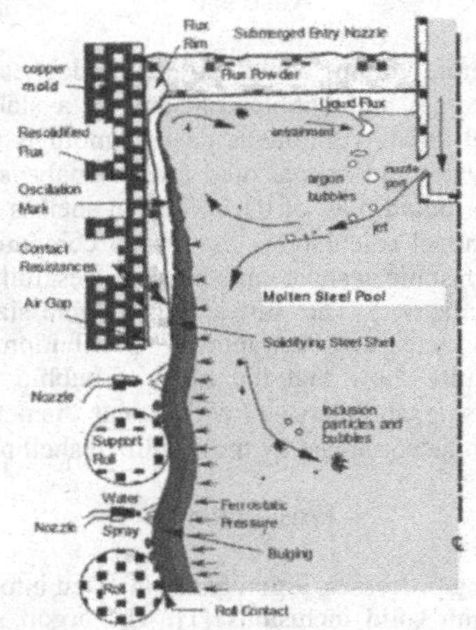

Figure 1. Complex Transport Phenomena in Continuous Casting Mold[1]

Model Formulation

Enthalpy-Porosity Model

Instead of tracking the liquid-solid front explicitly, an enthalpy-porosity formulation [11] is used to simulate the heat transfer (solidification of steel) in a slab continuous casting mold.

The enthalpy (H) of the material is computed as the sum of the sensible enthalpy, h , and the latent heat, ΔH :

$$H = h + \Delta H \tag{1}$$

$$h = h_{ref} + \int_{T_{ref}}^{T} c_p dT \tag{2}$$

where h_{ref} =reference enthalpy T_{ref} =reference temperature, C_p =specific heat at constant pressure.

The latent heat content can now be written in terms of the latent heat of the material, L :

$$\Delta H = \beta L \tag{3}$$

The liquid fraction, β , can be defined as

$$\begin{cases} \beta = 1, & if \;\; T > T_{liquidus} \\ \beta = \dfrac{T - T_{solidus}}{T_{liquidus} - T_{solidus}}, & if \;\; T_{solidus} < T < T_{liquidus} \\ \beta = 0 & , if \;\; T < T_{solidus} \end{cases} \tag{4}$$

The latent heat content can vary between zero (for a solid) and L (for a liquid).
For solidification/melting problems, the energy equation is written as

$$\frac{\partial(\rho H)}{\partial t} + \nabla \cdot (\rho \vec{v} H) = \nabla \cdot (k \nabla T) + S \tag{5}$$

where, ρ = density, $\vec{v}$ = fluid velocity, S = source term.
The momentum sink due to the reduced porosity in the mushy zone takes the following form:

$$S = \frac{(1-\beta)^2}{(\beta^3 + \varepsilon)} A_{much} \left(\vec{v} - \vec{v}_p \right) \tag{6}$$

Where, ε is a small number (0.001) to prevent division by zero, A_{much} is the mushy zone constant, and $\vec{v}_p$ is the solid velocity due to the pulling of solidified material out of the domain (also referred to as the pull velocity).

Large Eddy Simulation Model

Sinks are added to all of the turbulence equations in the mushy and solidified zones to account for the presence of solid matter.

$$S = \frac{(1-\beta)^2}{(\beta^3 + \varepsilon)} A_{much} \mu_t \tag{7}$$

where, μ_t represents the turbulence viscosity being solved using the large eddy simulation.

$$\mu_t = \rho L_s^2 |\bar{S}| \tag{8}$$

$$L_s = \min(\kappa d, C_s V^{1/3}) \tag{9}$$

Here, L_s is mixing length, $\bar{S}$ is strain rate tensor, κ is a constant (=0.42), d is the distance from nearest wall, C_s is Smagorinsky constant (=0.1), V is volume of cell.

Particle Transport Model

Consider a discrete particle traveling in a continuous fluid medium. The forces acting on the particle that affect the particle acceleration are due to the difference in velocity between the particle and fluid, as well as to the displacement of the fluid by the particle. The motion of inclusion particles can be simulated by integrating the following transport equation for each particle, which considers contributions from five different forces:

$$m_P \frac{dU_P}{dt} = F_D + F_L + F_{VM} + F_P + F_B + F_g \tag{10}$$

The terms on the right hand side of Eqs. (4) are gravitational force, drag force, buoyancy force, virtual mass force, and pressure gradient force. With further details presented elsewhere [12].

Boundary Conditions and Solution Method

The computational domain includes the entire submerged entry nozzle (SEN) and the complete mold region. In order to get the more real flow information, the continuous casting mold considered strand casters have a vertical part and a bending part, as exhibited in Figure 2. The inlet is the tundish bottom well. At the inlet, the velocity was fixed to have a normal component in the downward direction according to the casting speed. The

geometrical, physical properties and operating conditions used in numerical simulation model are shown in **Table 1**.

The effect of the air gap was neglected as it is confined to small region of the corner in this case and our study was focused on the fluid flow and the positions of bubbles and particles are entrapped on the solidifying shell. A uniform velocity for two phases is prescribed at the inlet opening based on the casting speed, and "No Perturbations" option inlet boundary conditions for the LES model. The thermophysical properties were assumed to be a constant in the solid and liquid phase and allowed to vary linearly with the liquid fraction in the two-phase mushy region.

Table 1 Comparison of process parameters between water model and actual continuous casting mold

Parameters	Water-Air System	Steel-Argon System	Parameters	Water-Air System	Steel-Argon System
Diameter of SEN (mm)	27	80	Length of vertical part(mm)	667	2000
Length of SEN (mm)	300	900	Length of bending part(mm)	1067	3200
Exit down angle of nozzle (°)	15	15	Radius of curvature (mm)	10250	10250
Height of SEN port (mm)	27	80	Fluid flow rate (L/min)	45.44	710
Width of SEN port (mm)	20	60	Gas flow rate, cold (L/min)	5	15.6
Submergence depth of SEN (mm)	100	300	Fluid density(kg/m^3)	1000	7020
Width of mold (mm)	579	1736	Gas density(kg/m^3)	1.29	0.56
Thickness of mold (mm)	76	228	Fluid viscosity (kg·m^{-1}·s^{-1})	0.001	0.0056
Length of real mold (mm)	300	900	Gas viscosity (kg·m^{-1}·s^{-1})	1.7×10^{-5}	7.42×10^{-5}

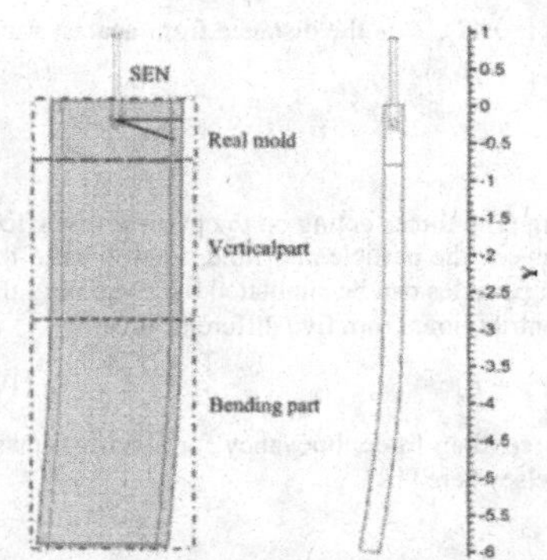

Figure 2. Schematic of the calculation model

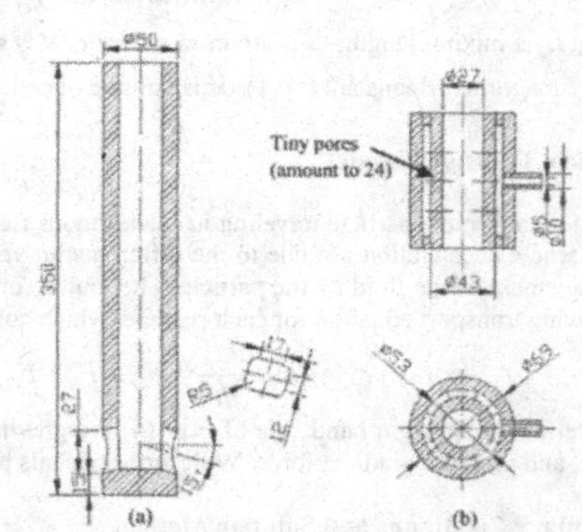

Figure 3. Schematic of submerged entry nozzle (a) and upper nozzle (b)

Model Validation

In order to visualize the flow pattern and validate the LES turbulence model, a one-third-scale water model was established, including tundish, upper nozzle, slide gate, submerged entry nozzle (SEN) and casting mold(considered strand casters have a vertical part and a bending part). The diameter of the tiny horizontal pores used in the upper nozzle of water model is 1.0mm. There are 24 pores around the upper nozzle. The air was injected through tiny horizontal pores into turbulent liquid flowing vertically down the wall, as shown in **Figure 3**.

Due to the fluctuating nature of turbulent flow, the asymmetric velocity field also provides asymmetric tracer dispersion patterns. But in most Reynolds-averaged simulations, symmetry is assumed between the flows in the two halves of the liquid pool. This assumption has been shown to be valid for long-term averages. **Figure 4** shows the fluid flow pattern inside the water model and the calculate mold. This snapshot was obtained by black-colored dye injection against a white background, as shown in **Figure 4(a)**. It can be noticed that the fluid jet flow is not symmetrical about the central plane. The jet flow on the left half descends in the downward direction, leading the fluid discharge through the left half of the mold. On the right half, there is a big colored domain. With this mathematical model, the flow pattern of molten steel inside the mold is given in **Figure 4(b)**. The flow pattern in the experiment and that predicted by the mathematical model are qualitatively very similar. The location of lower recirculation region eye was approximately the same in the two cases, which is around the interface of vertical part and bending part of the mold. So the LES model predictions match well with the water model results for asymmetry flow.

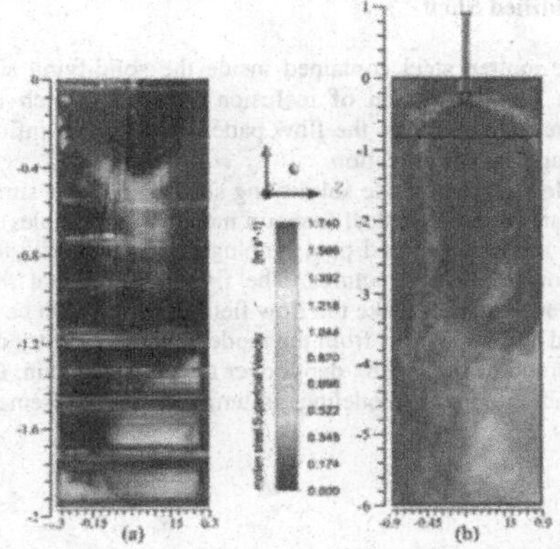

Figure 4. Fluid flow pattern obtain from the water model experiment and molten steel velocity contour obtain from LES model

The long-term research of molten steel flow in the continuous casting mold shows that the bubbles can't be removed by the slag layer if their ascent rate is no faster than casting speed. They would flow together with the molten steel, and finally stranded in slab to form slab defects. The bubble distribution in casting slabs was investigated by the ultrasonic testing (UT) method at the three cross sections of slab, as shown in **Figure 5**. The UT results revealed that bubble defects (many Al_2O_3 adhering to the surface of a bubble) present asymmetric distribution. They were mainly found at the 1/5, 1/4 width of the slab, and sometimes at the 1/2 width.

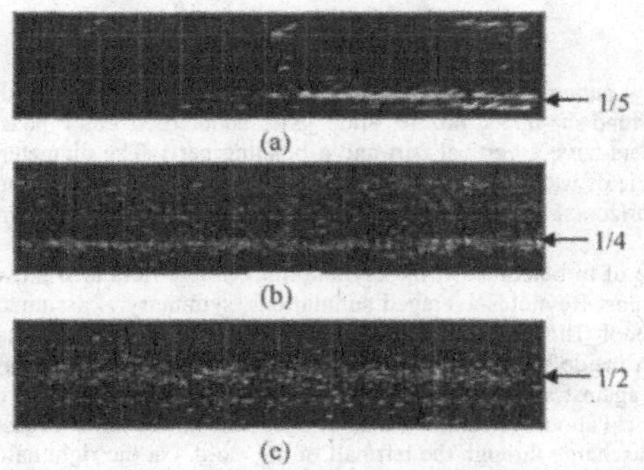

Figure 5. Inclusion (bubbles and solid particles) distribution in three cross sections of different slabs (the same continuous caster, different casting times)

Results and Discussion

Fluid Flow and Growth of Solidified Shell

The velocity distribution of the molten steel contained inside the solidifying shell of a continuous casting machine is very influential on the distribution of inclusion particles, which is important to the internal cleanliness and quality of the steel. In addition, the flow pattern has a great influence on heat transfer to the shell during the critical initial stages of solidification.

Figure 6 shows the predicted flow pattern inside solidifying shell using LES simulation, and shows a classic double-roll recirculating flow pattern. Molten steel (contain many argon bubbles) emerges from the inlet port as a jet, diffuses as it traverses across the liquid pool, impinges on the solidifying shell, and splits into two recirculation zones consisting of complex structures. The figure shows that the liquid pool is consist of multiple vortices. In addition, those vortexes make the flow field in the mold to be more complex.

Figure 7 compares the solidified shell thickness from the model and the industry monitoring value. The model reasonably predicts the overall trend of liquid flux depth over the entire domain. Considering the crude nature of the measuring apparatus and the many modeling assumption, the agreement between the model and experiment is significant.

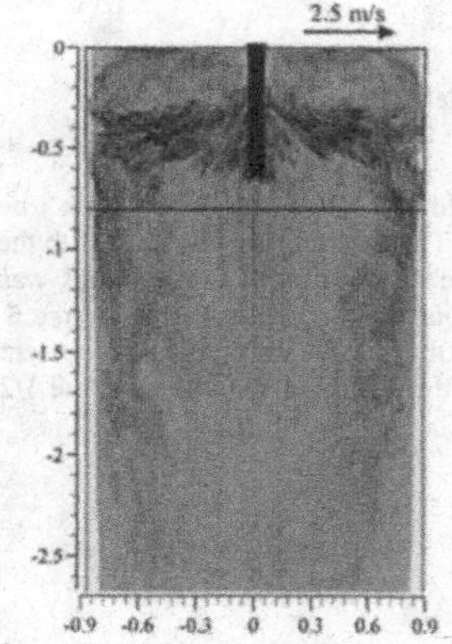

Figure 6. Molten steel velocity vector inside the solidified shell

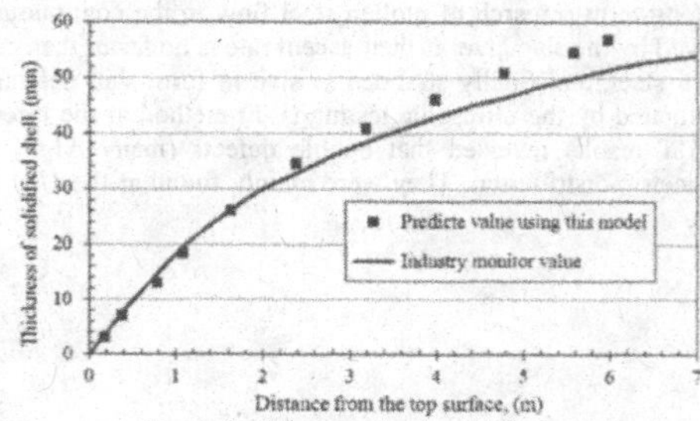

Figure 7. The growth of solidified shell

Bubbles Distribution and Removal Fraction

Figure 8 shows the distribution of bubbles inside the solidified shell with different sizes 0.1mm and 3mm. Larger bubbles float more easily and leave the mold quickly. They tend to release from the top surface near the SEN. But the bubble flotation (3mm) is difficult, because that they would adsorb on the nozzle wall for a long time. Smaller bubbles (0.1mm) penetrate deeper into the mold cavity, increasing their opportunity of entrapment into the solidifying shell.

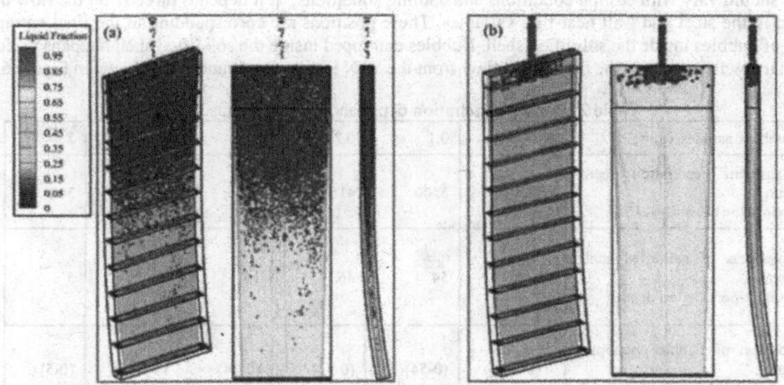

Figure 8. Effects of bubble size on bubble distribution in the mold (a) 0.1mm, (b) 3mm

The transport and capture of five groups of 10,000 bubbles, with five different sizes (0.05, 0.1, 0.2, 0.5, 1, and 3mm), are simulated in the mold. The destination of the bubbles, either to the top surface of the mold or entering the mold bottom and trapped by the solidified shell, were recorded. **Figure 9** shows the predicted removal fraction of bubbles from the liquid pool to the top surface of the mold. The results indicate that larger bubbles were easier to be removed; the smaller bubbles were easier to be entrapped by the solidified shell in the mold.

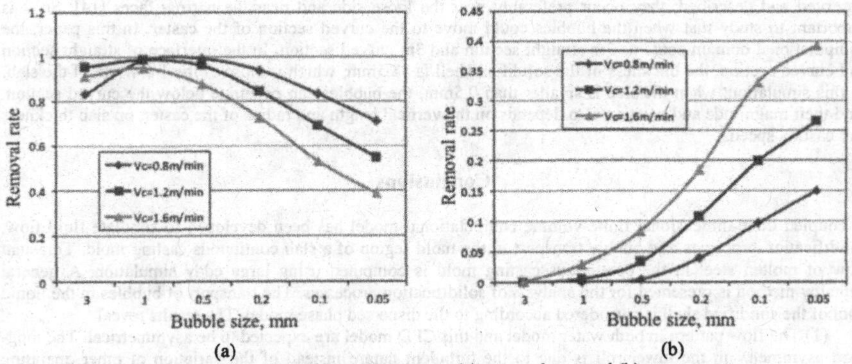

Figure 9. The bubble removal/entrapment rate, (a) removal rate and (b) entrapment rate

Entrapped Positions of Bubbles

The positions of the bubbles entrapped by the solidified shell can be divided in three categories [13]:

1) One part reaches the meniscus and form surface defects,

2) One part freezes with the molten steel in the vertical part of the mold, forming subskin blowhole defects,

3) The remaining bubbles freeze with the molten steel in the curved part of the mold, forming the inclusion band, which is internal hole defects.

The variation in thickness of the solidified shell around the mold perimeter is shown in Figure 8. This thickness is very important because it controls the positions of bubbles and particles are entrapped. The exact thickness should vary with casting conditions and cooling conditions, as it depends directly on the flow pattern developed in the steel and wall heat-flux variables. These positions are corresponding to the final entrapment positions of bubbles inside the solidified shell. Bubbles entrapped inside the solidified shell happens preferably near the narrow faces, where the downward flow from the SEN is most pronounced, as shown in Figure 6.

Table 2 Bubble penetration depth and entrapped position

Bubble diameter, (mm)	0.05	0.1	0.2	0.5	1	3
Maximum penetration depth, (mm) (below the top surface)	>6000	5500	4150	2350	1400	300
Thickness of solidified shell, (mm) (at max-penetration depth)	>56.5	54	46	34	23	5
Position of bubble entrapped, (mm) (below solidified shell surface)	> (0-56.5)	(0-54)	(0-46)	(0-34)	(0-23)	(0-5)

Table 2 shows the entrapped positions of bubbles inside the solidified shell. The entrapped positions of bubbles are corresponding to the bubble sizes. The penetration depth of large bubbles is smaller, and where the thickness of solidified shell is thinner, so the entrapped positions of large bubbles approach the surface of the slab. The smaller bubbles can be found closer to the slab center. The quality near the slab surface is worse than the slab center plane, and the inner quality of the slab is much better.

Entrapment under the solidification front in machines with a curved section leads to bubble accumulation bands by a mechanism similar to the formation of inclusion bands. Bubble accumulation bands have been observed and described; they occur preferably near the loose side and near the narrow faces [14]. So it is important to study that when the bubbles could move to the curved section of the caster. In this paper, the computational domain contains 3m straight section and 3m curved section. In the interface of straight section and curved section, the thickness of the solidified shell is 38.5mm, which is close to the 1/5 width of the slab. In this simulation, when the size is smaller than 0.5mm, the bubbles can penetrate below the curved section. And their magnitude and position also depends on the vertical length and radius of the caster, on slab thickness and casting speed.

Conclusions

A coupled three-dimensional finite-volume computational model has been developed to simulate fluid flow, solidification processes and bubble transport in the mold region of a slab continuous casting mold. Transient flow of molten steel in the continuous casting mold is computed using large eddy simulation. A general enthalpy method is presented for the analysis of solidification processes. The transport of bubbles in the liquid pool of the solidified shell is considered according to the dispersed phase model. The results reveal:

(1) The flow pattern in both water model and this CFD model are expected to be asymmetrical. The long-term asymmetry in the lower roll is due to the turbulent nature instead of the variation of other operating parameters.

(2) The vector of the molten steel contained inside the solidifying shell and the growth of the solidified shell are studied. And compared with the solidified shell thickness from the industry monitoring value, the model reasonably predicts the overall trend of liquid flux depth over the entire domain.

(3)The entrapped positions of bubbles are corresponding to the bubble sizes. The penetration depth of large bubbles is smaller, and where the thickness of solidified shell is thinner, so the entrapped positions of large bubbles approach the surface of the slab. The smaller bubbles can be found closer to the slab center. And their magnitude and position also depends on the vertical length and radius of the caster, on slab thickness and casting speed.

Acknowledgements

Authors are grateful to the National Natural Science Foundation of China for support of this research, Grant No. 50934008.

References

1. L.F. Zhang and B.G. Thomas, "State of the Art in Evaluation and Control of Steel Cleanliness", *ISIJ International*, 43(3), (2003), 271-291.
2. Y. Meng and B.G. Thomas, "Heat Transfer and Solidification Model of Continuous Slab Casting mon1d", *Metall. Mater. Trans. B*, 34(5), (2003), 685-705.
3. A.M. Eugene, "Mathematical Heat Transfer Model for Solidification of Continuously Cast Steel Slabs", *Transactions of the Metallurgical Society of Aime*, 239(11), (1967), 1747-1753.
4. B.G. Thomas, "Application of Mathematical Models to the Continuous Slab Casting Mold", 72nd Steelmaking Conference Proceedings, ISS, (1989).
5. B. Lally, L. Biegler and H. Henein, "Finite Difference Heat-Transfer Modeling for Continuous Casting", *Metall. Mater. Trans. B*, 21(8), (1990), 761-770.
6. V.R. Voller, C.R. Swaminathan and B.G. Thomas, "Fixed Grid Technique for Phase Change Problems: A Review", *Int. J. Num. Meth. Eng.*, 30, (1990), 875-898.
7. C.R. Swaminathan and V.R. Voller, "A General Enthalpy Method for Modeling Solidification Processes", *Metall. Mater. Trans. B*, 23(10), (1992), 651-664.
8. B.K. Li, Z.Q. Liu, F.S. Qi, F. Wang and G.D. Xu, "Large Eddy Simulation for Unsteady Turbulent Flow in Thin Slab Continuous Casting Mold", *Acta Metallurgica Sinica*, 48(1), (2012), 23-32.
9. Q. Yuan, B.G. Thomas and S.P. Vanka, "Study of Transient Flow and Particle Transport during Continuous Casting of Steel Slabs, Part 1. Fluid Flow", *Metal. Mater. Trans. B.*, 35 (4), (2004), 685-702.
10. Q. Yuan, B. G. Thomas, and S. P. Vanka, "Study of Transient Flow and Particle Transport during Continuous Casting of Steel Slabs, Part 2. Particle Transport", *Metal. Mater. Trans. B.*, 35(4), (2004), 703-714.
11. FLUENT 6.1 Manual. Fluent Inc., Lebanon, NH, 2003.
12. B.K Li, F. Tsukihashi, "Numerical Estimation of the Effect of the Magnetic Field Application on the Motion of Inclusion Particles in Slab Continuous Casting of Steel", *ISIJ International*, 43(6), (2003), 923-931.
13. M. Javurek, P. Gittler, R. Rossler, B. Kaufmann and H. Preblinger, "Simulation of Nonmetallic Inclusions in a Continuous Casting Strand", *Steel Research Int.*, 76(1), (2005), 64-70.
14. K.H. Tacke, "Overview of Particles and Bubbles in Continuously Cast Steel", *Journal of Iron and Steel Rsearch*, Int., 18, (2011), 211-219.

Materials Processing Fundamentals
Edited by: Lifeng Zhang, Antoine Allanore, Cong Wang, James A. Yurko, and Justin Crapps
TMS (The Minerals, Metals & Materials Society), 2013

Numerical Analysis of Coupled Fluid Flow, Heat Transfer and Solidification in Ultra-thick Slab Continuous Casting Mold

Xin Xie, Dengfu Chen*, Mujun Long, Leilei Zhang, Jialong Shen, Youguang Ma

Laboratory of Metallurgy and Materials, College of Materials Science and Engineering, Chongqing University, Chongqing 400030, China.

Keywords: Continuous casting, ultra-thick slab mold, numerical simulation, fluid field, temperature field, shell thickness.

Abstract

The fluid flow, temperature and solidification in ultra-thick slab continuous casting mold had been studied through comparing with conventional thick slab molds. The numerical simulation results shown the ultra-thick slab mold also had two recirculation regions. But its jet could not diffuse to the wide face owing to the thickness of mold. This leaded larger low velocity zone near the mold wall. And it had great influence to temperature distribution on the mold surface and shell thickness uniformity at the mold bottom section.

Introduction

Fluid flow in continuous casting (CC) slab mold is of great interest because it influences many important phenomena [1], including velocity, temperature, solidification and so on. A lot of researches [2, 3] have studied the conventional or thin slab mold though mathematics simulation, which was easier to visualize and quantify the fluid flow than water models and plant experiments. Thomas et al. [3] compared 4 different methods for studying fluid flow in slab mold, and proved that the popular standard K-ε model was able to simulate the time averaged velocity flied with almost equal accuracy to a large eddy simulation. Yang et al. [2] used the porous media theory to model the blockage of fluid flow by columnar dendrites in the mushy zone, and validated the calculated solid shell thickness and temperature distribution in liquid core with the plant results.

The ultra-thick slab is widely used in shipbuilding, armour, and nuclear facilities. The traditional technology to make ultra-thick slab is die casting, but this method has low productivity and high energy cost. So it's a great improvement to produce the ultra-thick slab though continuous casting due to its high efficiency and low power waste. As the development of technology and the increase of market demand to the thick rolling plates, the ultra-thick slab [4] continuous casting becomes an important invest direction of CC.

Although there are about ten ultra-thick slab casters in the world nowadays, a few references [5] can be found about the flow field in ultra-thick slab mold. Generally, the flow field of ultra-thick slab mold may be different from other slab mold owing to its thickness. And it is well known that the flow pattern of steel in the mold has an obvious effect to the manufacture and the quality of CC slab. So it is significant to research the flow field in the ultra-thick slab mold.

In the present study, a three-dimensional analysis of the fluid flow behavior in 300 mm, 360 mm conventional thick molds and 420 mm ultra-thick slab mold, are performed using the commercial

---

* Dengfu Chen is the professor of College of Materials Science and Engineering, Chongqing University of Chongqing, China. Contact e-mail: chendfu@cqu.edu.cn.

software FLUENT. And based on the corresponding results, the velocity, temperature and solidification shell distributions are investigated. The suppression effects of velocity in mush zone are considered depending on a porosity-enthalpy relationship.

Computational model

(a) Assumptions of model

The following assumptions for the molten steel flow in the mold were made in this model.

(1) The liquid steel in mold was assumed to be incompressible Newtonian fluid with steady flow, and the properties of steel were constants.

(2) The strand curvature, mold power and air gap were ignored. And only quarter of mold region was simulated.

(4) The radiative heat transfer and the heat transfer on the mold surface were neglected.

(5) The latent heat of solid phase transformation was considered to be negligible.

(b) Fluid and turbulence model

The fluid model was based on the solution of the continuity and momentum equations for incompressible viscous flow. The focus of this study was the flow field in different mold thickness. Therefore, the turbulence model was kept as simple as possible to save computing time. The K-ε equations were chosen, and the expressions could be found in other publications [2, 3].

(c) Solidification and heat transfer model

The enthalpy of the steel was computed as the sum of the sensible enthalpy, h, and the latent heat, ΔH:

$$H = h + \Delta H \tag{1}$$

$$h = h_{ref} + \int_{T_{ref}}^{T} C_p dT \tag{2}$$

h_{ref} was the reference enthalpy, T_{ref} was the reference temperature, and C_p was the specific heat at constant pressure.

The solidification profile was calculated based on an enthalpy method where the liquid fraction, f_l, was calculated by:

$$f_l = \begin{cases} 0 & T < T_{solidus} \\ \dfrac{T - T_{solidus}}{T_{liquidus} - T_{solidus}} & T_{solidus} < T < T_{liquidus} \\ 1 & T > T_{liquidus} \end{cases} \tag{3}$$

The latent heat content could be written in terms of the latent heat of the steel, L:

$$\Delta H = f_l L \tag{4}$$

So the energy equation was written as follows:

$$\nabla \cdot (\rho v H) = \nabla \cdot (K_{eff} \nabla T) \tag{5}$$

Where ρ was the density of steel, v was the fluid velocity, and K_{eff} was the thermal conductivity.

The enthalpy-porosity technique, that treats the mushy zone as a porous medium, was used which include the velocity sink resulting from the solidification. This sink term reduced the velocity depending on the solidification fraction at each cell as shown in Eq. (6).

$$S_{mom} = \frac{(1-f_l)^2}{(f_l^3+0.001)} A_{mush}(v - v_p) \tag{6}$$

Where A_{mush} was the constant of mushy zone [6] and v_p was the casting speed.

And Sinks were also added to two of the turbulence equations in the mushy and solidified zones to account for the presence of solid matter, which were shown in Eq. (7) and (8)

$$S_k = \frac{(1-f_l)^2}{(f_l^3+0.001)} A_{mush} k \tag{7}$$

$$S_\varepsilon = \frac{(1-f_l)^2}{(f_l^3+0.001)} A_{mush} \varepsilon \tag{8}$$

The S_k and S_ε are the sink terms added in the K and ε equations, respectively.

(d) Model description

A numerical model based on the commercial software FLUENT was created for different molds. Figure 1 shown the geometry of SEN. Figure 2 gave a part of the computational meshes for the mold and the SEN, and that has a total of 320 000 grid cells in a structured mesh. The inlet velocity was calculated to maintain the desired casting speed at the outlet. The turbulent kinetic energy, and the rate of turbulent energy dissipation at the inlet were estimated using the semi-empirical relations [2].

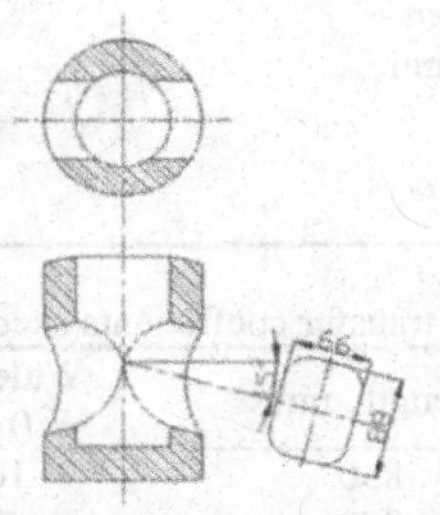

Figure 1. The geometry of SEN

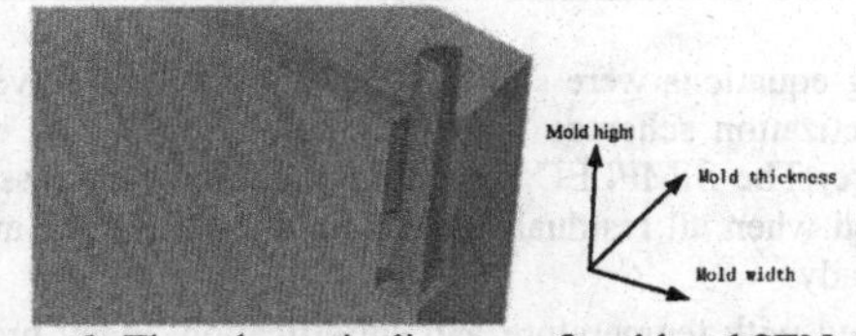

Figure 2. The schematic diagram of mold and SEN grids

Steel properties, the casting conditions and heat transfer conditions employed on the simulations were presented in the Table I, Table II, and Table III, respectively. Generally, heat flux [2, 7] and heat transfer coefficient [8, 9] were two ordinary heat transfer conditions used on the mold wall.

The heat flux could confirm from plant test or experimental equations, however the plant test got the temperature though the mold copper and considered the air gap and flux powder. So adding this heat flux directly on the slab without air gap and powder was inaccurate. What's more, it's obviously from the equation that the heat flux is only influenced by the position of grids; the heat transfer at the same position might be different for different operation condition. So one test or equation might not satisfy three different thickness molds in this research. On the other hand, the heat transfer coefficient considered the influence of the temperature gradient. Thus, heat transfer is good at comparing the simulation result of different conditions.

Table I. The physical parameters of steel

Physical Properties	Values
Density, kg/m^3	7020
Specific Heat, $J\cdot(kg\cdot K)^{-1}$	700
Thermal Coefficient, $W\cdot(m\cdot K)^{-1}$	28
Viscosity, $kg\cdot(m\cdot s)^{-1}$	0.0062
Latent Heat, J/kg	272000
Liquid Temperature, K	1790
Solid Temperature, K	1729

Table II. The Operating parameters of mold

Operating Parameters	Values
Mold section, mm× mm	(300, 360, 420)×2400
Mold Length, mm	850
Computer Length, mm	3000
Inside Diameter of SEN, mm	80
Outside Diameter of SEN, mm	132
Length of SEN, mm	850
Flow Rate, t/min	3.89
Casting Temperature, K	1815

Table III. The heat transfer coefficients used for the simulation

Simulation Regions	Length, mm	Wide Face, $W\cdot(m^2\cdot K)^{-1}$	Narrow Face, $W\cdot(m^2\cdot K)^{-1}$
Mold Region	850	1000	1400
Sub-mold Region I	850	700	800
Sub-mold Region II	1300	400	400

The non-lineal governing equations were solved using a segregated solver through an implicit, first order upwind discretization schemes for turbulence, moment and energy parameter, and PRESTO for the pressure; The SIMPLEC algorithm was used for pressure-velocity coupling. Convergence was reached when all residuals are below 1×10^{-3} and the average temperature on the outlet section was steady.

For the fluid flow coupled with temperature and solidification, many previous studies showed that the addition of the mush zone as a sink term in equations and solved by FLUENT gave results that agreed well to experimental measurements [10]. The current paper would not repeat the validation work. The FLUENT SOLIFICATION module was assumed to be accurate.

Results and discussion

Figure 3 shown half of the flow field of different mold thickness at the mid-plane between wide face, respectively. Comparing from the flow field, ultra-thick mold was similar with conventional mold. The jet left the nozzle first impacted on the narrow face, then divided into two parts which developed into upper and lower recirculation regions. However, as the increase of mold thickness, the lower recirculation region becomes larger. The lower recirculation depth of 300 mm, 360 mm and 420 mm mold are 1.94 m, 2.25 m and 2.26 m to meniscus, respectively. This might have great influence to the lower region of mold, especially the floating of inclusions.

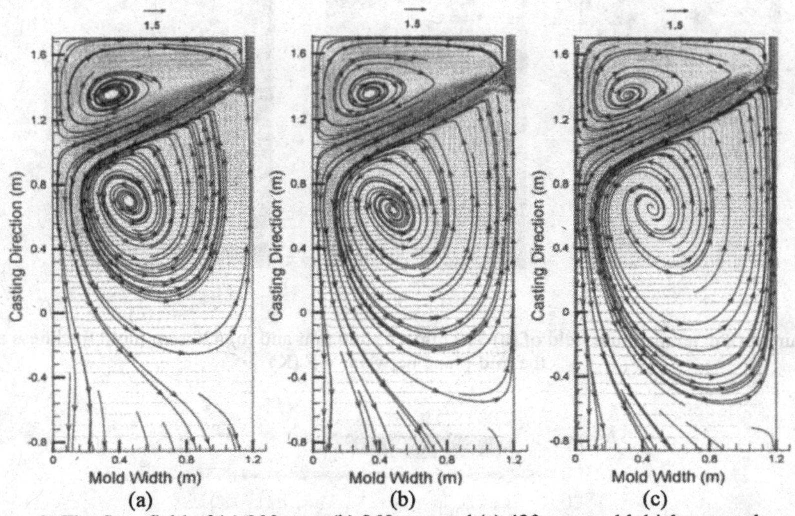

Figure 3. The flow field of (a) 300 mm, (b) 360 mm and (c) 420 mm mold thickness at the mid-plane between WF (m/s)

Figure 4 gave the temperature field of different mold thickness at the mid-plane between wide face. It was clearly that the temperature near mold surface was close to liquid line, and that may be influenced by the lower casting speed of ultra-thick mold. Because the ultra-thick slab needed thicker shell to bear the static pressure of liquid steel and avoid the bulging. On the other hand, it's hard to the heat transfer at the center of wide face due to the thickness of mold. As the increase of mold thickness, the temperature of upper recirculation decreased, especially at the region between nozzle and quarter of mold. The area near meniscus which temperature less than 1790.5 K was largest in 420 mm thickness mold. And the ultra-thick mold got higher temperature at the lower recirculation zone. Those were caused by the bigger lower recirculation zone of ultra-thick mold which could be obtained from figure 3

The temperature on the mold surface can reflect the upper region steel refreshing speed and the mold powder melting rate. Generally, the mold surface requires higher temperature and at the same time avoids the slag entrainment to stable the production and get high quality produce. It was readily apparent form Figure 5 that, as the increase of mold thickness, the higher temperature zone (larger than 1790.5 K) became smaller and the lower temperature zone (less

than 1790 K) near the mold wall became larger. So it's very important to take some methods to increase the temperature near wall in ultra-thick mold, avoiding the insufficient superheat transport to the meniscus.

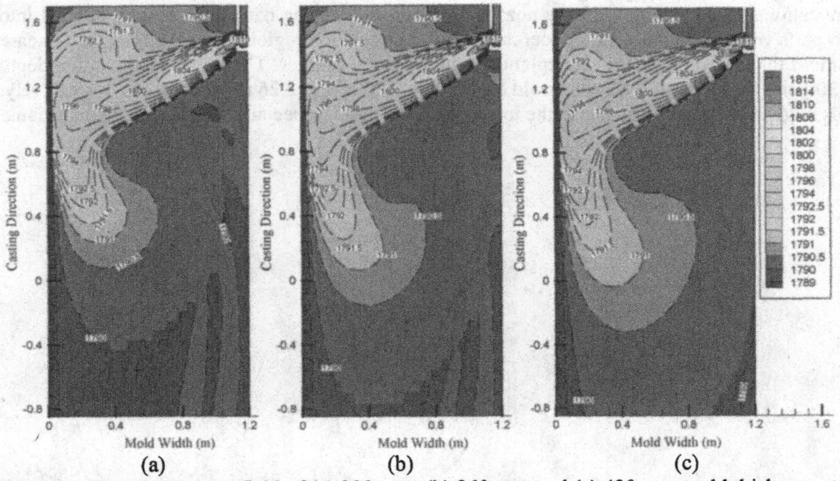

Figure 4. The temperature field of (a) 300 mm, (b) 360 mm and (c) 420 mm mold thickness at the mid-plane between WF (K)

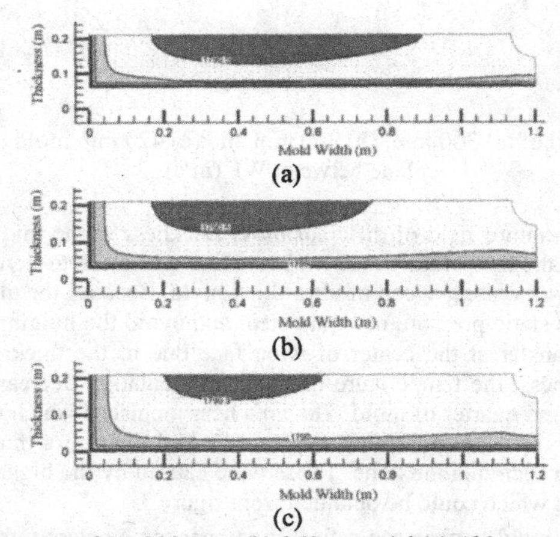

Figure 5. The temperature of (a) 300 mm, (b) 360 mm and (c) 420 mm mold thickness on the surface (K)

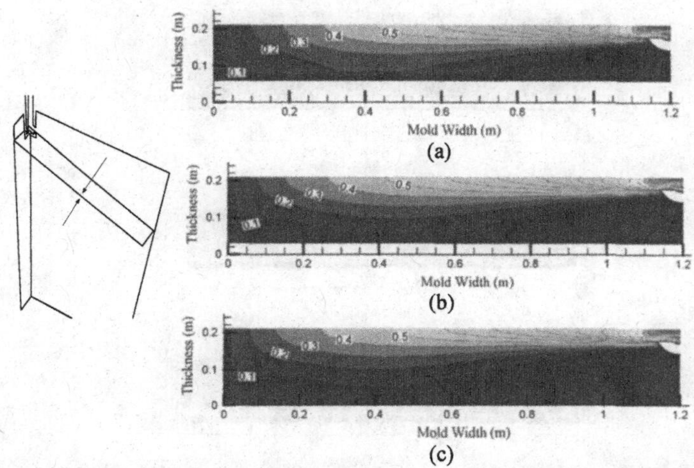

Figure 6. The flow field of (a) 300 mm, (b) 360 mm and (c) 420 mm mold thickness on the jet plane (m/s)

Basically, the jet properties directly influence the flow pattern in the mold. The jet plane was defined by the plane vertical to wide face and through the impact point on the narrow face and the nozzle port center. It presented in Figure 6 the thicker mold had a larger area of lower velocity region (less than 0.2 m/s) near the wide face. This indicated that the jet could not diffuse to the wide face when the thickness of mold was large. At the same time, the jet was less influenced by the mold wall when the thickness was larger. So at the same width of mold, the speed at the center of jet was larger. That meant the shell near the impact region would be shinning more by the higher speed liquid steel in thicker mold.

Figure 7 compared the liquid fraction on the section of mold bottom. And it's could get the shell thickness from the liquid fraction. When the solid fraction of the predicted shell was set to be 0.7, the shell thickness at the center of wide face of 300 mm, 360 mm and 420 mm mold are 18.4mm, 22.3 mm and 25.5 mm, respectively. The ultra-thick mold could get larger shell thickness on the wide face because the jet was hardly diffuse to the wide face which was shown in Figure 6. On the other hand, Figure 7 shown that the shell near impact region (except the corner of mold) was less than the wide face owing to the scour of high temperature liquid steel. And the uniform of shell in 420 mm thickness mold was the worst. This might cause long cracks and was great harmful to the slab quality

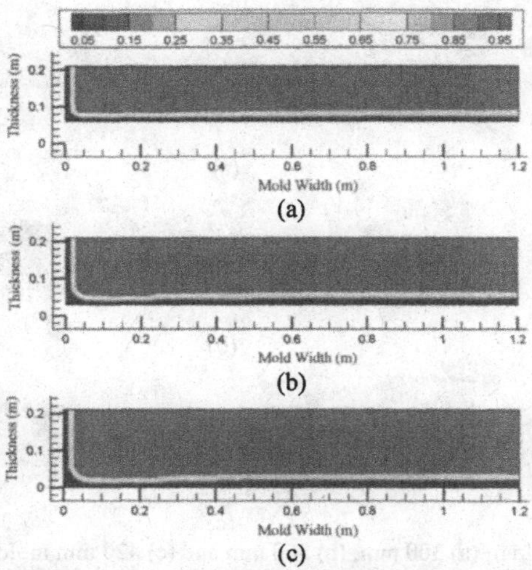

Figure 7. The liquid fraction of (a)300 mm, (b)360 mm and (c) 420 mm mold thickness at the mold exit

Conclusions

This paper had discussed the flow behavior difference of ultra-thick and conventional mold, and some conclusions could be found as follows:

(1) The flow field in ultra-thick mold was familiar with conventional mold. All of them were composed by upper and lower recirculation zone. However, the jet of ultra-thick mold was hardly diffuse to the wide face. So the liquid steel refreshing speed was lower and the shell thickness was larger at this region.

(2) Because of the lower casting speed and bad diffusing effect of jet, the temperature on the upper region of ultra-thick mold was lower. And the lower temperature zone was lager, especially near the wall of the mold surface.

(3) In order to maintain the process and improve the quality of ultra-thick slab continuous casting, the diffusing effect of jet and the refreshing speed of liquid steel on the upper region should be increased. And these requirements might be achieved though changing the structure and parameters of SEN or the operation conditions of caster.

Acknowledgements

The work was supported by "the Fundamental Research Funds for the Central Universities". The Project No. is CDJXS12132236.

Reference

1. B.G. Thomas and L.F. Zhang, "Mathematical modeling of fluid flow in continuous casting," *ISIJ International*, 41 (10) (2001), 1181-1193.
2. H.L. Yang, et al, "Mathematical simulation on coupled flow, heat, and solute transport in slab continuous casting process," *Metallurgical and Materials Transactions B-Process Metallurgy and Materials Processing Science*, 29 (6) (1998), 1345-1356.
3. B.G. Thomas, et al, "Comparison of four methods to evaluate fluid velocities in a continuous slab casting mold," *ISIJ* International, 41 (10) (2001), 1262-1271.
4. Mao Jinghua, "The characteristics of extra-thick slab caster mold," *Industrial Heating,* (3) (2010), 30-34. (in Chinese)
5. X. Xie, et al, "Effect of bottom structure of submerged entry nozzle on flow field in ultra-thick slab continuous casting mold," *Advanced Materials Research*, (154-155) (2011), 840-845.
6. A.D. Brent, V.R. Voller, and K.J. Reid, "Enthalpy-Porosity Technique for Modeling Convection-Diffusion Phase-Change - Application to the Melting of a Pure Metal," *Numerical Heat Transfer*, 13(3) (1988), 297-318.
7. A. Najera-Bastida, et al, "Shell Thinning Phenomena Affected by Heat Transfer, Nozzle Design and Flux Chemistry in Billets Moulds," *ISIJ International*, 50(6) (2010), 830-838.
8. M.R. Aboutalebi, M. Hasan, and R.I.L, "Guthrie. Coupled Turbulent-Flow, Heat, and Solute Transport in Continuous-Casting Processes," *Metallurgical and Materials Transactions B-Process Metallurgy and Materials Processing Science*, 26(4) (1995), 731-744.
9. S.H. Seyedein and M. Hasan, "A three-dimensional simulation of coupled turbulent flow and macroscopic solidification heat transfer for continuous slab casters," *International Journal of Heat and Mass Transfer*, 40(18) (1997), 4405-4423.
10. X.Y. Tian, et al, "Numerical Analysis of Coupled Fluid Flow, Heat Transfer and Macroscopic Solidification in the Thin Slab Funnel Shape Mold with a New Type EMBr," *Metallurgical and Materials Transactions B-Process Metallurgy and Materials Processing Science*, 41(1) (2010), 112-120.

Materials Processing Fundamentals
Edited by: Lifeng Zhang, Antoine Allanore, Cong Wang, James A. Yurko, and Justin Crapps
TMS (The Minerals, Metals & Materials Society), 2013

MEASUREMENT AND OBSERVATION OF THE FILLING PROCESS OF STEEL CASTINGS

Jinwu Kang [1], Gang Nie[1], Haimin Long[1], Rui You[1], Xiaokun Hao[1], Tianjiao Wang[1], Chengchun Zhang [2]

1）Department of Mechanical Engineering, Key Laboratory for Advanced Materials Processing Technology, Tsinghua University, Beijing 100084

2）Harbin Electric Machinery CO., LTD，Harbin 150040

Keywords： Steel casting, Mold filling, Filling time measurement, Wireless, Observation

Abstract

The filling process is significant for the quality of castings, and it has always been a hot but hard topic. The author develops a wireless measurement system for the filling process of casting based on contact time method and an observation system for the filling process of casting based on high speed camcorders working under high temperature. By using these two systems, the filling process of a turbine blade and a hub casting were measured and observed. And the filling situation of the castings with bottom gating system was obtained and the filling time of a number of typical positions was recorded. The result showed that liquid steel flows through the forward ingates at first. The velocity of the liquid steel in the mold was obtained by the calculation of the filling time. The study also shows that these two systems operate conveniently and reliably. They are effective tools for monitoring of the filling process of castings and future optimizing of gating system, and have broad prospect in casting production.

Introduction

The filling process of the melt into the mold cavity is the first step of the formation of a casting, which has significant effect on the quality of castings. Bad filling process may cause turbulence, mold wall erosion, cold shut, inclusions, misrun and etc. The research of the filling process has been the focus for many years. Physical measurement and numerical simulation are the two usual used methods. Numerical simulation has been widely used in research and production. However, the simulated results of castings have been always doubted due to no validation. Physical methods are rooted in the mind of researches, so，they have tried all kinds of methods to investigate the filling process. Among them，water analog was used for the investigation of gating system，cast specimens and simple castings[1-3], but it is hard to used for castings because the model of complicated casting was costly to make. High temperature resistant transparent window can be used in front of the mold cavity to observe the melt metal flow. Zhao [4] used transparent quartz pyrex window observed the filling process of aluminum. Khodai [5] used this method for the filling of aluminum, cast iron and steel during lost foam casting. However, this method is suitable for experimental study, and good for plate shaped cast specimen. In recent years, the development of high intensive X-ray provides a new method for the research of the filling process of test castings. Kashiwai[6] and Zhao[7] used in-situ x-ray examined the fluid flow for aluminum castings. Li [8] studied the filling process of turbine blade by cast iron replacement of Ni alloy. X-ray radiography method is suitable for experimental study, but, it is hard for the application in production because of its limit in size and thickness of castings, complex equipment and expense. Jong [9] proposed the contact time method, set of wires connected circuit to detect the melt flow, and took application of this method to aluminum castings successfully. Li[10] applied this method to study the filing process of an iron cast plate in lost foam casting and calculated the filling speed by the filling time and distance. However, this method hasn't yet used in steel castings and it is hard for production because of too many wires or cables on site which may cause trouble.

In this paper, a wireless filling process based on contact time method is proposed. And an observation system by high temperature resisted high speed camcorder is developed as well. These two systems are used to investigate the filling process of two real steel castings.

Development of the measurement method

The mold filling time measurement is performed by a wireless contact time measurement system, as shown in Fig. 1(a). The sensor is buried in the sand with the end exposed at the inner surface of the sand mold during molding, the emitter is placed outside the flask. Radio frequency signal is used for the emitter and receiver whose transmission distance is over 200m, enough for the onsite use. The receiver is connected to laptop computer. Software acquires the filing time of each measurement point, stores the data, and plots it on the casting CAD model in time for monitoring. The sensor for mold filling time measurement is an open circuit with a pair of electrode pointing into the inner surface of mold cavity to sense the filling of the liquid metal. As the melt flow through the pair of electrodes they are connected by the melt, the circuit is connected and there is electricity passing through the circuit. A signal emitter and a bulb are connected in the circuit for sending signal and illustration, as shown in Fig. 1(b). The developed sensor, emitter and receiver are shown in Fig. 2. One sensor is one circuit. One circuit can link an emitter, or a dozen of circuits connect a public emitter. As there is electricity, the emitter will send out a signal to a receiver which records the time.

Meanwhile, to better observe the filling process, another observation system is used with high temperature resistant camcorder used. The camcorder is buried in the sand mold with its lens exposed in the inner surface of sand mold. Each camcorder is connected to the computer to monitor the filing process. The number of camcorders can be installed at the necessary places. Therefore, the filling process of each area can be recorded. The observation system is also shown in Fig. 1(a).

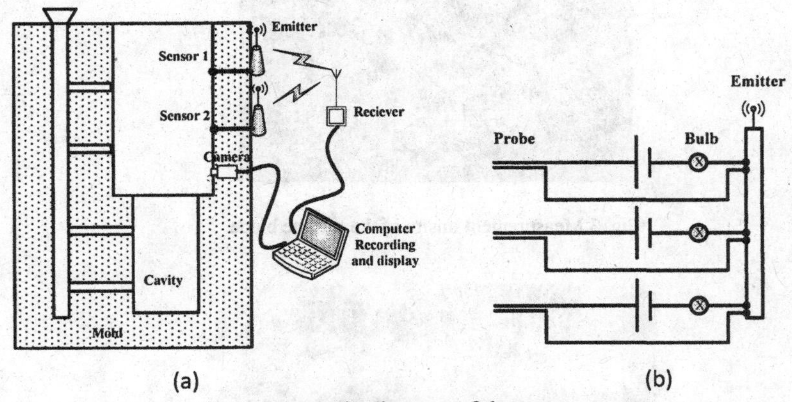

Fig.1 Schematic diagram of the test systems.
(a) Wireless measurement system for the filling process of casting based on contact time method and observation system for the filling process of casting by using high speed cameras working under high temperature, (b)the sensor circuit

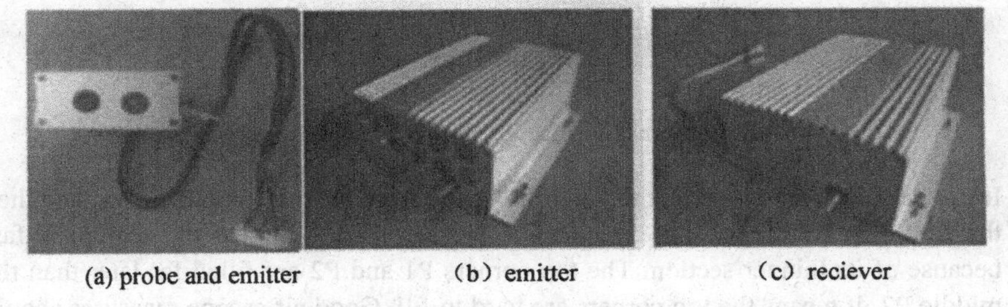

(a) probe and emitter (b) emitter (c) reciever

Fig. 2 Key parts of the wireless measurement system of mold filling time

Case studies

The developed system was used in the production of two steel castings, a hydro turbine blade and a hub casting produced in Harbin Electric Machinery Co., ltd.

Case study one: Filling process of a hydro turbine blade

The above developed system was used in the production of a hydro turbine blade produced in Harbin Electric Machinery Co., ltd. The casting is made of ZG0Cr13Ni4Mo, size 1460mm×1210mm×850mm and gross weight 0.9 tons. The gating system is of a bottom filling style. Ten measurement points were selected and one camcorder was placed. The onsite measurement is shown in Fig. 3. The measurement results are shown in Fig.4

Fig. 3 Measurement onsite of the turbine blade

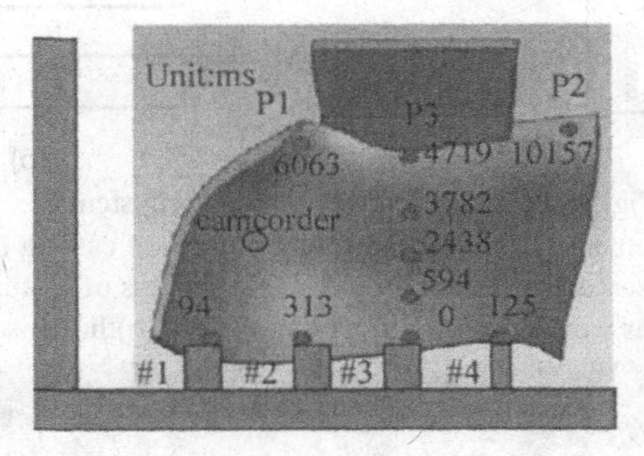

Fig. 4 Filling time measurement results of the turbine blade (unit: ms)

It can be seen from Fig. 4 that the melt flows out from the third ingate firstly, and then the first one, fourth one and second one the last. The filling of the bottom is fast because of its thinner section. The top corners P1 and P2 are filled far later than the middle P2, it means the top corners are hard to fill. Good air escape measures should be taken, such as ventilation holes. The observation of the camcorder is shown in Fig. 5. The melt comes out from the third ingate earlier than the second and the fourth ingates.

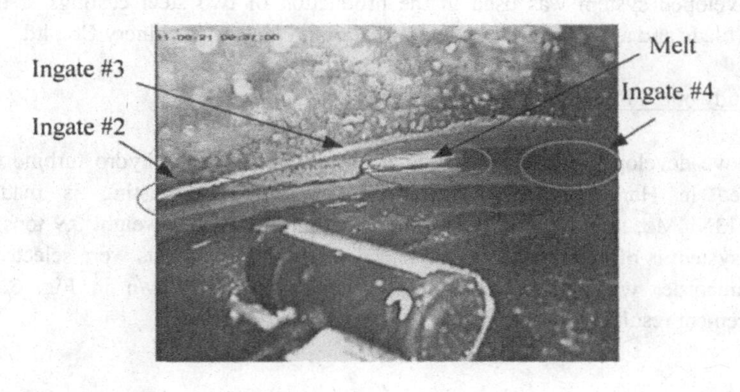

Fig. 5 In-mold Filling observation of the turbine blade at the beginning of pouring

Case study two: Filling process of a hub casting

The hub casting is made of B50E54D3, size 1460mm×1210mm×850mm and the melt weight 11.3 tons. The gating system is of a bottom filling style. Eighteen measurement points were selected at the ingates, the ribs and the top flank and two camcorder was placed on the top of the two risers. The measurement results are shown in Fig.6

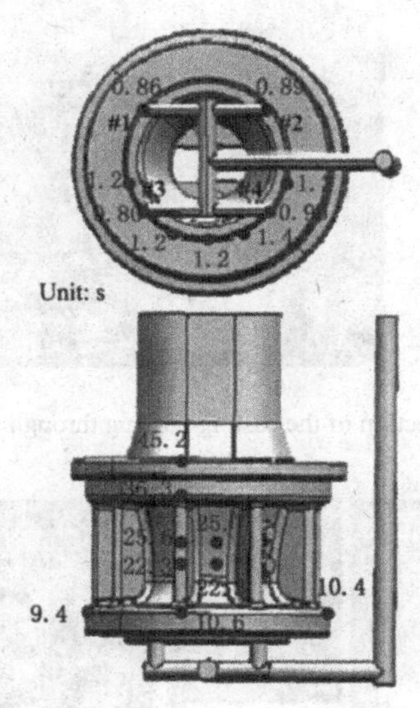

Fig. 6 Filling time measurement results of the hub casting (unit: s)

The four ingates fill the mold cavity almost at the same time, no big difference exist among them. But the two ingates #1 and #3 along the flow direction fills faster than the two backward ingates #2 and #4. That proves that the fluid flow always flow toward the forward direction first. For heavy castings, this effect can be neglected, while, for small casting it should be considered. By the filling times, the flow speed of the ingates can be calculated, it is 5m/s, which will result in flush of the flow and erosion of the sand roof it faces. The section view passing through a pair of ingates is shown in Fig. 7, it can be seen that there is still a certain height from the ingates to the roof sand, so, the opening of the ingate is correct. The filling of the bottom flank and the ribs is uniform. The melt level increasing speed of the bottom flank is 15mm/s, the ribs 30mm/s and the top flank 16mm/s.

The observed filling processing of the hub casting by the camcorders are merged, as shown in Fig. 8. It can be seen that there is splashing during the pouring beginning which will result in oxides and inclusions finally. The melt comes out from ingates #1 and #3 bigger (photo at 1s). At the beginning, there is overflow of the melt from the ingates, covering the whole bottom area, and then it flows back to the center. It can be seen from the photos at 1.09s, 2s, 3s, 4s, 5s and 7s. Later, the covered area of the melt on the bottom increases, as shown by the photos at 8s, 12s and 25s. Basically speaking, the filling process is uniform. There is smoke releasing from the core as the melt wraps the core and it becomes heavier as the filling proceeds. Slag is also found floating on the melt front the melt level gets into the riser.

Fig. 7 Section of the casting passing through a pair of ingates.

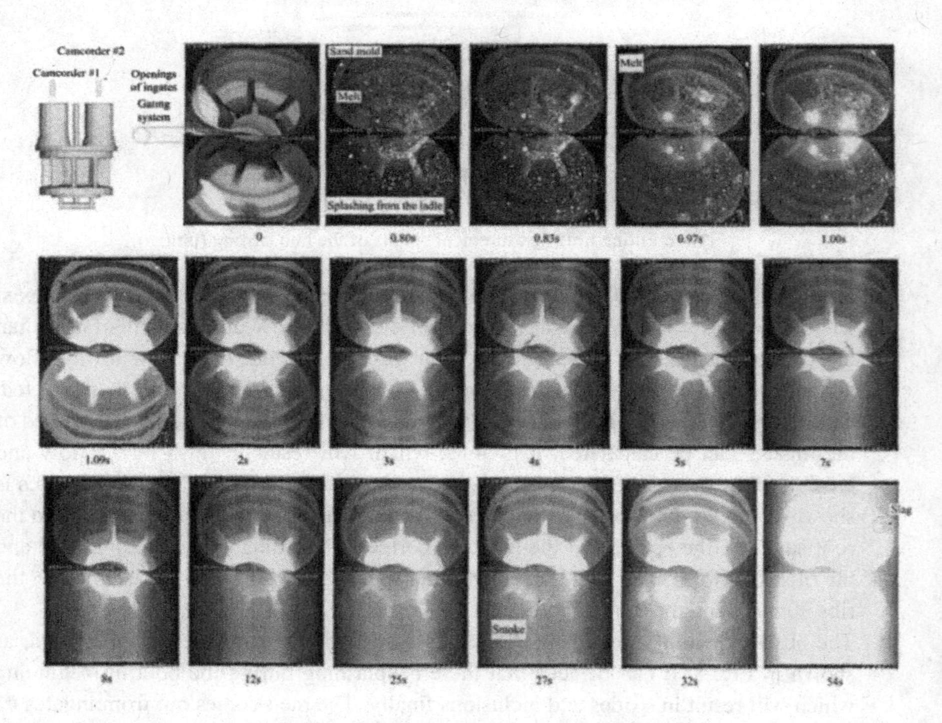

Fig. 8 Observed filling process of the hub casting

The successful application of these two systems reflects their effectiveness. The wireless system is easy to operate and doesn't interfere the production. These two methods can be used in the production.

Conclusions

A wireless measurement system for the filling process of casting based on contact time technique and an observation system for the filling process of casting based on high speed camcorder working under high temperature were developed.
By using these two systems, the filling process of a turbine blade and a hub casting were measured and observed. And the filling situation of these castings was obtained and the filling time of a number of typical positions was recorded. The result showed that liquid steel filled ingates with those along the direction firstly. The velocity of the liquid steel in the mold was obtained by the calculation based on filling times. The study also shows that these two systems operate conveniently and reliably. They are effective tools for monitoring the filling process of castings and optimizing gating system, and have broad prospect in casting production.

Acknowledgement

The project is funded by Major National Sci-Tech Project of China No 2011ZX04014-052 and National Basic Research Program of China (No. 2011CB012900).

References

1. F. Hou, S. Chen, X. Li, Q. Xu, The research on application of water simulation in permanent mold castings, Journal of Tianjin University, 1993, (4), 107-114.
2. L.N. Cai, X.J. Ma, B.J. Yang, J.Y. Su, Mold filling process examined using water as a substitute, Journal of Xi'An Jiaotong University, 1999,33(5),84-87.
3. Z.X. Zhao, S. T. Wang, S. L. Wen, Y.C. Zhai, Computer video technology of water simulation for aluminum liquid filling process, The Chinese Journal of Nonferrous Metals,2005,(8): 1263-1266.
4. Z.X. Zhao, S.L. Wen, S. T. Wang, Y.C. Zhai, Computer visualization study of molten aluminum filling mould process, Foundry, 2005, 54(10): 1010-1013.
5. M. Khodai, N. Parvin, Pressure Measurement and some observation in lost foam casting, Journal of materials processing technology, 2008, 206(1-3): 1-6.
6. S. Kashiwai, I. Ohnaka, A.K. Imatsuka, T. Kaneyoshi, T. Ohmichiand, J. Zhu. Numerical Simulation and x-ray direct observation of mould filling during vacuum suction casting, International Journal Of Cast Metals Research, 2005, 18(3): 144-148.
7. H.D Zhao, I. Ohnaka, J.D. Zhu, Modeling of mold filling of al gravity casting and validation with x-ray in-situ observation, Applied Mathematical Modelling, 2008, 32: 185-194

8. D.Z. Li, J. Campbell, Y.Y. Li, Filling system for investment cast ni-base turbine blades. journal of materials processing technology, 2004,148(3): 310-316

9. S.I. Jong, W.S. Hwang, Measurement and visualization of the filling pattern or molten metal in. actual industrial castings. AFS Transactions, 1992: 489-497

10. F.J. Li, H. F. Shen, B.C. Liu, D.D. Zhang, Mold filling velocity during investment casting, Material Science And Technology,2003,11(3): 222-229

MATERIALS PROCESSING FUNDAMENTALS

Physical Metallurgy of Metals

Session Chair
Cong Wang

Materials Processing Fundamentals
Edited by: Lifeng Zhang, Antoine Allanore, Cong Wang, James A. Yurko, and Justin Crapps
TMS (The Minerals, Metals & Materials Society), 2013

INFLUENCE OF LOAD PATHS AND BAKE HARDENING CONDITIONS ON THE MECHANICAL PROPERTIES OF DUAL PHASE STEEL

Mehdi Asadi[1], Heinz Palkowski[2]

[1]Benteler Automotive, An der Talle 27-35, 33102 Paderborn, Germany
[2]Technical University of Clausthal, Institute of Metallurgy, Metal Forming and Processing
Robert-Koch-Str. 42, Clausthal-Zellerfeld, Germany

Keywords: Dual phase steel, load paths, bake hardening

Abstract

In recent years high strength hot rolled dual phase (DP) steels became available. This multiphase steel type shows a strong Bake Hardening (BH) potential being of importance for shaping of body structures. Normally the BH effect is assessed using the uniaxial tension test. In reality shaped parts show a variety of stress paths, starting from deep drawing over uniaxial, plane strain to biaxial conditions. This is the special aspect of the analysis in combination with BH conditions on the evolution of strength parameters. DP steels samples were pre-strained under uniaxial, biaxial and plane strain conditions with different degrees of pre-strain. Subsequently, samples were undertaken a baking treatment at different temperatures. The mechanical properties were analysed particularly with regard to temperature and pre-strain. Pre-straining the samples with defined degrees of deformation and a subsequent aging treatment leads to enhanced material's strengthening.

Introduction

DP steels are characterized by a good formability, high strength and a good compromise between strength and ductility [1]. Moreover, the DP steels exhibit a continuous yielding behaviour, low yield point and a high strain-hardening coefficient [2]. Furthermore, the DP steels often show a large potential for bake hardening (BH). BH refers to the increase in yield strength as a result of the paint baking treatment of the shaped auto-body parts. The primary mechanism that causes the additional strengthening is the immobilization of dislocations by the segregation of interstitial atoms, known as classical static strain aging [3]. The increase of strength thus achieved allows a further reduction of sheet thickness and improves the crash safety and the dent resistance. The BH of special steel qualities is technically used in DP, where e.g. the increase in strength is realized in the final heat treatment [4]. Previous own investigations [5-6] state that the BH effect of DP is much stronger than that one for conventional BH steels.
Strength and the amplitude of the deformability due to aging as well as BH effects obviously depend on the degree of pre-strain and on the load path history [7,8]. If a structural element is inhomogeneously deformed, the activation of this effect through a final heat treatment also results in different strength values. One aim of this research project is to examine the deformation behavior of multiphase steels. The automotive industry in particular is responding with growing interest in the aging effect. The aging procedure is based on the effect that in a shaped construction unit, dislocations are pinned by carbon and/or nitrogen atoms in solid solution following thermal treatment. This lack of dislocation mobility increases the unit's strength. In the automotive production process, the thermal treatment is carried out at the end of the production line by means of the paint baking process [9]. The combination of prestrain, baking temperature and aging time results in an increase in the material's yield strength. If it is possible to compute this increase in yield strength, then a construction unit can be optimised for beneficial weight or crash characteristics. This requires an examination of the material specific shaping and heat treating processes since the BH effect in complex shaped construction units (e.g. tailored blanks, B-pillar or a crash protection unit), results in locally different strength behaviors following the thermal treatment of the units [10]. For this reason, the mechanical-technological properties were analysed particularly with regard to temperature, holding time and pre-strain. Under these aspects two basic questions have to be answered:
Where is the maximum increase of yield strength and tensile strength at a given pre-strain and what influence does the load path (kind of pre-strain) have?
When does the yield strength reach a maximum as a function of pre-strain, temperature and holding time and can this effect be computed?
Considering an actual shaped construction unit it becomes clear that the deformation behaviour of the material cannot be completely described by using uniaxial tension tests [7,11]. On this basis, the influence of type of pre-strain on the aging effect was investigated.

Material and Experiments

The DP steel used was delivered as a hot rolled sheet with a thickness of 1,9 mm. Its chemical composition is listed in Table I.

Table I. Chemical composition of the steel (wt%)

Steel	DIN 10336	C	Si	Mn	Cr	Mo	Nb	P	N	Al
DP W 600	HDT580X	0.06	0.10	1.30	0.60	0.005	0.002	0.04	0.006	0.035

In order to generate different strain paths to cover the Forming Limit Diagram (FLD), uniaxial, biaxial and plane strain conditions were used.

Uniaxial tests were conducted using a universal tensile machine UTS 250 kN. The samples were stretched to specified pre-strains (Table II) and subsequently aged for 20 min at 100 °C, 170 °C and 240 °C. Finally, the tensile test was performed on the samples to determine the mechanical properties.

Considering the different choices to generate biaxial stress, a Marciniak forming tool (Figure 1) with a diameter of 250 mm was chosen, mounted on a 2500 kN hydraulic press. Thus it is possible to produce biaxial pre-strained specimens of large area. For biaxial pre-straining the specimens were cut into 400 x 400 mm square sheets. The specimens are strained by a flat-bottomed cylindrical punch. Between the punch and the specimen is a steel driver with a central hole. By means of this, frictionless deformation of the sheet in the hollow area takes place. When a large width specimen is used, the external forces acting on the sample are greater than the internal forces; this results in biaxial stretching [12].

To control the plastic deformation and adjustment, the non-contact photogrammatical measuring system ARGUS (GOMmbH) was used. This system is well suited for measuring three dimensional deformation and strain distributions in components subject to static or dynamic load, especially for small ranges of elongation.

The 240 mm long tensile specimens were taken from the stretched drawn regions of the sheets (Figure 1). The tensile specimens were subsequently age treated under the same conditions as uniaxial pre-strained specimens. Finally, the tensile tests were performed on the samples to determine the mechanical properties of the pre-strained areas and thus ascertain the effect of pre-straining on the strengthening.

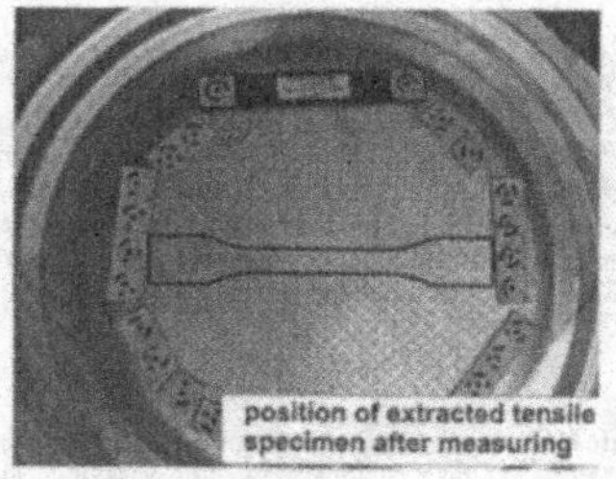

Figure 1. Marciniak tool to produce the biaxial strain condition. The specimens were cut into 400 x 400 mm sheets

The model developed by Vegter [13] was used for the plane strain test. A special specimen has been developed to create plane strain conditions, see Figure 2 [14,15]. The advantage of this specimen is that a plane strain test can be performed using an ordinary tensile testing machine. One of the difficulties is to achieve a uniform plane strain state across the specimen, as is shown in Figure 3. Due to the existence of a uniaxial stress state at the edges of the specimen, deviations from a plane strain state always occur in this type of specimen. A correction has to be made to subtract this edge effect from the test results. A simple geometric average stress - strain curve over the whole width of the specimen would result in a 2 % lower value of the plane strain stress. Even such a small difference will lead to significant differences in the prediction of FLD's and differences in the prediction of strain distribution for critical sheet formed products [16].

The plane strain specimens having specified pre-strains were stretched using a MAN 1000 kN hydrostatic universal testing machine. The generated strain in the specimen was controlled photogrammetrically. The 210 mm long tensile specimens were taken from the pre-stretched regions of the sheets (Figure 4) and underwent the same conditions as the uniaxial pre-strained specimens. Finally, the tensile test was performed on the samples to determine the mechanical properties of the pre-strained regions and thus ascertain the effect of pre-straining on strengthening.

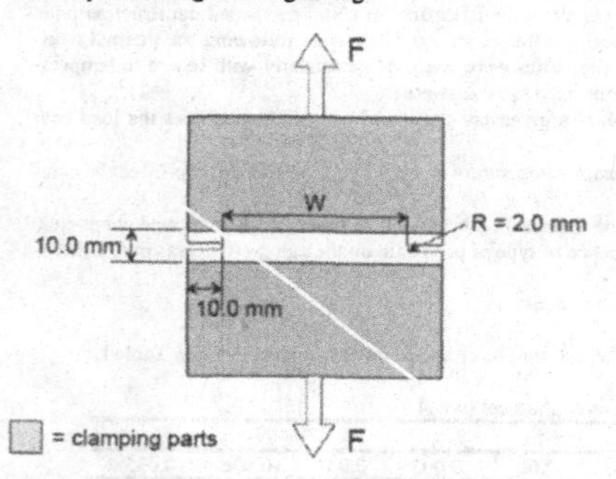

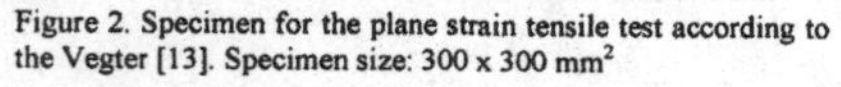

Figure 2. Specimen for the plane strain tensile test according to the Vegter [13]. Specimen size: 300 x 300 mm^2

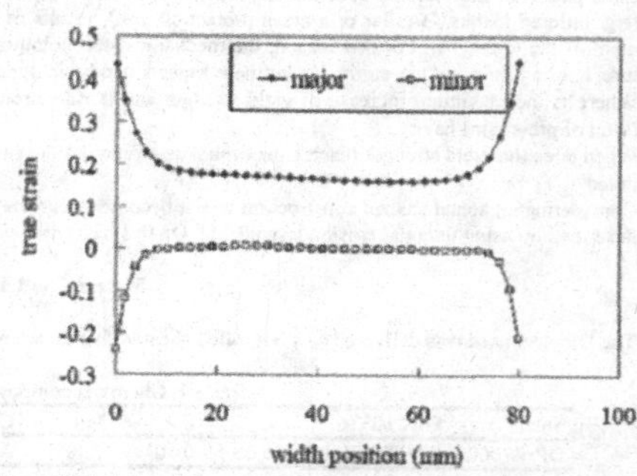

Figure 3. Axial and transverse strain distribution in the plane strain test

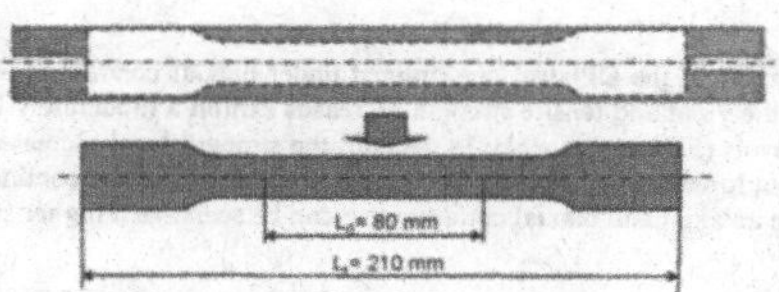

Figure 4. Tensile specimen taken from pre-strained region having a plane strain state

The change in length and the corresponding degree of pre-strain for the DP-steel for each type of strain path are listed in Table II.

Table II. The change in length and the corresponding degree of pre-strain for the DP steel for different types of load path

Type of load path	Change in length [mm]	Degree of pre-strain ε [%]
Uniaxial	-	2, 5, 10
Biaxial	5, 10, 15, 20	2, 3, 4, 5, 7
Plane-strain	6, 7, 8, 9, 9.5	0.50, 0.75, 1.50, 2.25, 4.00

Results and Discussion

Uniaxial condition

Figure 5 shows the mechanical properties of DP-W 600 conditioned with uniaxial pre-strains of up to 10 % and aging temperatures of 100 °C, 170 °C and 240 °C with a holding time of 20 min.. The strengthening effect is stronger with increasing aging temperatures which improve diffusion conditions. The pre-straining influences the strengthening values such that the strength increase becomes significantly higher for a higher degree of pre-straining (Figure 5 (a)-(b)). It can be seen that for all aging temperatures the yield and tensile strength increase exhibit a moderately linear dependence on the degree of pre-strain. Total elongation decreases with increasing pre-strain and aging temperature (Figure 5 (c)).

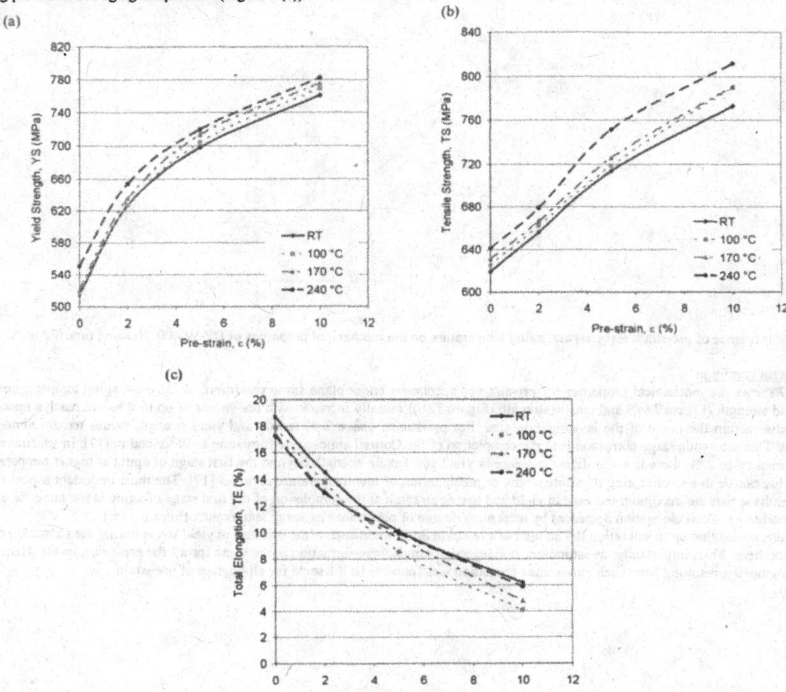

Figure 5. Influence of pre-strain (uni-axial) and aging temperature on the mechanical properties of DP-W 600. Holding time: 20 min

Biaxial condition

Figure 6 shows the mechanical behavior of the DP-steel pre-strained under biaxial conditions with different degrees of pre-strain. Similar to the pre-strain conditions, the yield and tensile strength increases exhibit a moderately linear dependence on the degree of pre-strain at all aging temperature levels (Figure 6 (a)-(b)). In general, the strength level increases with increasing of pre-strain and temperature. The highest strength, but lowest ductility (Figure 6 (c)) was observed for the condition with the highest degree of pre-strain aged at 240 °C. Comparing the uniaxial and biaxial conditions, it can be seen that a higher strength level with lower ductility is found for the biaxial condition.

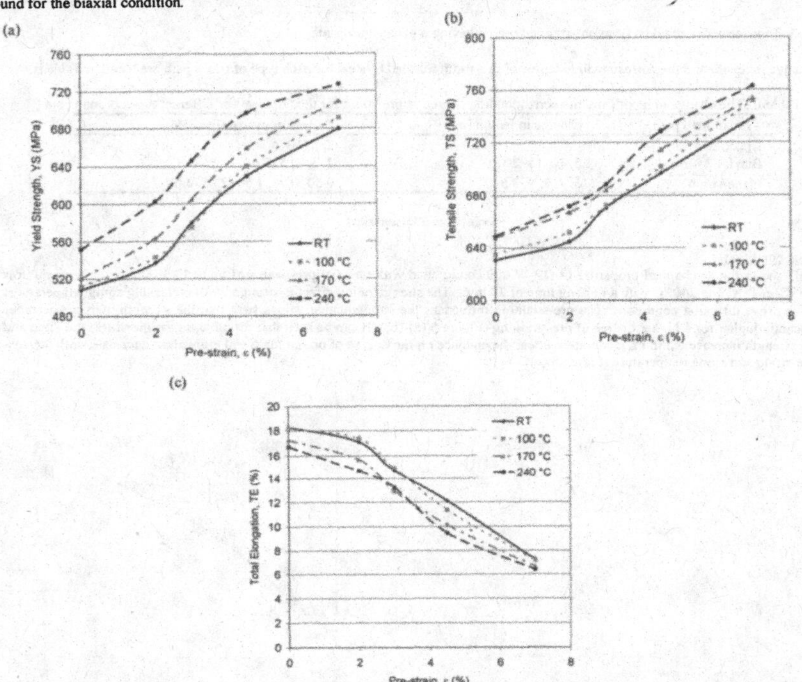

Figure 6. Influence of pre-strain (biaxial) and aging temperature on the mechanical properties of DP-W 600. Holding time: 20 min.

Plane strain condition

Figure 7 shows the mechanical properties for pre-strained specimens under plane strain conditions at different aging temperatures. The yield strength (Figure 7 (a)) and tensile strength (Figure 7 (b)) initially increase with pre-strains of up to 2 % and reach a maximum value within the range of the investigation area. For pre-strains above 2 % tensile and yield strength values remain almost constant. This saturation stage corresponds to the completion of the Cottrell atmosphere formation as described in [17]. In specimens pre-strained up to 2 %, there is a significant increase in yield and tensile strength beyond the first stage of aging at higher temperatures. This can be due to clustering of solute atoms or precipitation of low temperature carbides [17]. The most important aspect of these results is that the maximum increase in yield and tensile strength at the completion of the first stage of aging is the same for all aging conditions. Total elongation decreases by increasing degree of pre-strain and aging temperature (Figure 7 (c)).

The results reveal that up to saturation the amount of pre-strain does not influence the increase in yield stress during the formation of the atmosphere. Most importantly, at saturation, the degree of atmosphere formation is the same for all the pre-strain levels. Hence the strengthening resulting from such atmosphere formation was found to be the same for all degrees of pre-strain.

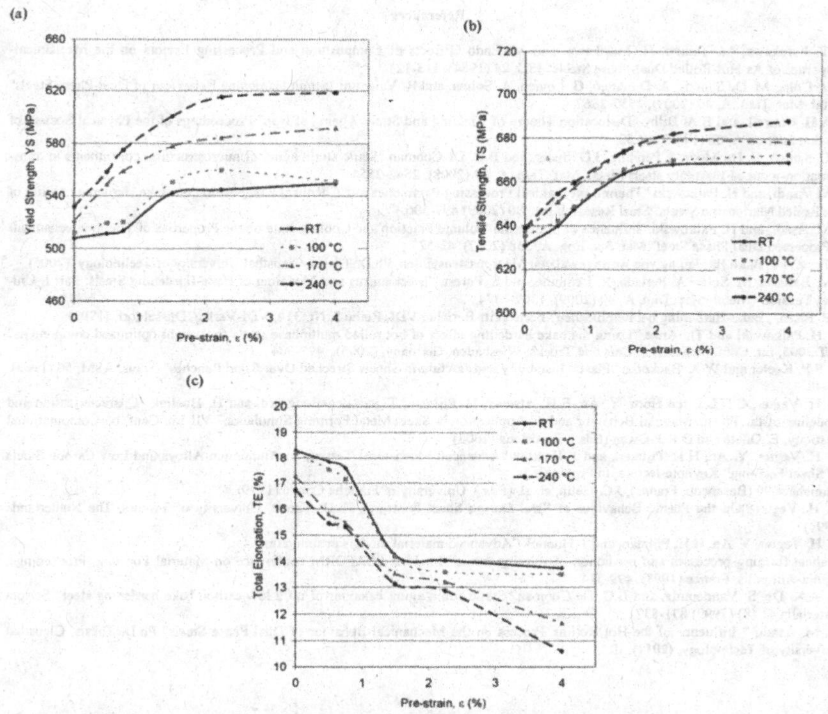

Figure 7. Influence of plane strain pre-straining and aging temperature on mechanical properties of DP-W 600. Holding time: 20 min.

For all pre-straining conditions, an aging treatment at 100 °C exhibited no influence on the mechanical properties. Aging at temperatures of 170 °C and 240 °C led to an increase of the yield and tensile strengths and a decrease of the total elongation. This can be due to clustering of solute atoms or precipitation carbides at high temperature [17].

Conclusions and Outlook

From the present investigation the following conclusions can be drawn:

1- The local mechanical strength behavior regarding the BH effect depend on the type of load path, degree of pre-strain and bake hardening conditions.
2- Considering this effect and the differences of uniaxial, biaxial and plane strain pre-strain, the type of the load path is of importance. A higher strength level can be observed for biaxial condition at the same degree of pre-strain.
3- For all load conditions, a noticeable higher strength increment is observed after baking at 170 and 240 °C for 20 min. This increment is higher for the baking condition with 240 °C and 20 min.
The results give the chance to develop a formula for different combinations of basic load paths to be able to calculate the mechanical properties under these specific conditions.

Acknowledgement

The authors would like to thank the DFG "Deutsche Forschungsgemeinschaft", Bonn, Germany, for their financial support of the project as part of the Collaborative Research Center 675 named "Creation of high strength metallic structures and joints by setting up scaled local material properties". The authors are also grateful to Salzgitter Flachstahl GmbH for providing the steels.

References

1. T. Furukawa, M. Tanino, H. Morikawa, and M. Endo "Effects of Composition and Processing Factors on the Mechanical-Properties of As-Hot-Rolled Dual-Phase Steels" ISIJ, 24 (1984), 113-121.
2. V. Colla, M. De Sanctis, A. Dimatteo, G. Lovicu, A. Solina, and R. Valentini "Strain Hardening Behaviour of Dual-Phase Steels" Metal. Mat. Tran. A, 40 (2009), 2557-2567.
3. A.H. Cottrell, and B.A. Bilby "Dislocation Theory of Yielding and Strain Ageing of Iron" Proceedings of the Physical Society of London Section A, 62 (1949), 49-62.
4. L. Samek, E. De Moor, J. Penning, J.G. Speer, and B.C. De Cooman "Static strain aging of microstructural constituents in transformation-induced-plasticity steel" Metal. Mat. Tran. A, 39 (2008), 2542-2554.
5. M. Asadi, and H. Palkowski "Thermo-mechanical Processing Parameters and Chemical Composition on Bake Hardening ability of Hot Rolled Martensitic Steels" Steel Research Int., 80 (2009) 499-506.
6. M. Asadi, and H. Palkowski "Influence of Martensite Volume Fraction and Cooling Rate on the Properties of Thermomechanically Processed Dual Phase Steel" Mat. Sci. Eng. A, 538 (2012), 42-52.
7. Th. Anke: Bake Hardening von warmgewalzten Mehrphastenstählen. Ph.D. Thesis, Clausthal University of Technology, (2005).
8. V. Ballarin, M .Soler, A. Perlade, X. Lemoine, and S. Forest, "Mechanisms and Modeling of Bake-Hardening Steels: Part I. Uniaxial Tension" Metal. Mat. Tran. A, 40 (2009), 1367-1374.
9. P. Elsen, „Bake-Hardening bei Feinblechen" Fortschritt-Berichte VDI, Reihe 5, Nr. 314, VDI-Verlag, Düsseldorf, (1993).
10. H. Palkowski and Th. Anke "Using the bake hardening effect of hot rolled multiphase steels for weight optimized components" SCT 2005, Int. Conf. on Steels in Cars and Trucks, Wiesbaden, Germany, (2005), 497-504
11. S.P. Keeler and W.A. Backofen "Plastic Instability and fracture in Sheets Streched Over Rigid Punches" Trans. ASM, 56 (1963), 25–48
12. H. Vegter, C.H.L.J. ten Horn, Y. An, E.H. Atzema, H. Pijlman, T. van den Boogaard, and H. Huétink "Characterization and Modeling of the Plastic Material Behavior and its Application in Sheet Metal Forming Simulation" VII Int. Conf. on Computational Plasticity, E. Oñate and D.R.J. Owen (Eds.), Barcelona (2003).
13. H. Vegter, Y. An, H.H. Pijlman, and J. Huétink "Advanced Mechanical Testing on Aluminium Alloys and Low Carbon Steels for Sheet Forming" Keynote-lecture, in Proceedings
Numisheet'99 (Besançon, France), J.C. Gelin, et. al. (Eds.), University of Franche Comté (1999).
14. H. Vegter "On the Plastic Behaviour of Steel During Sheet Forming" Ph.D. Thesis, University of Twente, The Netherlands (1991).
15. H. Vegter, Y. An, H.H. Pijlman, and J. Huétink "Advanced material models in simulation
of sheet forming processes and prediction of forming limits" In: First ESAFORM conference on Material Forming, Proceedings, Sophia-Antipolis, France (1998), 499-514.
16. A.K. De, S. Vandeputte, and B.C. De Cooman "Static strain aging behavior of ultra low carbon bake hardening steel" Scripta Materialia 41 (8) (1999) 831-837.
17. M. Asadi, " Influence of the Hot Rolling Process on the Mechanical Behavior of Dual Phase Steels" Ph.D. Thesis, Clausthal University of Technology, (2011).

Materials Processing Fundamentals
Edited by: Lifeng Zhang, Antoine Allanore, Cong Wang, James A. Yurko, and Justin Crapps
TMS (The Minerals, Metals & Materials Society), 2013

THE EFFECT OF PHOSPHORUS AND SULFUR ON THE CRACK SUSCEPTIBILITY OF CONTINUOUS CASTING STEEL

Weiling Wang, Sen Luo, Zhaozhen Cai, Miaoyong Zhu*

School of Materials and Metallurgy, Northeastern University, Shenyang 110819, Liaoning, China

Keywords: Crack susceptibility; Peritectic reaction zone, Brittle temperature range, Thermal strain, Difference of deformation energy

Abstract

Based on the coupled macro-heat transfer and micro-segregation model for continuous casting wide-thick slab, the effect of phosphorus and sulfur on crack susceptibility was investigated from the peritectic reaction zone, the brittle temperature range, the thermal strain and the difference of deformation energy. The results show that the crack sensibility of continuous casting steel, especially the hypo-peritectic steel, intensifies with the improvement of phosphorus and sulfur in steel matrix. Compared with phosphorus, sulfur does a more powerful effect on the crack susceptibility of steel. The peritectic reaction zone, the brittle temperature range and the thermal strain are capable to characterize the influence of phosphorus and sulfur on the crack susceptibility of steel. However, the calculated difference of deformation energy decreases with increasing phosphorus and sulfur, which is inconsistent with the experiment results. Therefore the difference of deformation energy is inappropriate to be applied to characterize the impact of phosphorus and sulfur to crack susceptibility of steel.

Introduction

The continuous casting famous for high efficiency, high metal yield and low energy consumption has gradually become the main process for steel casting since 1950s. However, the continuous casting steel suffers from serious quality problems, many of which originate from the solidification of steel during that process and cannot be eliminated in the following rolling process, and therefore affect the performance of final products. The quality problems of continuous casting steel are classified into surface ones and inner ones[1]. Among those quality problems, the surface longitudinal cracks and the inner cracks are known to originate from the mushy zone and be affected by the micro-segregation behavior of solute elements in steel matrix[2,3]. Many researchers[2-17] have carried out the related experiments and mathematical models to describe the generation criteria and distribution in steel strand of cracks.

In order to evaluate the characteristic temperature such as zero strength temperature (ZST), zero ductile temperature (ZDT), liquid impenetrable temperature (LIT) and temperature range of strength generation (TRSG), as well as the critical strain and stress for cracks, the bending tests and the tensile tests within solidification temperature range were implemented[6-9]. Suzuki et al[4] suggested that both ZST and ZDT were proportional to solidus temperature according the results from Gleeble tests. Matsumiya et al[5] measured the critical strain of inner cracks in continuous casting slabs with different carbon content from bending tests and proposed that the critical strain decrease with the increase of solidification temperature range. Whereas, Shin et al[6], Nakagawa et al[7], Seol et al[8] and Guo et al[9] obtained the specifically corresponding solidification fraction and

* Corresponding author: myzhu@mail.neu.edu.cn

temperature to ZST, ZDT and TRSG from tensile tests. Meanwhile, Wolfgang et al[10] and Won et al[11] calculated ZST, ZDT and LIT with an analytical model (Scheil-Gulliver model) and a numerical micro-segregation model, respectively. Yamanaka et al[12] proposed that the inner cracks occurred when the total strain was higher than 1.6% and was regardless of the deformation mode. Kim et al[2] proposed that thermal strain within the brittle temperature range from ZDT to LIT was appropriate to describe the longitudinal crack susceptibility of steel and investigated the influence of carbon and sulfur utilizing a micro-segregation model. Won et al[3] took both brittle and ductile fracture into consideration and developed a new criterion called difference of deformation energy to predict the inner cracks sensibility of steel, which showed a good performance to reveal the effect of initial carbon composition in steel matrix. Won et al[13] introduced the specific crack susceptibility from a two dimensional thermo-mechanical model to describe the possibility and distribution of inner cracks in steel strand, with a special consideration of casting conditions. Han et al[14] compared the tensile strain at solidification front with the critical strain to investigate the occurrence of inner cracks, with a full consideration of the effect of bulging, unbending and misalignment of supporting rolls. Recently, Cai et al[15] investigated the inner crack susceptibility by the maximum principal stress from a two-dimension transient thermo-mechanical model, which presented a sound performance. Some researchers[16, 17] took the effect of the precipitation of MnS inclusion on the inner crack susceptibility into consideration and proposed that high Mn/S ratio is helpful to reduce the occurrence of inner cracks.
Among the main five elements in common steel, phosphorus and sulfur exhibit significant influences on the micro-segregation of steel, characteristic temperature and cracks observed in steel works. However, the comprehensive investigation of effect of phosphorus and sulfur on the crack susceptibility of continuous casting steel is seldom reported. Therefore, based on a mathematical model[18] that combined the macro-heat transfer and micro-segregation, present paper investigates the influence of phosphorus and sulfur on the crack sensibility of continuous casting steel in terms of the peritectic reaction zone, the brittle temperature range, the thermal strain and the difference of deformation energy.

Characteristic Parameters

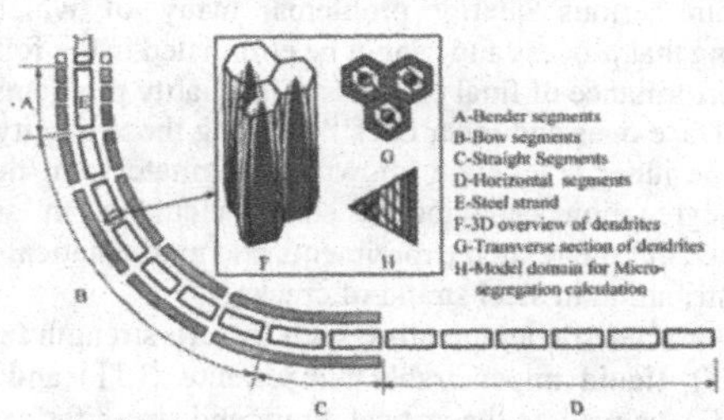

Figure 1. Schematic diagram of the solidification for continuous casting wide and thick slab[18]

The solidification of peritectic steel which contains 0.132C, 0.26Si, 1.475Mn, 0.017P and 0.008S united in wt% during wide and thick slab continuous casting is of great interest, as shown in Figure 1. The macro-heat transfer model and the micro-segregation model for the peritectic steel were described in detail in [18]. The two models are combined in such a mechanism. The macro-heat transfer model provides the effective cooling rate to the micro-segregation calculation, while the later model feedbacks the liquidus and the solidus temperature to the former one. The calculated effective cooling rate equals to 1.78 °C/s, which is used to determine the dendrite arm spacing for the description of micro-segregation behavior in [18], as well as the investigation of the impact of phosphorus and sulfur on crack susceptibility of continuous casting steel with the single variable approach in present work.

The brittle temperature range, ΔT_B, is defined as the difference between LIT and ZDT, according to Kim et al[2] and Won et al[3]. In present study, LIT and ZDT correspond to the temperature at solidification of 0.9 and 1.0, respectively. And thus, ΔT_B is expressed as follow:

$$\Delta T_{\mathrm{B}} = \mathrm{LIT} - \mathrm{ZDT} = T(f_s = 0.9) - T(f_s = 1.0) \tag{1}$$

The thermal strain in the brittle temperature range is mainly caused by variation of steel volume and simply treated as a function of steel density, which is expressed as follow[19, 20]:

$$\varepsilon^{\mathrm{th}}(T) = \sqrt[3]{\frac{\rho(T_{\mathrm{LIT}})}{\rho(T_{\mathrm{ZDT}})}} - 1 \tag{2}$$

Where $\rho(T)$ is steel density, $\mathrm{kg\cdot m^{-3}}$, dependent on the phase density[19, 21] and fraction. f_δ, f_γ and f_l are phase fraction of δ, γ and L phase respectively. The related equations are expressed as follows:

$$\rho(T) = f_\delta \cdot \rho_\delta(T) + f_\gamma \cdot \rho_\gamma(T) + f_l \cdot \rho_l(T) \tag{3}$$

$$\rho_\delta(T) = \frac{100(8106 - 0.51T)}{(100 - w_{\mathrm{C}})(1 + 0.008 w_{\mathrm{C}}^3)} \tag{4}$$

$$\rho_\gamma(T) = \frac{100(8111 - 0.47T)}{(100 - w_{\mathrm{C}})(1 + 0.013 w_{\mathrm{C}}^3)} \tag{5}$$

$$\rho_l = 7100 - 73 w_{\mathrm{C}} - (0.8 - 0.09 w_{\mathrm{C}})(T - 1550) \tag{6}$$

$$f_\delta + f_\gamma + f_l = 1 \tag{7}$$

The difference of deformation energy, W_B between ductile fracture and brittle fracture proposed by Won et al[3] is dependent on the critical strain in the brittle temperature range, which is expressed as follows:

$$W_{\mathrm{B}} = \left(k_\delta \frac{\varepsilon_c^{n_\delta+1}}{n_\delta+1} + k_\gamma \frac{\varepsilon_c^{n_\gamma+1}}{n_\gamma+1} \right)\Bigg|_{\mathrm{ZDT}} - \left(k_\delta \frac{\varepsilon_c^{n_\delta+1}}{n_\delta+1} + k_\gamma \frac{\varepsilon_c^{n_\gamma+1}}{n_\gamma+1} \right)\Bigg|_{\mathrm{LIT}} \tag{8}$$

$$\varepsilon_{\mathrm{c}} = \frac{\varphi}{\dot{\varepsilon}^{m'} \Delta T_{\mathrm{B}}^{n'}} \tag{9}$$

$$k_\delta = \left(\frac{f_s - {}^c f_s}{1 - {}^c f_s} \right) \frac{{}^\delta f_s}{\beta_\delta} \sinh^{-1}\left[\frac{\dot{\varepsilon}}{A_\delta} \cdot \exp\left(\frac{Q_\delta}{RT} \right) \right]^{m_\delta} \tag{10}$$

$$k_\gamma = \left(\frac{f_s - {}^c f_s}{1 - {}^c f_s} \right) \frac{{}^\gamma f_s}{\beta_\gamma} \sinh^{-1}\left[\frac{\dot{\varepsilon}}{A_\gamma} \cdot \exp\left(\frac{Q_\gamma}{RT} \right) \right]^{m_\gamma} \tag{11}$$

$$\sigma_{\mathrm{c}} = \left(\frac{f_s - {}^c f_s}{1 - {}^c f_s} \right) \left({}^\delta f_s \sigma_{\mathrm{c}}^\delta + {}^\gamma f_s \sigma_{\mathrm{c}}^\gamma \right) \tag{12}$$

$$f_s = f_\delta + f_\gamma \quad {}^\delta f_s = f_\delta / f_s \quad {}^\gamma f_s = f_\gamma / f_s \tag{13}$$

$$\sigma_{\mathrm{c}}^\delta = k_\delta \varepsilon_{\mathrm{c}}^{n_\delta} \quad \sigma_{\mathrm{c}}^\gamma = k_\gamma \varepsilon_{\mathrm{c}}^{n_\gamma} \tag{14}$$

Where ε_c is the critical strain, σ_c is the critical fracture stress, $\dot{\varepsilon}$ is strain rate equaling to 0.01 $\mathrm{s^{-1}}$[3],

${}^{\delta}f_s$ and ${}^{\gamma}f_s$ is the relative fraction of δ and γ in solid phase[22], Q is active energy for deformation, R is gas constant. φ, m^* and n^* are fitted parameters, equaling to 0.02821, 0.3131, 0.8638, respectively, β, n, A and m are constant, which are listed in table I[3].

Table I. Parameters for the critical stress calculation

Phase	β MPa	n	A s^{-1}	m	Q $kJ \cdot mol^{-1}$
δ	0.0933	0.0379	6.754×10^{8}	0.1028	216.9
γ	0.0381	0.2100	1.192×10^{10}	0.2363	373.4

Results and discussion

In order to reveal the effect mechanism of phosphorous and sulfur on the crack sensibility of continuous casting steel, present study carried out the analysis from peritectic reaction zone, brittle temperature range, thermal strain and difference of deformation energy based on the developed micro-segregation model. Meanwhile, the ability of those characteristic parameters in characterizing the crack susceptibility of steel was discussed, especially with consideration of the influence of phosphorous and sulfur.

Parameter Validations

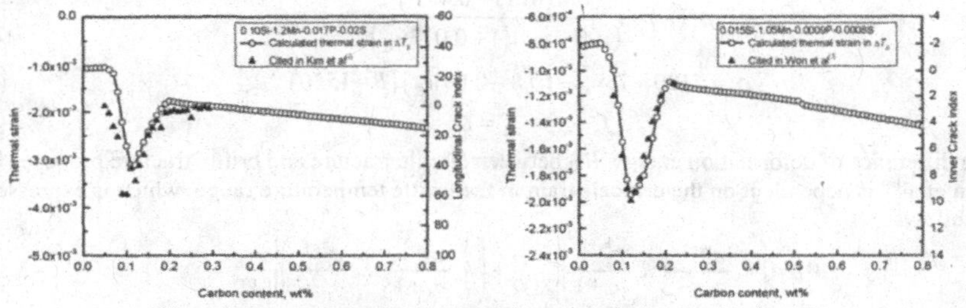

Figure 2. Comparison of tendency of thermal strain in ΔT_B with that of (a) longitudinal and (b) inner crack index

Figures 2 (a) and (b) show the comparison of the tendency of thermal strain in brittle temperature range, ΔT_B, simply "thermal strain" for short, with the measured occurrence of longitudinal and inner cracks within the whole carbon range. Within the low carbon range, the thermal strain is lower than that in other carbon range and decreases slightly, which does not agree with Cai and Zhu[20], mainly due to different carbon segregation behavior caused by neglecting the effect of carbon content on the dendrite arming space[18]. In the hypo-peritectic range, the thermal strain increases dramatically due the peritectic reaction at the end of solidification. In the hyper-peritectic range, the peritectic reaction ends before arriving at the end of solidification, which causes the thermal strain decreasing with the increase of initial carbon content. Then the thermal strain increases linearly in the medium and high carbon range. Therefore there exists a peak for the thermal strain, which corresponds to the turning point from hypo to hyper-peritectic range, that is, the peritectic point. The tendency of calculated thermal strain in the peritectic reaction range agrees well with the occurrence of longitudinal and inner cracks. Therefore, thermal strain in ΔT_B

is capable of characterizing the crack susceptibility in the peritectic reaction range[2]. Figure 3 shows the comparison between the calculated critical fracture stresses, σ_c with the measured values. The calculated critical fracture stresses agree well with the experimental parameters, as the carbon content varies from low carbon range to the high one. In the low carbon range, the critical fracture stress varies little with the change of initial carbon content. In the peritectic reaction zone, the critical fracture stress dramatically increases first and then decreases a little, causing a peak in the hyper-peritectic range. And then the critical fracture stress keeps linearly increasing in the medium and high carbon range. Moreover, the critical fracture stress for low carbon steel is much lower than that of medium and high carbon steel, indicating a higher sensibility to inner crack, as reported by Won et al[3].

Figure 4 shows the comparison of the tendency of the calculated difference of deformation energy, W_B with the measured inner crack index. The difference of deformation energy presents a different behavior compared with other characteristic parameters within the whole carbon range. In the low carbon range, W_B drops sharply with the increase of initial carbon content. In the hypo-peritectic range, W_B increases dramatically to a maximum value near the peritectic point. In the hyper-peritectic range, W_B declines dramatically again. Subsequently, in the medium and high carbon range, W_B decrease gently. The tendency of W_B in peritectic reaction range agrees well with the incidence of measured inner cracks. Moreover, W_B for ultra-low-carbon steel is much lower than that of medium and high carbon steel, accounting for a more susceptibility to inner cracks[3].The difference of deformation energy is reliable to characterize the inner crack in the whole carbon range, as proposed by Won et al[3].

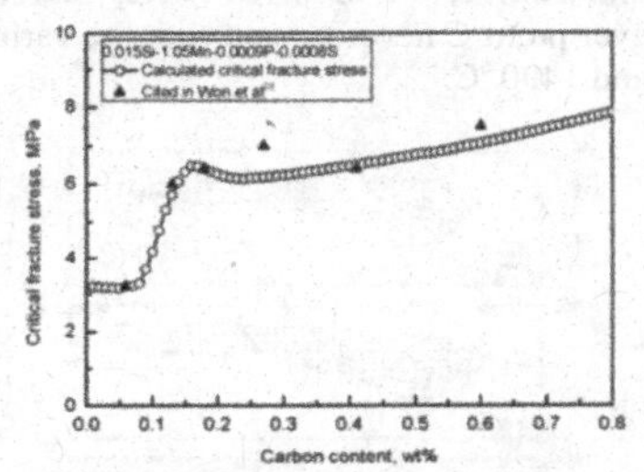

Figure 3. Comparison between calculated σ_c and measured values

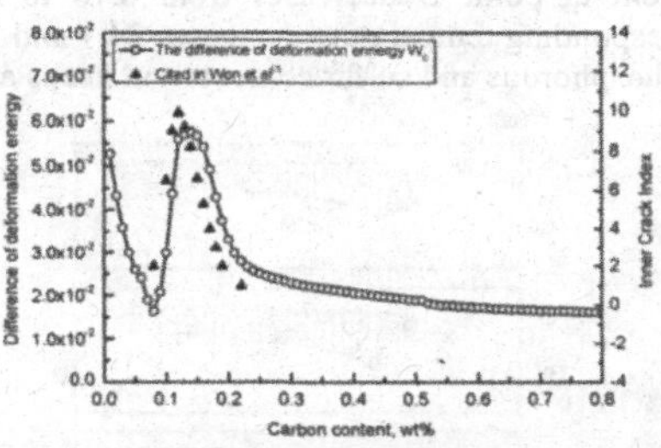

Figure 4. Comparison between the tendency of W_B and measured inner crack index

Micro-segregation Behavior

Based on the calculated effective cooling rate in mushy zone and the micro-segregation analysis, the effect of phosphorous and sulfur content on the micro-segregation behavior of steel is investigated. Figures 5(a) and (b) show variation of solidus and liquidus temperature of continuous casting steel with the increases of phosphorous and sulfur content. In order to illustrate the effect of phosphorous and sulfur, their initial contents in steel matrix are set extremely high, which are seldom produced in steel works With the increase of phosphorous content from 0.007 to 0.054 wt% and sulfur content from 0.016 to 0.048 wt%, the liquidus temperature decreases 1.62 and 1.22 °C in the whole carbon range, respectively. However, the solidus temperature of steel exhibits a different behavior as phosphorous and sulfur content increase. The decrease effect within low carbon range is weaker than that at medium and high carbon range, due to easier distribution of phosphorous and sulfur from γ phase to L phase. Moreover, the decrease effect enlarges with the increase of carbon content, which is consistent with Kim et al[2].

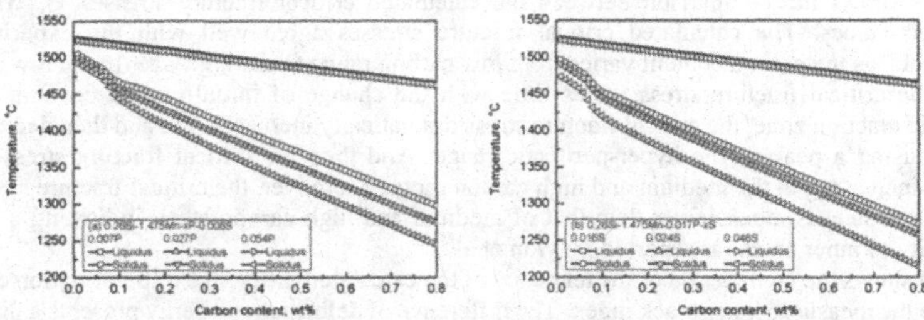

Figure 5. Influence of (a) phosphorus and (b) sulfur on the liquidus and solidus temperature of continuous casting steel

Figures 6 and 7 show the change of peritectic zone with increasing phosphorous and sulfur content. The first peritectic reaction starting before complete solidification, peritectic point and first direct austenization point are noted as point A, B and C for analysis in this paper. According to the scale in figures, the area of peritectic zone extends 1.28 and 1.21 times as phosphorous content increases 0.047 wt% and phosphorous content increases 0.032 wt%. The carbon content at point A keeps constant at 0.06 wt% with the increases of phosphorous and sulfur content, however the corresponding temperature decreases from 15.0 and 20.7 °C, respectively. With increasing phosphorous and sulfur content, the carbon content at point B decreases from 0.16 to 0.136 wt% and from 0.144 to 0.109 wt%, moreover, corresponding temperature decreases 24.7 and 22.3 °C. However point C never moves with the variation of phosphorous and sulfur contents and keeps at 0.46 wt%C and 1490 °C.

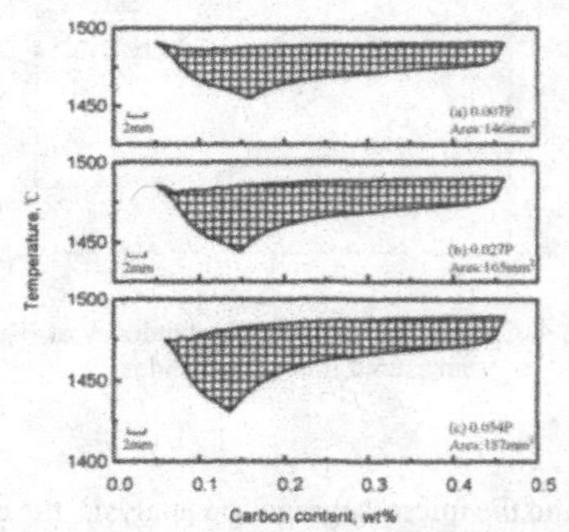

Figure 6. Influence of phosphorous on the peritectic reaction during steel solidification, (a) 0.007, (b) 0.017, (c) 0.027 wt%

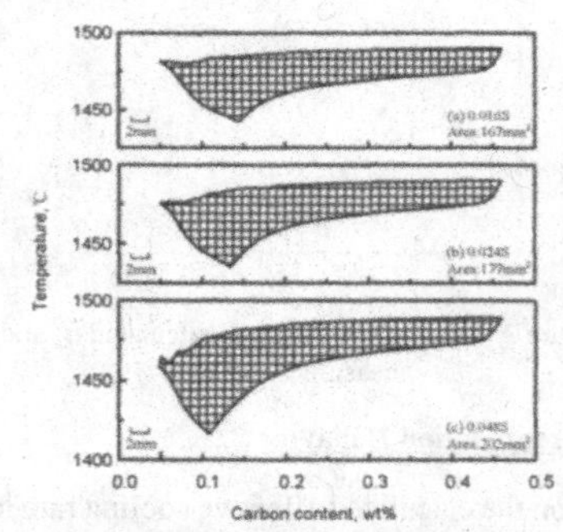

Figure 7. Influence of sulfur on the peritectic reaction during steel solidification, (a) 0.016, (b) 0.024, (c) 0.048 wt%

The Brittle Temperature Range

Figures 8(a) and (b) show the variation of the brittle temperature range, ΔT_B with the increase of phosphorus and sulfur. ΔT_B increases with the increase of initial carbon content within the whole carbon range. Taking the case of 0.017 wt% phosphorus and 0.008 wt% sulfur for instance, the distribution of ΔT_B

the whole carbon range is divided into five parts. In the low carbon range from 0.01 to 0.06 wt%, ΔT_B increases linearly with the increase of initial carbon content, due to the linear effect of carbon on LIT and ZDT. In the carbon range from 0.07 to 0.11 wt%, the nonlinear effect of carbon on ZDT due to the peritectic reaction at the end of solidification dramatically extends the brittle temperature range. In the carbon range from 0.12 to 0.15 wt%, both LIT and ZDT fall into the peritectic reaction zone, and thus change nonlinearly causing a slight drop of ΔT_B. In the carbon range from 0.16 to 0.24 wt%, peritectic reaction ends before reaching ZDT, causing a linear increase of ΔT_B. Then in the carbon range from 0.25 to 0.80 wt%, peritectic reaction ends before reaching LIT, also causing a linear increase of ΔT_B.

Meanwhile, the increment of ΔT_B gradually increases with the increase of phosphorus and sulfur, especially when the carbon content is higher than 0.11 wt%. As phosphorus increases 0.02 wt% and sulfur increases 0.008 wt%, the brittle temperature range increases 7.6 and 6.0 °C at carbon content of 0.11 wt%, while 18.9 and 12.2 °C at carbon content of 0.80 wt%. Therefore as the increase of phosphorus and sulfur, the continuous casting steel becomes more susceptible to cracks, as proposed by Kim et al[2]. Moreover, the impact of sulfur to the brittle temperature range is more powerful than that of phosphorus, similar to Won et al[3] and Cai et al[20].

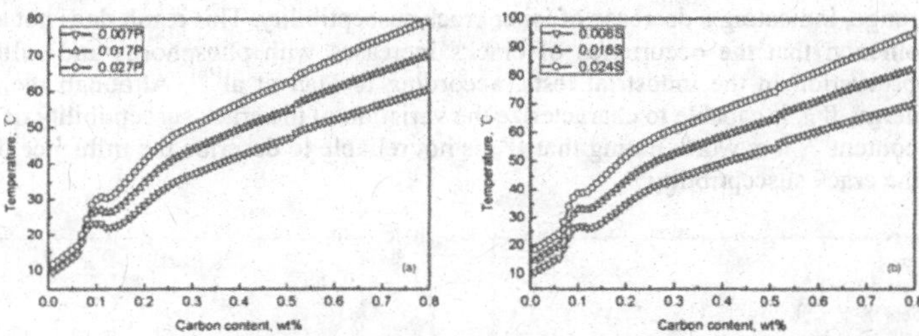

Figure 8. Variation of the brittle temperature range, T_B with (a) phosphorus and (b) sulfur within the whole carbon range

The Thermal Strain

Figures 9(a) and (b) show the variation of the thermal strain, ε^{th} with the increase of phosphorus and sulfur. ε^{th} increases with the increase of phosphorus and sulfur within the whole carbon range. The increment of εth keeps constant at low carbon range, and dramatically increases in the hypo-peritectic range, then gradually decreases in the hyper-peritectic range, and gradually increases in the medium and high carbon range. Consequently, the hypo-peritectic steel is more sensible than other steel. Moreover, the increment of εth in hypo-peritectic range with the increase of sulfur is much higher that caused by phosphorus. For example, the maximum increment near the peritectic point reaches 4.7×10^{-4} and 4.2×10^{-4}, since phosphorus increases 0.02 wt% and sulfur increases 0.008 wt% respectively. Therefore, sulfur gives a more significant influence on the crack susceptibility of continuous casting steel. Meanwhile, with the increase of phosphorus and sulfur, the carbon content corresponding to the maximum thermal strain slightly moves to low carbon range, as reported by Kim et al[2] and Cai et al[20].

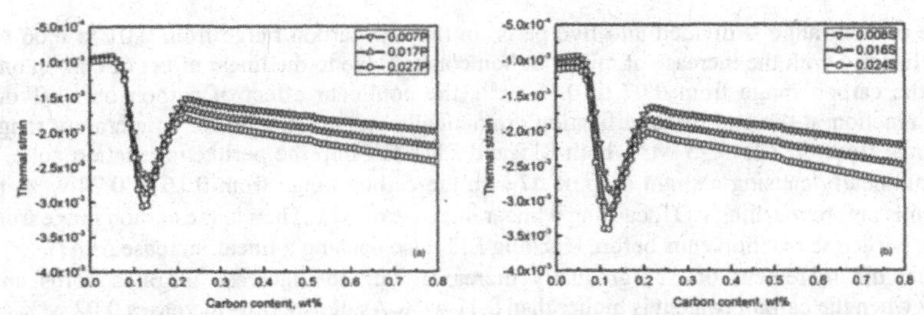

Figure 9. Variation of thermal strain, ε^{th} with (a) phosphorus and (b) sulfur within the whole carbon range

The Difference of Deformation Energy

Figures 10(a) and (b) show the variation of difference of deformation energy, W_B with the increase of phosphorus and sulfur. With the increase of phosphorus and sulfur, W_B generally decreases within the whole carbon range, indicating a decrease of inner crack susceptibility. This result does not agree with the realistic phenomenon that the occurrence of cracks increases with phosphorus and sulfur increasing through the observation in the industrial tests, according to Han et al[14]. Although the difference of deformation energy, W_B, is capable to characterize the variation of the crack susceptibility of steel with the initial carbon content[3], it is worth noting that W_B is not reliable to describe the influence of phosphorus and sulfur on the crack susceptibility.

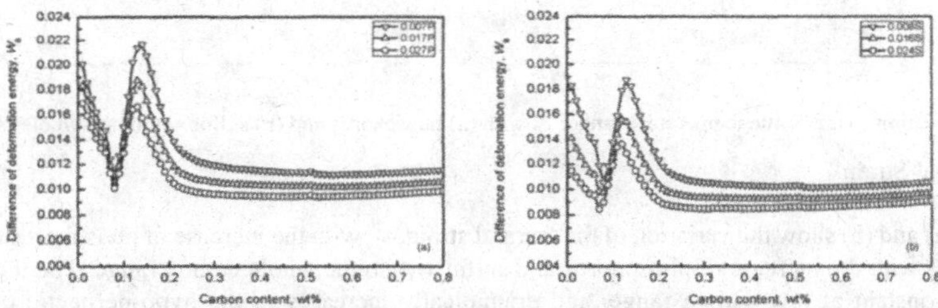

Fig. 10 Variation of the difference of deformation energy, W_B with (a) phosphorus and (b) sulfur within the whole carbon range

Conclusions

This paper has systematically analyzed the effect of phosphorus and sulfur on the pritectic reaction zone, the brittle temperature range, the thermal strain and the difference of deformation energy to investigate their impacts to the crack susceptibility of continuous casting steel, which are built on the developed micro-segregation model. Meanwhile, the capacity of those characteristic parameters to account for the effect of phosphorus and sulfur is discussed. The main conclusions are listed as follows:

(1) In the peritectic range except for the vicinity of peritectic point, the gradients of the brittle temperature range, the thermal strain and the difference of deformation energy to the initial carbon content is higher than that in other carbon range. Meanwhile, both the thermal strain and the difference of deformation energy increase in hypo-peritectic range, decrease in the hyper-peritectic range and reach a peak near the peritectic point. Although the brittle temperature range presents a different behavior in the peritectic range,

it increases in hypo-peritectic range. Therefore, the hypo-peritectic steel is more sensible to cracks than other steels.
(2) With the increase of phosphorus and sulfur, the peritectic reaction zone extents, meanwhile the brittle temperature range and the thermal strain increase within the whole carbon range, indicating an increase of crack susceptibility. Moreover, sulfur does a more powerful effect on the crack susceptibility than phosphorus. For instance, the peritectic reaction zone extends 1.13 and 1.06 times, the brittle temperature range at carbon content of 0.11 wt% increases 7.6 and 6.0 °C, the maximum thermal strain increases 4.7×10^{-4} and 4.2×10^{-4}, respectively, when phosphorus increases 0.02 wt% and sulfur increases 0.008 wt%.
(3) The thermal strain is capable to account for both longitudinal and inner crack susceptibility in the peritectic zone and the impact of phosphorus and sulfur. Whereas, the difference of deformation energy is limited to describe the variation of inner crack susceptibility with the initial carbon content. The brittle temperature range and the peritectic reaction zone are capable to explain the influence of phosphorus and sulfur on crack susceptibility to some extent.

Acknowledgement

The authors sincerely acknowledge the financial support from National Natural Science Foundation of China No. 50925415 and Fundamental Research Funds for Central University of China No. N100102001.

References

1. J. K. Brimacombe, and K. Sorimachi, "Crack Formation in the Continuous Casting of Steel," *Metallurgical and Materials Transactions B*, 8 (2) (1977), 489-505.
2. K. H. Kim et al., "Effect of Carbon and Sulfur in Continuously Cast Strand on Longitudinal Surface Cracks," *ISIJ International*, 36 (3) (1996), 284-289.
3. Y. M. Won et al., "A new Criterion for Internal Crack Formation in Continuously Cast Steels," *Metallurgical and Materials Transactions B*, 31 (4) (2000), 779-794.
4. H. G. Suzuki, S. Nishimura, and Y. Nakamura, "Improvement of Hot Ductility of Continuously Cast Carbon Steels," *Transactions of the Iron and Steel Institute of Japan*, 24 (1) (1984), 54-59.
5. T. Matsumiya et al., "An Evaluation of Critical Strain for Internal Crack Formation in Continuously Cast Slabs," *Transactions of the Iron and Steel Institute of Japan*, 26 (6) (1986), 540-546.
6. G. Shin et al., "Mechanical Properties of Carbon Steels during Solidification," *Tetsu-to-Hagane*, 78 (4) (1992), 587-593.
7. T. Nakagawa et al., "Deformation Behavior during Solidification of Steels," *ISIJ International*, 35 (6) (1995), 723-729.
8. D. J. Seol et al., "Mechanical Behavior of Carbon Steels in the Temperature Range of Mushy Zone," *ISIJ International*, 40 (4) (2000), 356-363.
9. L. Guo et al., "High Temperature Mechanical Properties of Micro-alloyed Carbon Steel in the Mushy Zone," *Steel Research International*, 81 (5) (2010), 387-393.
10. R. Wolfgang, K. Ernst, and B. Bruno, "Computer Simulation of the Brittle-temperature-range (BTR) for Hot Cracking in Steels," *Steel Research*, 71 (11) 2000, 460-465.
11. Y. M. Won et al, "Effect of Cooling Rate on ZST, LIT and ZDT of Carbon Steels near Melting Point," *ISIJ International*, 38 (10) (1998), 1093-1099.
12. A. Yamanaka, K. Nakajima, and K. Okamura, "Critical Strain for Internal Crack Formation in Continuous Casting" *Ironmaking and steelmaking*, 22 (6) (1995), 508-512.
13. Y. M. Won et al., "Analysis of Solidification Cracking Using the Specific Crack Susceptibility," *ISIJ International*, 40 (2) (2000), 129-136.

14. Z. Q. Han, K. K.Cai, and B. C. Liu, "Prediction and Analysis on Formation of Internal Cracks in Continuously Cast Slabs by Mathematical Models," *ISIJ International*, 41 (12) (2001), 1473-1480.
15. Z. Z. Cai, W. L. Wang, and M. Y. Zhu, "Analysis of Shell Crack Susceptibility in Slab Continuous Casting Mold," (Paper presented at Materials Science & Technology 2012, Pittsburgh, Pennsylvania, 07-11 October 2012), 61-72.
16. G. Alvarez De Toledo, O. Campo, and E. Lainez, "Influence of Sulfur and Mn/S Ratio on the Hot Ductility of Steels During Continuous Casting," *Steel Research*, 64 (6) (1993), 292-299.
17. K. Wünnenberg, and R. Flender, "Investigation of Internal Crack Formation in Continuous Casting Using a Hot Model," *Ironmaking and Steelmaking*, 12 (1) (1985), 22-29.
18. W. L. Wang et al., "Micro-Segregation Behavior of Solute Elements in the Mushy Zone of Continuous Casting Wide-Thick Slab," *Steel Research International*, 83 (12) (2012), 1152-1162.
19. I. Jimbo, and A. W. Cramb "The Density of Liquid Iron-carbon Alloys," *Metallurgical and Materials Transactions B*, 24 (1) (1993), 5-10.
20. Z. Z. Cai, and M. Y. Zhu, "Microsegregation of Solute Elements in Solidifying Mushy Zone of Steel and Its Effect on Longitudinal Surface Cracks of Continuous Casting Strand," *Acta Metallurgica Sinica*, 45 (8) (2009), 949-955.
21. Y. Meng, and B. G. Thomas, "Heat-transfer and Solidification Model of Continuous Slab Casting: CON1D," *Metallurgical and Materials Transactions B*, 34 (5) (2003), 685-705.
22. K. H. Kim, K. H. Oh, and D. N. Lee, "Mechanical behavior of Carbon Steels during Continuous Casting," *Scripta Materialia*, 34 (2) (1996), 301-307.

Materials Processing Fundamentals
Edited by: Lifeng Zhang, Antoine Allanore, Cong Wang, James A. Yurko, and Justin Crapps
TMS (The Minerals, Metals & Materials Society), 2013

MATHEMATICAL MODELING OF HEAT TRANSFER AND THERMAL BEHAVIOUR OF TOOL STEEL H13 IN MOLTEN ALUMINUM ALLOY A380

Ting Ding, Junfeng Su, Henry Hu, Xueyuan Nie and Ronald M Barron

Department of Mechanical, Automotive and Materials Engineering
University of Windsor, 401 Sunset Ave., Windsor, Ontario, Canada N9B 3P4
E-mail: ding@uwindsor.ca, huh@uwindsor.ca

Abstract

In high pressure die casting, the die tooling suffers from soldering, erosion and thermal fatigue, which result in heat checking and cracking. To understand the thermal resistance between molten aluminum and tool steel against thermal cracking, a 2D numerical model was adapted to predict heat transfer and thermal behaviour of a cylindrical tool steel (H13) pin in the bath of molten aluminum alloy A380. The established solid model with the Ansys Fluent CFD code was carried out for the simulation. The simulation results show the shell formation of the aluminum alloy upon immersion of the steel pin into the bath of the molten aluminum. As the time increases, the melting of the shell takes place. The thickness and formation progression as well as the melting speed of the shell are dependent on the superheat of the bath. The influence of molten aluminum temperatures on the thermal resistance of the shell was also analyzed based on the results of predicted temperatures at the center of the steel pin as well at the original interface between the steel cylinder and molten aluminum. The verification of the predictions was also carried out experimentally in a laboratory setup.

1. Introduction

Erosion of steel mould due to solidification of liquid aluminum and heat transfer between aluminum and tool steel often takes place in high pressure die casting (HPDC) since the iron and/or the other alloying elements in the die steel are dissolvable in hot liquid aluminum [1, 2]. The similar phenomenon of shell formation on alloying addition such as Ti, Cr, Si, and Ni etc., also occur when iron and/or steel are processed. This is because the solidification of iron or steel at the interface between solid additions and liquid iron or steel owing to the chill effect of additions leads to the formation of a shell of iron or steel around additions. Researchers in the iron and steel industries have made considerable efforts in developing an in-depth understanding of heat transfer and solidification phenomena during shell formation [3-9]. The kinetics of the shell formation and the thermal resistance between the steel shell and the alloying addition has been estimated by using an inverse technique [4-8].

Although solidifications mechanism in the shell formation is quite similar to that present in the erosion of steel mould in HPDC, studies on heat transfer and solidifications phenomena taking place as the steel mould is eroded by liquid aluminum is limited. In this study, a 2D heat transfer and fluid flow model has been developed in an effort to predict the phenomena of heat transfer and solidification occurring in the process of die erosion. The model considering the immersion of the steel pin into liquid aluminum bath is capable of predicting the solidification process of the shell growth as well as the melting process of the shell after reaching its maximum thickness. The rapid heat transfer was simulated from the liquid aluminum to the cylindrical steel pin after immersion by computing the temperature rise in the centre of the pin.

2. Theoretical Considerations and Mathematical Formulation

The analysis of the physical phenomena involved indicates a moving boundary problem, including a heat source (i.e., release of latent heat due to solidification) during the formation of a thin shell, which is further complicated by the presence of a heat sink (i.e. latent heat) as the shell melts back. In an attempt to develop a mathematical formulation for such a complex system, a cylindrical steel pin was immersed into a cylindrical container with liquid aluminum. Figure 1 shows schematically the details of the physical system used in this study.

In order to develop the mathematical formulation for this complex case in the presence of natural convection and heat source, certain simplifying assumptions were made: (i) the fluid motion of the liquid aluminum is a two-dimensional axisymmetric laminar incompressible flow; (ii) the volume variation associated with phase change and the viscous dissipation is ignored; (iii) the liquid aluminum is a Newtonian fluid; (iv) the immersion of the steel bar into the liquid aluminum is assumed to be instantaneous.

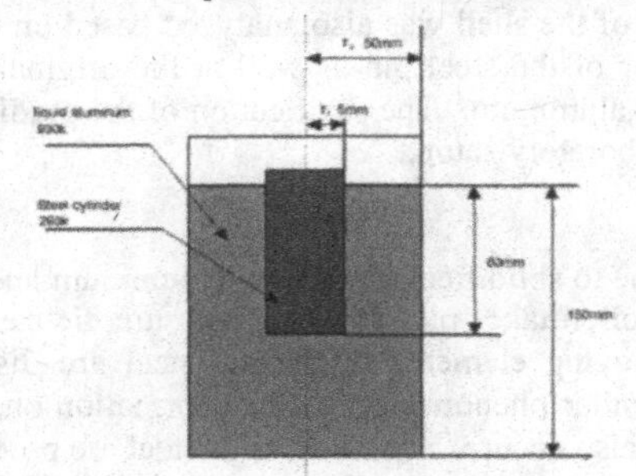

Figure 1. Schematic illustration of the physical system for establishing the model.

The governing differential equations for the laminar and incompressible flow with no viscous dissipation, in a cylindrical polar coordinate system can be represented as:

Continuity Equation:

$$\frac{\partial \rho}{\partial t}+\frac{1}{r}\frac{\partial}{\partial r}(\rho r v_r)+\frac{\partial}{\partial z}(\rho v_z)=0 \quad (1)$$

Momentum Equation:
r - component:

$$\frac{\partial}{\partial t}(\rho v_r)+v_r\frac{\partial}{\partial r}(\rho v_r)+v_z\frac{\partial}{\partial z}(\rho v_r)=$$

$$-\frac{\partial p}{\partial r}+\left[\frac{1}{r}\frac{\partial}{\partial r}(\mu r\frac{\partial v_r}{\partial r})+\frac{\partial}{\partial z}(\mu\frac{\partial v_r}{\partial z})\right]+(-\frac{\mu}{r^2})v_r \quad (2)$$

z - component:

$$\frac{\partial}{\partial t}(\rho v_z) + v_r \frac{\partial}{\partial r}(\rho v_z) + v_z \frac{\partial}{\partial z}(\rho v_z) =$$

$$-\frac{\partial p}{\partial z} + \left[\frac{1}{r}\frac{\partial}{\partial r}(\mu r \frac{\partial v_z}{\partial r}) + \frac{\partial}{\partial z}(\mu \frac{\partial v_z}{\partial z})\right] + \rho g_z \quad (3)$$

Energy Equation:

$$\frac{\partial}{\partial t}(\rho H) + v_r \frac{\partial}{\partial r}(\rho H) + v_z \frac{\partial}{\partial z}(\rho H)$$

$$= \left[\frac{1}{r}\frac{\partial}{\partial r}(kr\frac{\partial T}{\partial r}) + \frac{\partial}{\partial z}(k\frac{\partial T}{\partial z})\right] \quad (4)$$

Since the energy equation consists of both transient and conductive terms in which two dependent variables are included, temperature T and enthalpy H, an enthalpy technique [9-12] which accounts for the phase change is used to convert the energy equation into a non-linear equation with only one dependent variable, enthalpy. Thus, the existing control-volume algorithm can easily be implemented. Temperature and enthalpy are related via the following state equation based on thermodynamics:

$$\frac{dH}{dT} = C(T) \quad (5)$$

If a constant specific heat for each phase is considered and $H = 0$ is chosen to correspond to AZ91D at its solidus temperature, the relation between temperature and enthalpy becomes

$$T = \begin{cases} T_s + H/C_s \; ; & H \le 0 \\ T_s + H \times [\Delta T/(C_m \Delta T + H_f)] \; ; & 0 < H < H_f + C_m \Delta T \\ T_s + [H - H_f - (C_m - C_l)\Delta T]/C_l \; ; & H \ge H_f + C_m \Delta T \end{cases} \quad (6)$$

where $\Delta T = T_l - T_s$ is the solidification range.

The energy equation can be rewritten as follows:

$$\frac{\partial}{\partial t}(\rho H) = [\frac{1}{r}\frac{\partial}{\partial r}(\gamma r \frac{\partial H}{\partial r}) + \frac{\partial}{\partial z}(\gamma \frac{\partial H}{\partial z})] + [\frac{1}{r}\frac{\partial}{\partial r}(r\frac{\partial S}{\partial r}) + \frac{\partial^2 S}{\partial z^2}] \quad (7)$$

$$\begin{cases} \text{Solid Phase} \quad H \le 0; & \gamma = k_s/C_s; \quad S = 0; \\ \text{Mushy Phase} \quad 0 < H < H_f + C_m \Delta T; & \gamma = k_m \Delta T/(C_m \Delta T + H_m); \; S = 0; \\ \text{Liquid Phase} \quad H \ge H_f + C_m \Delta T; & \gamma = k_l/C_l; \quad S = -\frac{k_l}{C_l}[H_f + (C_m - C_l)\Delta T] \end{cases}$$

$$k_m = \lambda k_l + (1-\lambda)k_s \; ; \; C_m = \lambda C_l + (1-\lambda)C_s \; ;$$

$$\rho_m = \lambda \rho_l + (1-\lambda)\rho_s \; ; \; \lambda = \frac{H}{H_f}$$

Initial conditions:

In the domain of calculation, the velocity for the momentum equations was set to zero; the temperature in the steel was set at 293 K; the temperature of the liquid aluminum was specified to be a desired temperature. Mathematically, these are expressed as:

In the steel pin,
$v_r = 0$, $v_z = 0$, T = 293 K, for t = 0, $0<r<r_1$,
In the liquid aluminum,
$v_r = 0$, $v_z = 0$, T = 870, 890, 910 or 930 K, for t = 0, $r_1 < r < r_2$.

Boundary conditions:
The free surface condition was applied to the top surface of the steel. The radial velocity gradient at the free surface and the vertical velocity gradient on the plane of symmetry are zero. The velocity was set to zero at all solid walls. A zero flux condition was used for the concentration and the temperature is kept constant at all the solid walls. An adiabatic condition was assumed for the temperature gradient and the concentration gradient at the top surface due to minor changes in temperature during jmmersion. They can be represented in mathematical forms as follows:

At the axis of symmetry,

$$v_r = 0, \frac{\partial v_z}{\partial r} = 0 \frac{\partial T}{\partial r} = 0,$$

for r = 0, $0 < z < z_2$;

At the free surface,

$$\frac{\partial v_r}{\partial z} = 0, \quad v_z = 0, \quad \frac{\partial T}{\partial z} = 0,$$

for $0 < r < r_2$, $z = z_2$;

At the bottom wall and side wall,
$v_r = 0$, $v_z = 0$, T = 870, or 890, 910, 930 K,
for $0 < r < r_2$, z = 0, and $r = r_2$, $0 < z < z_2$;

For transient simulations, the governing equations must be discretized in both space and time. The spatial discretization for the time-dependent equations is identical to the steady-state case. Temporal discretization involves the integration of every term in the differential equations over a time step. After the discretization of the governing equations, the discretized equations were resolved by the Ansys Fluent CFD code with the control volume implicit finite difference procedure. The thermophysical properties of alloy A380 and tool steel H13, which were assumed constant, are given in Table 1.

Table 1. Thermophysical properties of aluminum alloy A380 and tool steel H13

	Aluminum A380	Steel H13
C_p (J/(kg-K))	963	500
Latent heat (J/kg)	321000	
Thermal conductivity (W/(m-K))	96.4	30
Viscosity (kg/(m-s))	1.8e-5	
Density (kg/m^3)	2.74	7.75
Solidus temperature (K)	813	1448
Liquidus temperature (K)	868	1526

3. Results and Discussion

3.1. Temperature Distribution and Solidification Front

The model was used to determine temperature profiles and the movement of the solidification front of the aluminum shell formed on the surface of the cylindrical steel pin. Figure 2 shows the temperature iso-contours at different instants of time after the immersion of the steel pin into the aluminum bath (870 K). As can be seen from the predictions, the temperature profiles are vertically symmetric and horizontally uniform at the beginning of solidification upon the immersion of the steel pin. Due to the chill effect of the pin which conducts heat rapidly away from aluminum adjacent to the pin, aluminum in contact with the pin solidifies to form a shell around the pin. As the aluminum solidification proceeds, the temperature of the steel pin tends to increase as the shell grows and becomes thicker and thicker. Four seconds after the immersion, the thickness of the shell has increased to its maximum of 4 mm. Meanwhile, the rising temperature at the center of the pin reaches a plateau of 825 K. After that, the continuing heat transfer from the bath of liquid aluminum to the pin causes the meltback of the shell. As soon as the meltback starts, the shape of the moving interface between the shell and liquid aluminum progresses quickly into nonuniform contours along the vertical symmetrical line. Due to the effect of the two-dimensional heat conduction and convection in both radial and longitudinal directions, the bottom half of the shell melts relatively faster than its top portion. Also, heat conduction away from the top section of the pin extracts additional heat to the surrounding media. As a result, the aluminum shell takes the shape of a cone cup.

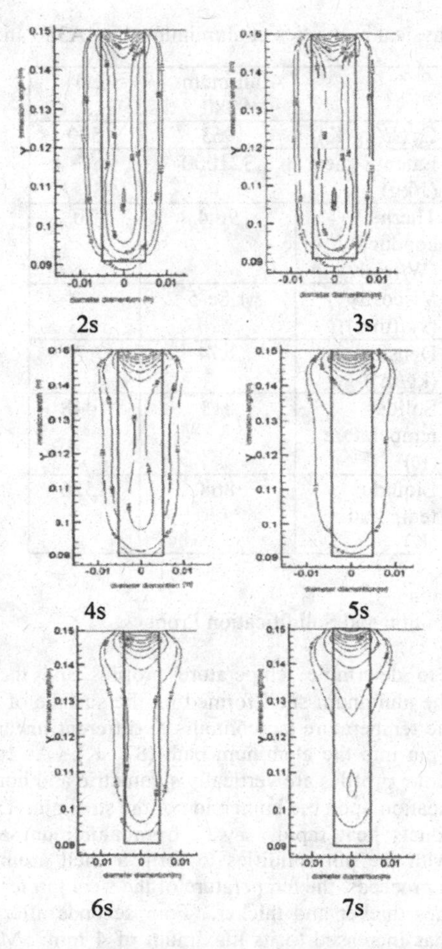

Figure 2. Steel H13 pin in liquid aluminum at various times after immersion: isotherms and shell formation, (a) 2, (b) 3, (c) 4, (d) 6 and (e) 7 s, respectively.

Figure 3 displays the variation of the aluminum shell thicknesses with immersion times at different superheats of the molten bath. As seen from Figure 3, the aluminum shell experiences two periods of the formation process, i.e., the growth and the meltback. In the first period, upon the immersion, the aluminum shell grows very fast. After a certain period, the shell reaches its maximum thickness at a peak value. The maximum thickness of the shell is influenced by the bath superheat. As the superheat is increased from 2 K (870 K) to 52 K (930 K), the peak value of the shell thickness rises from 2.7 mm to 4.2 mm (Figure 4), which leads to 50% increase in the maximum thickness. The lack of heat capacity in the bath with a low superheat should be responsible for the formation of a large aluminum shell around the steel pin. It is interesting to see from Figure 3 that the reduction in superheat postponed the arrival of the peak value of the shell formation. Once passing its growth peak, the shell entered its second period, i.e, the meltback. Due to continuous heat transfer from the bath to the steel pin, the shell thickness decreases after it starts melting back. The rate of the shell meltback is also dependent on the bath superheat. The higher the superheat, the faster the meltback took place. It is worthwhile noting that a decrease in the superheat from 52 K (930 K) to 2 K (870 K) extended the entire shell period including the growth and the meltback from 5.1 to 7.5 seconds.

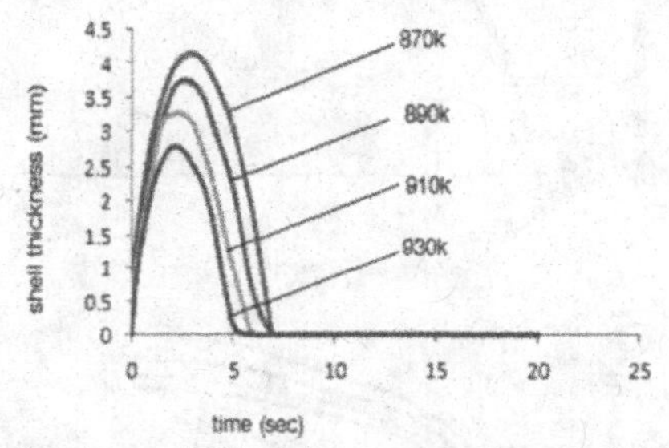

Figure 3. Thickness of aluminum shell vs immersion time at different superheats of the bath.

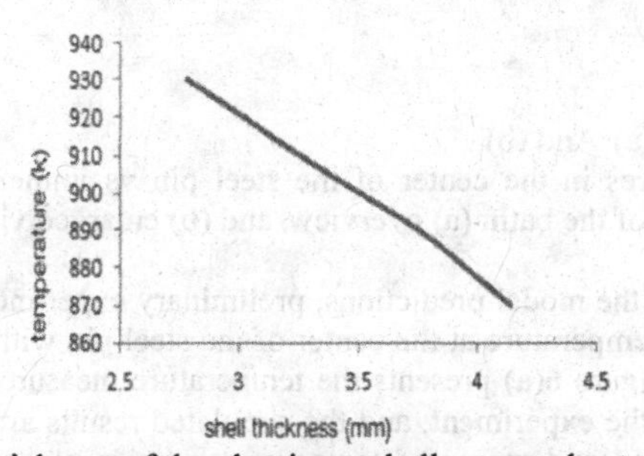

Figure 4. Maximum thickness of the aluminum shell vs superheats of the bath.

Figure 5 illustrates temperatures in the center of the steel pin vs immersion time at different superheats of the bath. It can be seen from Figure 5 that, upon the immersion, the temperature at the center of the steel pin rises very rapidly owing to the presence of the large temperature gradient at the interface between the pin and the molten aluminum bath. However, the shell formation generates a thermal resistance between the pin and the molten aluminum bath. The generated thermal resistance reduces the inward heat flux from the bath to the inner pin. The increase in temperature at the center of the steel pin is slowed down. Even after the shell melts back, which eliminates the thermal resistance, the temperature at the center of the steel pin keeps increasing, but at a very slow pace. The slow increase in the central temperature should be attributed to the drop in the temperature gradient between the pin and the molten aluminum bath following an extended period of immersion, which in turn reduces the inward heat flux in the radial direction to the pin. The effect of the bath superheat on the temperature variation at the center of the steel pin appeared less significant than on the maximum thickness of the shell. The reason for this difference is due to the fact that the variation of the central temperature in the pin is primarily influenced by the temperature gradient between the pin and the bath while the shell formation is mainly determined by the superheat.

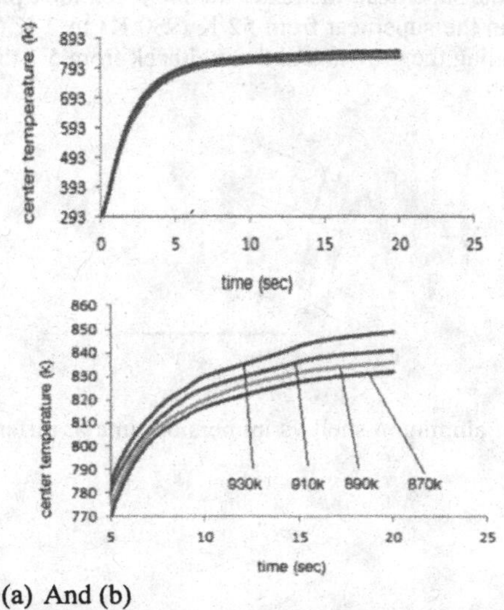

(a) And (b)

Figure 5. Temperatures in the center of the steel pin vs immersion time at different superheats of the bath; (a) overview, and (b) enlarged view.

In an attempt to verify the model predictions, preliminary experimental work was carried out by measuring the temperature at the center of the steel pin with the bath superheat of 52 K (930 K) [13]. Figure 6(a) presents the temperature measurements at the center of the pin obtained from the experiment, and the simulated results are given in Figure 6(b). In general, the predicted and measured temperatures are in close agreement during the entire immersion period. However, there are some unavoidable discrepancies between the numerical and experimental results because of the very complicated nature of the problem. It can be seen from Figure 6(a) that, 10 seconds after immersion, the central temperature has risen from 293 to 788 K. Meanwhile, the predicted central temperature becomes 829 K. This minor discrepancy (5%) might, at least in part, be due to the inaccuracy of the thermophysical data used for the computation, and the uncertainty of the temperature measurements. Further experimental investigation on the change of shell thickness with immersion times by using load cell techniques is assuredly needed, which could give valuable detailed information on the mechanism of the shell formation.

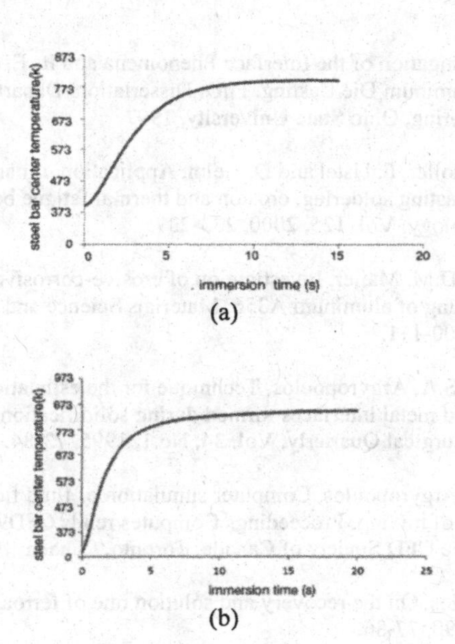

Figure 6. Comparison of the experimental (a) and simulated (b) results of the temperature in the centre of the steel pin.

4. Conclusions

A mathematical model has been developed based on the finite volume technique and the enthalpy method to predict heat transfer and thermal behaviour of a cylindrical tool steel (H13) pin in the bath of molten aluminum alloy A380. The model predicts the temperature distributions, the heating curves, and the shape and position of the aluminum shell formed during the immersion of the pin into the aluminum bath.

The predictions indicate that the shell formation process consisted of two periods, i.e., the growth and the meltback. The simulated results suggest the presence of a relationship between the bath superheat and the maximum shell thickness, as well as the exact time corresponding to the arrival of the maximum shell thickness. It was also predicted that the temperature at the center of the steel pin varied with the immersion time, which was influenced by the presence of the thermal resistance after the formation of the aluminum shell. Comparison also revealed reasonable agreement between computational results and the preliminary experimental measurements.

ACKNOWLEDGEMENTS

The authors would like to express their appreciations to the Natural Sciences and Engineering Research Council of Canada and the University of Windsor for supporting this work.

References

1. Y.-L. Chu, Investigation of the Interface Phenomena and its Effect on Erosion and Corrosion in Aluminum Die Casting, PhD. Dissertation, Department of Industrial and Systems Engineering, Ohio State University, 1997.

2. C. Mitterer, F. Holler, F. Ustel and D. Heim, Application of hard coatings in aluminium die-casting soldering, erosion and thermal fatigue behaviour. Surface and Coatings Technology, Vol. 125, 2000, 233-239.

3. A.E. Miller and D.M. Maijer, Investigation of erosive-corrosive wear in the low pressure die casting of aluminum A356. Materials Science and Engineering A, Vol. 435-436 2006, 100-111.

4. N.J.Goudie and S.A. Argyropoulos, Technique for the estimation of thermal resistance at solid metal interfaces formed during solidification and melting, Canadian Metallurgical Quarterly, Vol. 34, No.1, 1995, 73-84.

5. H. Hu and S.A. Argyropoulos, Computer simulation of fluid flow associated with an exothermic heat of mixing. Proceedings Computes rends CFD94, Second Annual Conference of the CFD Society of Canada, Toronto, Canada, 1994, 559-566.

6. S.A. Argyropoulos, On the recovery and solution rate of ferroalloys. Transactions of the ISS, May 1990, 77-86.

7. S.A. Argyropoulos and R.I.L. Guthrie, The dissolution of titanium in liquid steel, Canadian Metallurgical Quarterly, Vol. 18, 1979, 267-281.

8. S.A. Argyropoulos and P. G. Sismanis, The mass transfer kinetics of niobium solution into liquid steel. Metallurgical Transactions B, Vol. 22B, August 1991, 417-427.

9. J. Crank, Free and Moving Boundary Problems. Oxford Science Publications, 1984.

10. V. Alexiades and A.D. Solomon, Mathematical Modelling of Melting and Freezing Processes, Hemisphere Publishing Corporation, 1993.

11. H. Hu and S.A. Argyropoulos, Mathematical simulation and experimental verification of melting resulting from the coupled effect of natural convection and exothermic heat of mixing. Metallurgical and Materials Transactions B, Vol. 28B, 1997, 135-148.

12. H. Hu and S.A. Argyropoulos, Mathematical modeling and experimental measurements of moving boundary problems associated with exothermic heat of mixing. International Journal of Heat and Mass Transfer, Vol. 39, No.5, 1995, 1005-1021.

13. H. Hu, Unpublished Internal Research, University of Windsor, Windsor, Ontario, Canada, 2012.

Materials Processing Fundamentals
Edited by: Lifeng Zhang, Antoine Allanore, Cong Wang, James A. Yurko, and Justin Crapps
TMS (The Minerals, Metals & Materials Society), 2013

Optimization Investigation on the Soft Reduction Parameters of Medium Carbon Microalloy Steel

Chao Xiao[1,2], Jiongming Zhang[1,2], Yanzhao Luo[1,2], Lian Wu[1,2], Shunxi Wang[1,2]
[1]State Key Laboratory of Advanced Metallurgy, University of Science and Technology Beijing, Beijing 100083, China
[2]School of Metallurgical and Ecological Engineering, University of Science and Technology Beijing, Beijing 100083, China

Key words: medium carbon microalloy steel, soft reduction, central segregation, internal crack

Abstract

In current paper, medium carbon microalloy steel is selected for systematically carrying out the soft reduction tests, a set of conventional process which doesn't apply soft reduction technology is arranged to compare. Taking into account of both the etched results and the carbon segregation distributions along the thickness direction, it is found that influence of soft reduction process is so effective that central segregation of strands significantly decrease. Internal defects including central segregation, porosity and internal crack are comprehensively considered, the optimum soft reduction parameters are the total reduction amount (TRA) of 5mm, the solid fractions between 0.2 and 0.5, and the corresponding reduction zone of 6~7segments.

Introduction

Internal defects such as central segregation, central porosity and so on occurred during the continuous casting process, have an extremely adverse effect on the properties of the finished product. There have been many attempts to solve these defects by the low temperature casting[1], the final electromagnetic stirring[2], the continuous forging technology[3], the soft reduction technology[4] et al. Among these various methods the soft reduction technology has been widely used in global steel mills due to their effectiveness in reducing macrosegregation of strands[5].

The basic idea of any kind of soft reduction is to suppress the formation of central macrosegregation and porosity by compensation of the solidification shrinkage and interrupt the suction flow of the residual melt, at the same time, the corresponding operation is to carry out an suitable reduction amount at an appropriate mushy zone of the final solidification stage by using the pinch rolls or other specialized equipments. However, the dynamic soft reduction is that it can online track the thermal state of strands and thus adjust the reduction positions rapidly based on the real-time solidifaication condition. At present, dynamic soft reduction is greatly developing in the worldwide context and becoming the most important continuous casting technology for any steel plant, because of its unique advantages of improving internal quality of strands[6-9], and this technology has be one of the key technologies which can represent the level of the continuous casting machine[10].

Since the 2# continuous caster of a steel plant put into production, there is no attemp to optimize the soft reduction parameters suggested by VAI, the drawn result is that the internal quality of produced slabs is very poor. In current paper, medium carbon microalloy steel is selected for systematically carrying out the soft reduction experiments, and a set of conventional process which doesn't apply soft reduction technology is arranged to compare. Considering both the assessed results of eched samples and carbon segregation distributions along the thickness direction, the optimum soft reduction parameters are obtained.

Experimental Scheme

The main soft reduction parameters are the total reduction amonut (TRA), reduction positions and reduction rate. Among these paremeters, the reduction rate is equal to the product of the reduction amount (its unit is mm/m) and the casting speed, if the reduction amount or casting speed changes, accordingly, the reduction rate will have an change, however, the casting speed is almost a constant during the casting process, so the reduction rate will not be taken into account in the following analysis, only online tracking and recording the parameter of casting speed.

In order to find out the optimal soft reduction parameters as soon as possible, soft reduction trials are carried out in two campaigns on slab caster 2#. The objective of the first campaign is to determine the most favourable reduction range from a metallurgical point of view, by setting the TRA of 4mm and 5mm used commonly in normal production and the reduction positions of fs = 0.1~0.5, 0.0~0.2, 0.2~0.5, 0.2~1.0. In the second campaign, four-group soft parameters of TRA of 6mm and reduction zone of fs=0.0~0.20, 3mm and fs=0.0~0.20 and 0.1~0.7, 3.5mm and fs= 0.5~0.95 are set in terms of the results of previous soft reduction trials using the parameters given by VAI, the main purpose is to determine the optimum TRA, in addition, a set of conventional process which doesn't apply soft reduction technology is arranged to compare. After these soft reduction trials are performed, the optimum soft reduction parameters will be got easily.The regarded steel grade is the medium carbon microalloy steel with the transverse section of 220mm×2080mm, the soft reduction parameters and the basic casting parameters for each trial are shown in Table I.

Table I. Soft Reduction Parameters and Basic Casting Parameters

Number	Steel Grade	Casting speed m/min	Tundish temperature,°C	TRA,mm	Reduction zone fs	Correspondi-ng segments
1	Q345-T	1.15	1530	4	0.1~0.5	Seg 6~7
2	Q345-T	1.15	1537	4	0.0~0.2	Seg 5~6
3	Q345-T	1.15	1532	4	0.2~0.5	Seg 6~7
4	Q345	1.2	1518	4	0.2~1.0	Seg 7~10
5	A32	1.1	1526	5	0.1~0.5	Seg 6~7
6	A32	1.15	1527	5	0.0~0.2	Seg 5~6
7	A32	1.1	1529	5	0.2~0.5	Seg 6~7
8	Q345	1.2	1528	5	0.2~1.0	Seg 7~10
9	Q345R	1.15	1522	6	0.0~0.2	Seg 5~6
10	Q345R	1.15	1524	0	0	
11	A32	1.15	1531	3	0.4~0.95	Seg 7~9
12	Q345R	1.1	1529	3	0.1~0.7	Seg 6~8
13	Q345qD	1.05	1525	3.5	0.5~0.95	Seg 7~8

Evaluation Methods

Soft reduction trials are carried out in accordance with the above scheme. When the casting parameters of each trial keep stable about 30min, a slab sample with the cross section of 220mm×2080mm is taken by flame cutting, its thickness along the casting direction is 80mm or so, shown in Figure 1. In the following, after the slab samples are etched with 50% hydrochloric acid-water solution for revealing the macrostructure, they are placed under the magnascope to examine the structure and all visible macrostructural features such as nature and area fraction of chill, columnar and equiaxed zones in the cast structure are recorded by the picture. Furthermore use the standard diagrams to assess their internal qualities and thus get the corresponding degree, for evaluating the effectiveness of various soft reduction parameters.

After the etched experiment is completed, take a sample with the width of 50mm at the half of the slab width to analyze the carbon segregation distribution along the slab thickness direction, as presented in Figure 2. Besides, the etched experiment may influence the element concentration, so the etched surface of sample is treated by machining, and the amount is not less than 2mm.

After the samples for the segregation distribution analysis are machined, use the shaper to get the steel chips, then the chips are analyzed to obtain the carbon concentration using the high frequency infrared carbon and sulfur analyzer CS-3000, the thickness for each group of steel chips is 0.5mm, clearly shown in Figure 3. For convenience of shaping steel chips and carbon segregation distribution analysis, a coordinate axis whose coordinate origin is the center of slab thickness is established, and the inner arc side is defined as the positive semiaxis, the outer arc side is the negative semiaxis, so the coordinates of the center point for each group of steel chips are as follows: 105, 90, 75, 60, 45, 35, 25, 15, 12, 10, 8, 6, 4, 2, 0, -2, -4, -6, -8, -10, altogether twenty groups of steel chips are taken for each sample.

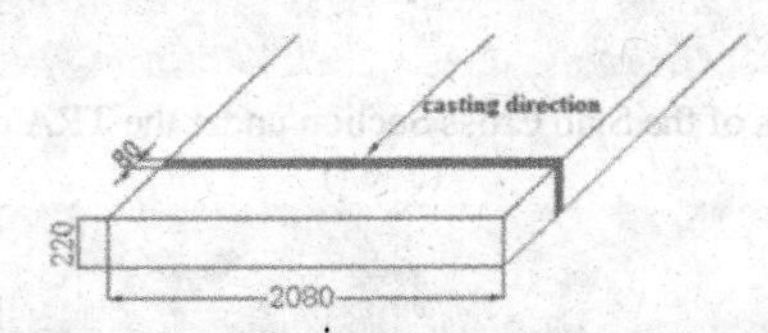

Figure 1. Schematic Diagram of Taking the Slab Samples for Etched Experiment

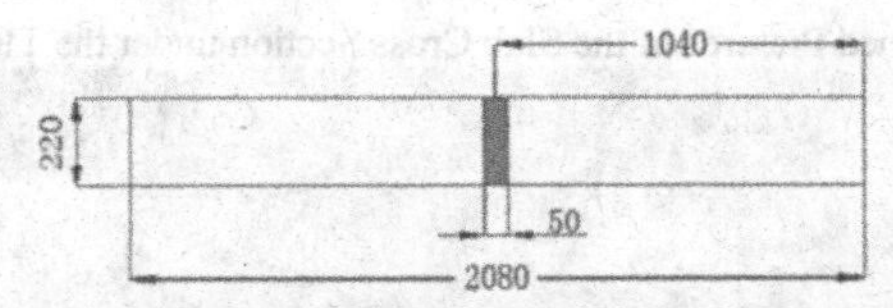

Figure 2. Schematic Diagram of Taking the Samples for Analyzing the Carbon Segregation Distribution

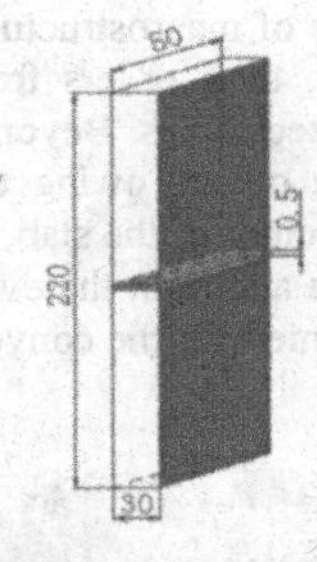

Figure 3. Schematic Diagram of Taking Steel Chips

Results and Discussion

Results of Etched Experiment

On the basis of the difference of the TRA, the pictures of macrostructure inspection is divided into five parts, after the etched experiment is completed, represented in Fig.4 to Fig.8, respectively, in particular, the photo of the sample without soft reduction is acted as a single part.

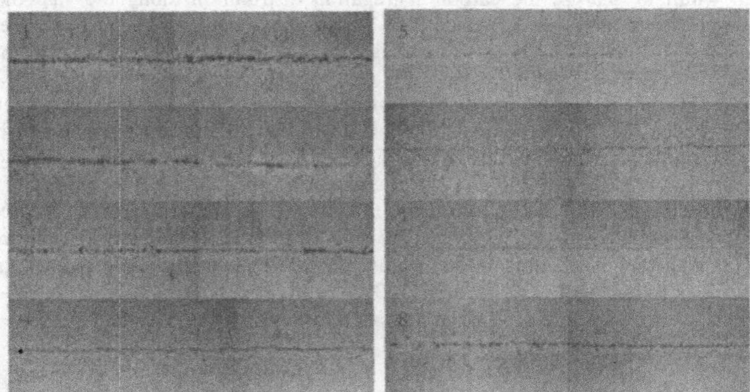

Figure 4. Etched Pictures of the Slab Cross Section under the TRA of 4mm (1~4#) and 5mm (5~8#)

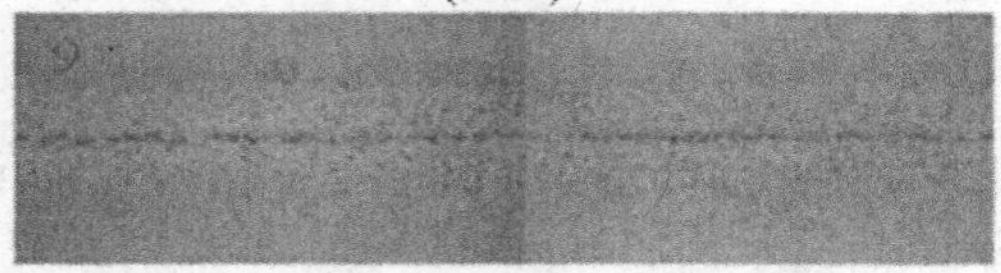

Figure 5. Etched Pictures of the Slab Cross Section under the TRA of 6mm

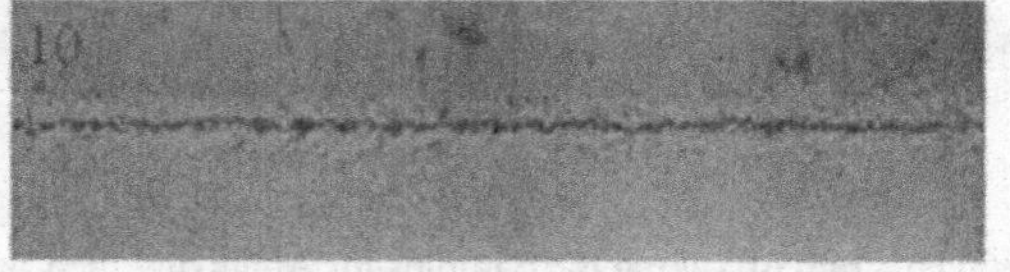

Figure 6. Etched Pictures of the Slab Cross Section without Soft Reduction

It can be found from the above pictures of macrostructure inspection that the macrostructure of the slab is divided into three parts, their orders from the surface to the center are as follows: the chill, columnar and equiaxed zones. Beyond a short distance from the surface, rapid coarsening of columnar dendrite occurs owing to reduced heat extraction from the surface, and they almost extend to the centre of the slab, so the chill and equiaxed zones only occupies a quite small area fraction. In addition, the evaluated degrees of the defects of the etched samples are listed in Table II, aimed for the convenience to assess the effectiveness of different soft reduction parameters.

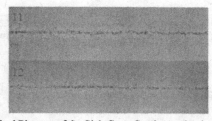

Figure 7. Etched Pictures of the Slab Cross Section under the TRA of 3mm

Figure 8. Etched Pictures of the Slab Cross Section under the TRA of 3.5m

Table II. Assessed Results of the Etched Slabs using Standard Diagrams

Numbers	Central segregation	Central porosity	Intermediate crack
1	B1.5	1.5	1.5
2	B1.5	1.0	1.0
3	B1.0	1.5	1.5
4	C1.5	1.0	0.5
5	B1.0	1.0	1.0
6	C1.0	1.0	0.5
7	C1.0	1.0	0.5
8	B1.5	1.0	1.5
9	C1.5	1.0	1.0
10	B2.0	1.5	2.0
11	B1.5	1.0	1.0
12	B1.5	1.0	0.5
13	B1.5	1.0	0.5

Table II describes that each of the thirteen slab smaples has the intermediate crack occuring, but for the No.10 sample its crack is the most serious, the degree reaches to 2.0. The samples with the segregation level of B1.5 have a relatively high proportion, nearly occupying a half, particularly, the No.10 sample has the severest central segregation degree of B2.0. It can be seen that the No.7 sample have the best internal quality, at the same time, the internal quality of the No.10 sample without soft reduction is the poorest, clearly, the soft reduction technology could obviously improve the internal quality of slab.

Taking into account of the macrostructure inspection of Figure 4 to Figure 8 and the assessed results of Table II, the findings resulting from the trials can be summarized as follows: among the results of different TRA, the trials with the TRA of 5mm receive the optimum internal quaity, however, for the reduction zone, the most favourable is solidification fraction range of fs = 0.2~0.5, meanwhile, for the No.7 sample with the TRA of 5mm and the reduction positions of fs = 0.2~0.5, its internal quality is the best from the metallurgical point of view, therefore, the optimal soft reduction parameters preliminarily determined are the TRA of 5mm and the solid fraction range of fs = 0.2~0.5.

Carbon Segregation Distribution

The obtained steel chips are detected to get the carbon concentration by the high frequency infrared carbon and sulfur analyzer, at the same time, the segregation index is used to evaluate the carbon segregation distribution for convenience. Each slab sample takes twenty groups of steel chips, so it would get the twenty carbon contents, taking the average value of these 20 datas is acted as the initial carbon content $C_0 = \sum_{i=1}^{20} C_i / 20$, thus easily get the ratio of each carbon centent to the initial centent C_0, it means that the segregation index $\rho_i = {C_i}/{C_0}$ for any position of taking steel chips is presented, where $\rho_i > 1$ is the positive segregation, $\rho_i < 1$ is the negative segregation.

Determination of the Optimum Reduction Zone.Figure 9 shows that for the reduction ranges of fs=0.2~0.5 and 0.2~1.0, their maximum segregation indexes are: 1.396 and 1.291, respectively, when the TRA is 4mm, and they are much smaller than the results of the other reduction positions. Therefore, the reduction ranges of fs=0.2~0.5, 0.2~1.0 are relatively suitable for the TRA of 4mm.

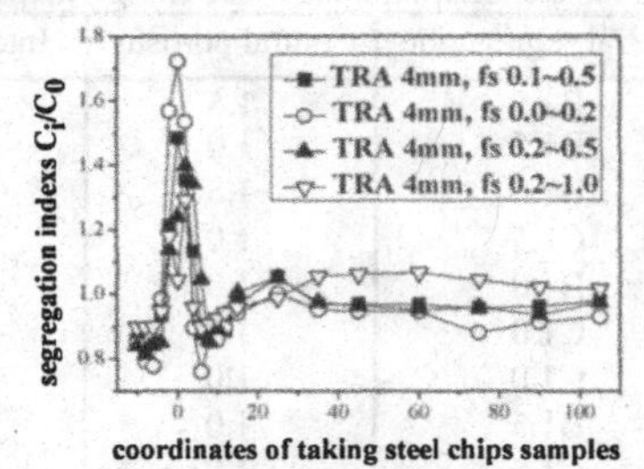

Figure 9. Carbon Segregation Distribution along the Thickness Direction of Slab under the TRA of 4mm

As shown in Figure 10, when the TRA is 5mm, the segregation indexes of the reduction zones of fs=0.1~0.5 and 0.2~0.5 are 1.143 and 1.142, respectively, and they are much smaller than the indexes of the other two reduction ranges, so the more favourable reductionzones are the solid fraction ranges of fs=0.1~0.5 and 0.2~0.5, for the TRA of 5mm.

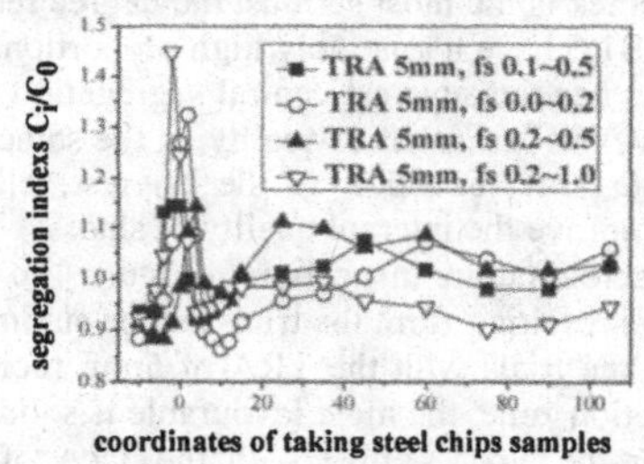

Figure 10. Carbon Segregation Distribution along the Thickness Direction of Slab under the TRA of 5mm

From the carbon segregation distribution of the TRA of 4mm (Figure 9) and of the TRA of 5mm (Figure 10), the conclusions can be drawn that the optimal reduction range is the solid fraction range of fs =0.2~0.5.

Determination of the Optimum TRA.In terms of the above analysis, the best result can be obtained if the soft reduction trials with the reduction zone of solid fraction range of fs=0.2~0.5 are carried out. Due to the crack susceptibility of the medium carbon microalloy steel, it often has the intermediate crack occurring during the continuous casting process, and the occurring probability could greatly increase when the TRA is relatively large during the soft reduction process, on the other hand, from the viewpoint of the previous researcher, if the TRA is too large, the risk of occurring internal crack will have a sharp increase[11]. As presented in Figure 11, when the soft reduction parameters are the TRA of 5mm and the reduction zone of fs=0.2~0.5, the maximum carbon segregation is controlled to 1.142, the minimum segregation index is only 0.883, and except the area near the center point of the slab, the segregation indexes of the rest positions only have a quite small fluctuation between 1.0 and 1.1, these results are much better as compared to the segregation distribution of the TRA of 4mm, therefore, the TRA of 5mm is temporarily intended to be optimum.

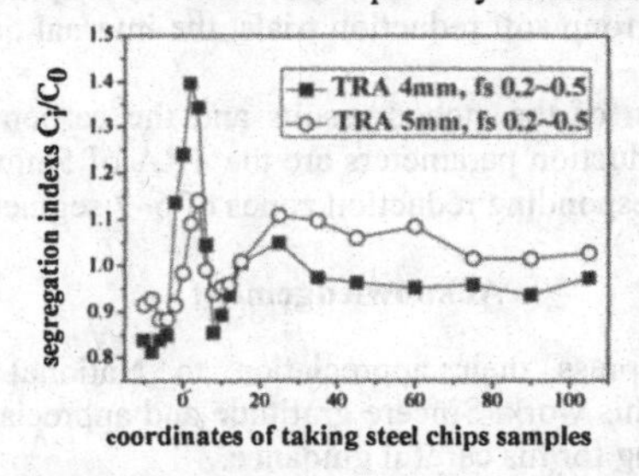

Figure 11. Carbon Segregation Distribution along the Thickness Direction of Slab under the Reduction Zone of Solid Fraction fs=0.2~0.5

Determination of the Optimum Reduction Parameters. From the 8-group soft reduction experiments for the TRA of 4mm and 5mm, the best soft reduction parameters are: the TRA of 5mm, reduction solid fraction range of fs=0.2~0.5, and in the following analysis, the results drawn by this group parameters are acted as the basis to compare with carbon segregation distribution of the other 5-group trials, in order to determine the optimal soft reduction parameters.

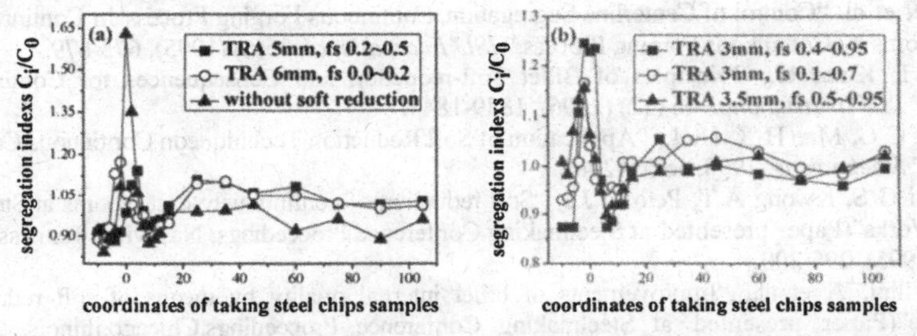

Figure 12. Carbon Segregation Distribution along the Thickness Direction of Slab under the TRA of 3, 3.5, 5, 6mm and without Soft Reduction

As shown in Figure 12, the maximum segregation index is controlled to 1.173, when the trial with the TRA of 6mm and the solid fraction range of fs= 0.0~0.2 is performed, however, at the same reduction zone, the maximum segregation indexes for the TRA of 4mm and 5mm are only controlled to 1.722 and 1.323, respectively, these segregation levels are relatively high, moreover, the controlled result of internal cracks must be considered, when the reduction range of fs=0.0~0.2 is applied, the degrees of intermediate crack for the TRA of 4, 5, 6mm are in order of 1.0, 0.5, 1.0, and these 3-group soft reduction trials are all carried out in the segments 5~6, if the above three soft reduction process is proceeded at the straightening segments 7~8, the risk of the intermediate crack occurring must have an gteat increase, therefore, the optimal soft reduction parameters are the TRA of 5mm, the solid fraction range of fs=0.2~0.5, and the corresponding reduction zones of segments 6~7, after considering the carbon segregation distributions of six-group soft reduction parameters, represented in Figure 12.

Conclusions

(1) The continuous casting soft reduction could remarkably improve the internal quality of the slab, among the the 13-group soft reduction trials, the internal quality of the slab without soft reduction is the poorest.
(2) Tking into consideration of the etched results and the carbon segregation distribution analysis, the optimal soft reduction parameters are the TRA of 5mm, the solid fraction range of fs = 0.2~0.5, and the corresponding reduction zones of 6~7 segments.

Acknowledgement

The authors gratefully express their appreciation to National Natural Science Fund (51174024) for sponsoring this work. Sincere gratitude and appreciation should be expressed to Professor Jiongming Zhang for his careful guidance.

References

1. A. Shiraishi et al. ,"Improvement of Internal Quality by Stable Casting Under Low Teeming Temperature Using Tundish Induction Heater" (Paper presented at The Sixth International Iron and Steel Congress, Nagoya, Japan, 21-26 October 1990), 264-270.
2. Kyung S O, Young W C, "Macrosegregation Behavior in Continuously Cast High Carbon Steel Blooms and Billets at the Final Stage of Solidification in Combination Stirring,"*ISIJ International*, 35(7) (1995), 866-875.
3. Seiji N et al. "Control of Centerline Segregation Continuous Forging Process In Continuously Cast Blooms by Continuous Forging Process," *ISIJ International*, 35(6) (1995), 673-679.
4. Ralf T, Klaus H, "Principles of Billet Soft-reduction and Consequences for Continuous Casting," *ISIJ International*, 46 (12) (2006), 1839-1844.
5. Cheng C G, Mao H X, Hu L, "Application of Soft Reduction Techniqueon Continuous Casting Process," *Steelmaking*, 1998, no.5: 42-45.
6. Sakaki G S, Kwong A T, Petozzi J J, "Soft reduction of continuously-cast blooms at Stelco's Hilton Works"(Paper presented at Steelmaking Conference Proceedings, Nashville, Tennessee, 2-5 April 1995), 295-300.
7. Cristallini. A et al., "Improvements of billet internal quality by means of soft reduction technique"(Paper presented at Steelmaking Conference Proceedings,Chicago,illinois, 20-23 March 1994), 309-315.
8. Itakurav T et al., "High carbon tool steel casting by a vertical bending caster with soft reduction"(Paper presented at Steelmaking Conference Proceedings,Chicago,illinois, 20-23 March 1994), 291-295.
9. Sivesson P, Örtlund T, Widell B, "Improvement of inner quality in continuously cast billets through thermal soft reduction and use of multivariate analysis of saved process variables," *Ironmaking and Steelmaking*, 23(6) (1996), 504-511.
10. Ji Cheng, Zhu Miaoyong, Cheng Nailiang, "Design and realization of dynamic soft reduction process control system for slab continuous casting machine,"*Metallurgical Industry Automation*, 2007, no.1: 51-55.
11. Akihiro Y, Kazuo O, Takashi K, "Mechanism of Internal Cracking in Continuous Casting," *Tetsu-to-Hagané*, 82(12) (1996), 35-40.

Materials Processing Fundamentals
Edited by: Lifeng Zhang, Antoine Allanore, Cong Wang, James A. Yurko, and Justin Crapps
TMS (The Minerals, Metals & Materials Society), 2013

Effect of Microstructure Evolution on Hot Cracks of HSLA Steel during Hot Charge Process

Jiang Li[1], Qian Wang[1], Yongjian Lu[1], Banglun Wang[1], Shaoda Zhang[1]
[1]Chongqing Key Laboratory of Metallurgical Engineering, College of Materials Science and Engineering, Chongqing University, Chongqing 400030, China

Keywords: Nb-containing steel, Hot charge, Surface cracks, Microstructure evolution

Abstract

The microstructure evolution under different hot charge treatments has been studied in the paper. It was found that the microstructure morphologies of specimens charged at 750°C and 700°C respectively was lath martensite, banite, polygonal ferrite and the film-like ferrite along the prior austenite grain boundaries. As the specimen was charged at 700°C and reheated to 800°C~1150°C, it can be obviously observed that cracks both distributed on the film-like ferrite matrix of water-quenched specimens and furnace-cooling specimens, by using optical microscopy and SEM observation. For further understanding the formation mechanism of hot cracks, the volume fraction of phases was determined by using IPP statistical software, and the results show that the film-like ferrite was 3.6% of all and 9.5% of the ferrite. In combination of the inhomogeneous microstructure and precipitates at the prior austenite grain boundaries during reheating process, cracks formed as specimen charged from 700°C to 800°C.

Introduction

The manufacturing process of steel products has undergone profound changes from the standpoint of saving energy, and recently hot charging rolling process (HCR) has been interested. Hot charging rolling process is a process, connecting the rolling process with the continuous casting process. In this process, continuous cast slabs are directly charged to furnace for rolling without cooling to room temperature [1,2] However, the HCR process has not been successfully implement in the production of HSLA medium and thick plates, which are usually alloyed with small quantities of strong carbide-forming elements such as Nb、V and Ti in order to improve the strength of steels, due to the formation of hot cracks during hot charge process [3,4]. In the study by Zhu et al [5], half of slabs were charged at the charging temperature of ~700℃ to reheating furnace. It was found that the incidence of slab surface cracks via the hot charge process was 6.55% while only 0.39% via conventional cold charge process. This result indicated that the hot cracks incidence of low alloyed steels procuced in hot charge process is greater than that of C-Mn steels due to the affection of micro-alloy elements on the microstructure evolution. It is important to investigate the microstructure of the HSLA steel during hot charging process to reveal the cause of surface cracks on the HSLA plates.

In present study, the microstructure of Nb containing steel has been studied by simulating in laboratory. The specimens were simulated to charge with different temperatures, especially for the charging temperature within the two-phase region. The key objective of this study was to discuss the effect of microstructure evolution on hot cracks of Nb containing steel during hot charge process.

Experimental procedures

The chemical composition(mass%) of the specimen used in this study is 0.14C, 0.4Si, 1.46Mn, 0.014P, 0.004S, 0.04Al, 0.02Nb, 0.002V, 0.018Ti and 0.0027N(mass-%). Samples were cut from a 250 mm-thick continuously cast slab in the position of about 10mm under the slab surface and machined into 45 mm in length and 10 mm in diameter. The dilatometric experiments and formula suggested by Mintz were used to determine phase transformation temperatures and the results were as follows: A_{r3}=782□, A_{r1}=671□, A_{c3}=830□, A_{c1}=690□. A_{r3} and A_{r1} were measured at a constant cooling rate of 0.2□·S^{-1} while A_{c3} and A_{c1} were calculated at constant heating rate of 0.2□·S^{-1}.

In the thermal history □, the specimen was heated to 1300°C and held for 15min to dissolve the microalloying and to produce a coarse grain size similar to the as-cast grain size. Then the specimen was cooled at a rate of 6.7°C·min^{-1} to the test temperatures ranging from 750□ to 550□, then quenched with water and the microstructure at different hot charge temperature was detected. The thermal history □ was designed to simulate the hot charge process of Nb containing steel, and the specimen was cooled from the solution temperature to charge temperatures and subsequently reheated to 1150□, Three typical charging temperatures were designed: 650□ (charging temperature below the Ar1 temperature), 700□ and 750□ (charging temperature within the austenite + ferrite two-phase region). Other samples were subjected to furnace cooling under the same thermo treatments.

In order to investigate the relationship between the formation temperature of hot cracks and the microstructure evolution of Nb containing steel during hot charge process, the specimens were charged from 700□ to different reheating temperature, then quenched with water and the microstructure evolution of the steel were studied.

The exact control of the thermal conditions during the text is a prerequisite for the analysis of the results. For this purpose, a pair of Ni-Cr/Ni-Si thermocouple wires was clamped inside the surface center of each specimen in order to measure the temperature.

The quenched specimens were polished and etched by 4%nital acid to reveal the microstructure. The microstructure was identified by an optical microscope and SEM.

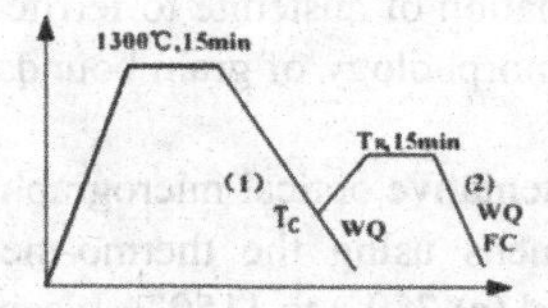

Figure 1. Illustration of thermal simulation test in different thermal histories

Results and Discussion

Micro-cracks were observed on the ferrite matrix of furnace-cooling specimens under reheating from 700□ to 1150□, as is shown in Figure 2. Numerous studies have shown that microstructure evolution is a key factor for the cracks formation. For understanding the formation of cracks, which were observed under the reheating process, microstructure of specimens at different charge temperature was firstly investigated. Meanwhile, in order to identify whether cracks was formed from the reheating process, other samples were subjected to furnace cooling under the same thermo treatments.

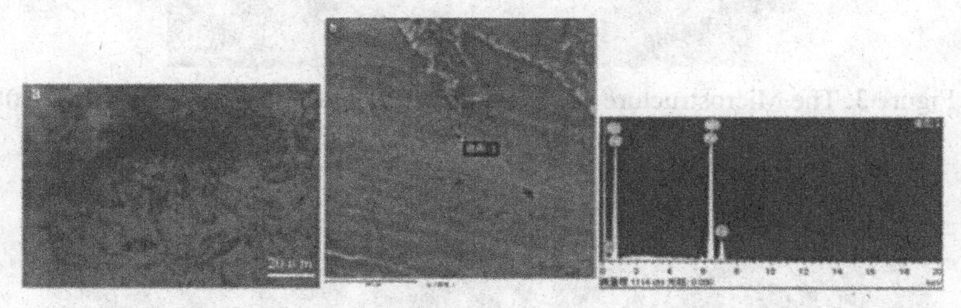

Figure 2. Cracks on the matrix of furnace cooling specimens reheated from 700°C to 1150°C.(a) Metallographic picture; (b) Representative SEM images and EDS analysis.

<u>Microstructure of specimens at different temperature</u>

Figure 3 shows the microstructures of water quenched specimens at (a) 550□, (b) 650□, (c) 700□, (d) 750□. The microstructure of specimens at 650□ and 550□ showed the ferrite and lamellar pearlite, attribute to the finished transformation $\gamma \rightarrow \alpha + Fe_3C$. It is found that the microstructure of the specimen at 700°C and 750°C exhibited lath martensite and ferrite, and the morphology of ferrite were polygon ferrite and the film-like ferrite along the prior austenite grain boundaries. Furthermore, the width of the film-like ferrite of specimen at 750□ was range from 8μm to 25μm. These values are in rather good agreement to the date reported in by J.M. Gregg et al [6].

According to the comparison of phases between 750°C and 700°C, a complex structure including austenite, polygon ferrite and the film-like ferrite was observed at 700°C by the decrease of charge temperature. The reason for this phenomenon can be explained that the transformation of austenite to ferrite increase and the ferrite grains continue to grow, then the morphology of grain boundary ferrite change into polygon ferrite.

Figure 4 shows some representative optical micrographs of the martensite structure in the water-quenched specimens using the thermo-mechanical conditions: reheated from (a) 650□, (b) 700□ and (c) 750□ to 1150□. It can be observed that micro cracks were distributed on the matrix of specimens reheated from 700□ to 1150□.

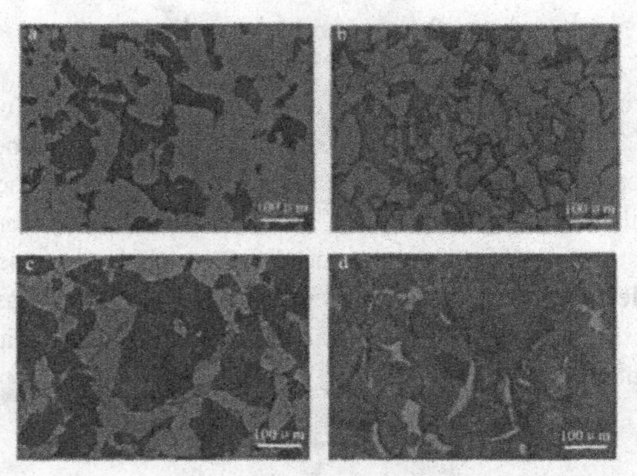

Figure 3. The Microstructure of water-quenched specimens: (a) 550□, (b) 650□, (c) 700□, (d) 750□.

Figure 4. The microstructure of water-quenched specimens reheated from (a) 650°C, (b) 700°C,(c) 750°C to 1150°C, respectively.

<u>Microstructure of specimens charged from 700□ to different reheating temperature</u>

In order to understand the relationship between the formation temperature of hot cracks and the microstructure evolution of Nb containing steel during hot charge process, microstructure examination was carried out and phase volume fraction was determined by Image-Pro-Plus statistics software.

Figure 5 shows the microstructure of water-quenched specimens at charge temperature 700°C. It can be found that the prior austenite grain was extremely coarse and severe mixed grain structure was apparent when the reheated temperature ranging from 725°C to 800°C. The reasons for these phenomena in reheating process are as follows: part of the austenite grain has transformed to ferrite before the specimen is reheated, thus, the ferrite grain which was transformed from the austenite will become a new nucleus of austenite grain, and however, the original austenitic grains will continue to grow with the temperature increase. Therefore, the mixed grain size structure was appeared when the primary austenite and secondary austenite coexist in the matrix of microstructure. Yongjian Lu [7] reported that the hot ductility behavior of Nb containing steel is greatly influenced by the microstructures constituents and the worst ductility of specimens charged at ($\gamma+\alpha$) range temperature caused by a combination of microstructure and the precipitation of Nb at austenite grain boundary.

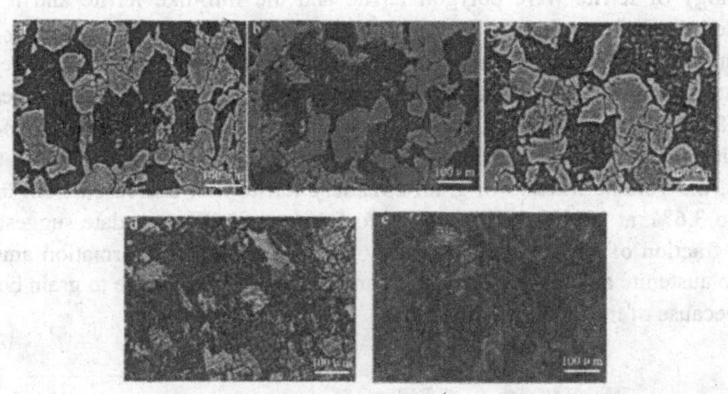

Figure 5. The microstructure of water-quenched specimens reheated from 700°C to (a) 725°C,(b) 750°C,(c) 800°C,(d) 825°C,(e) 1150°C, respectively.

Figure 6. Cracks on the matrix of water-quenched specimens reheated from 700°C to (a) 800°C, (b) 825°C and (c) 900°C.

Figure 7. Cracks on the matrix of furnace cooling specimens reheated from 700°C to (a) 800°C, (b) 825°C and (c) 900°C.

Figure 6 shows that micro cracks were distributed on the ferrite matrix of water-quenched specimens. Meanwhile, by comparison of micrographs of specimens using furnace cooling, as shown in Figure 7, micro cracks were also observed. Therefore, it can be concluded that cracks were mainly formed in reheating process.

Study of the relationship between the formation temperature of hot cracks and the phase volume fraction

For further understanding the relationship between the formation temperature of hot cracks and the microstructure evolution of Nb containing steel during hot charge process, microstructure examination was carried out and phase volume fraction was determined by Image-Pro-Plus statistics software [8]. As mentioned above, the morphology of ferrite were polygon ferrite and the film-like ferrite and it can be concluded that micro cracks were mainly distributed on the grain boundary ferrite. The number histogram of the measured ferrite distribution over the reheating temperature and ferrite volume of specimens charged at 700°C and reheated to 725°C~825°C were shown in Figure 8 (a). With the increase of reheated temperature, the ferrite volume fraction decreased, due to the transformation of $\alpha+Fe_3C$ to γ. In contrast, the volume fraction of grain boundary ferrite first decreased from 6.4% at 725℃ to 3.6% at 800℃, then increased to 5.6% at 825℃. The date suggesting the volume fraction of phases is explained by the increase of transformation amount of ferrite to austenite and the morphology changes of polygonal ferrite to grain boundary ferrite because of the diffusion of carbon.

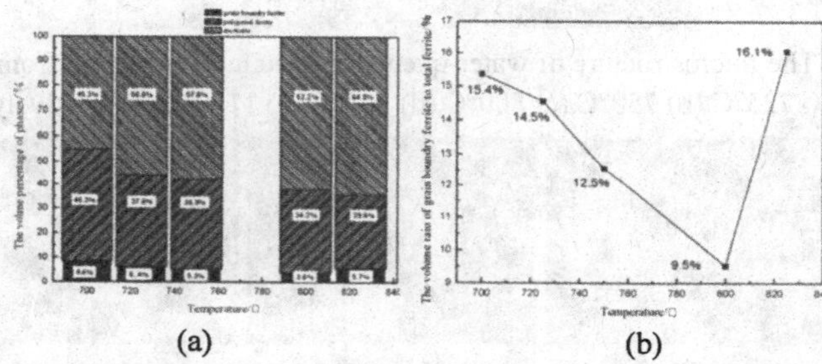

(a) (b)

Figure 8. (a) The ferrite volume fraction of specimens charged at 700°C and reheated to 725°C~825°C; (b) The ferrite volume fraction of grain boundary ferrite to ferrite.

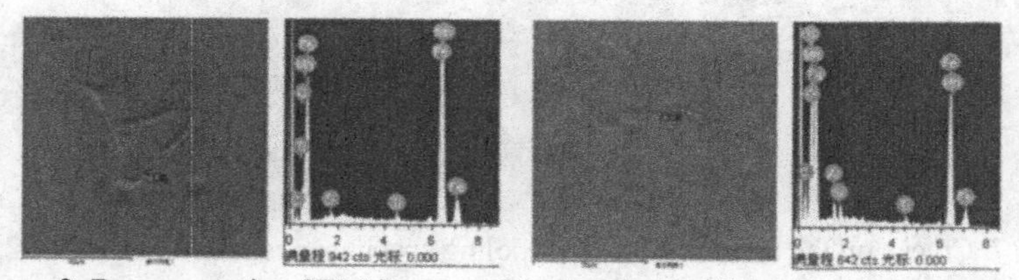

Figure 9. Representative SEM images of the cracks distribution and the EDS analysis for water-quenched specimens charged at 700°C, then reheated to (a) 800°C, (b) 825°C.

It was established that the film-like ferrite represents 9.5% of the area, whereas the polygon ferrite fraction accounts for 91% of the area, as shown in Figure 8 (b). Numerous studies [9-11] reported that the state of niobium (in solution or in precipitate), determined by the reheating temperature, can affect the recrystallization, the grain growth and the γ to α transformation of austenite. And the solubility of precipitation into ferrite is much worse than into austenite. Combined with coarse Nb(CN) particles can be preferred sites for ferrite nucleation. Therefore, the larger Nb(CN) or NbC existing on the ferrite matrix weakens the continuity of the base and accelerated the forming and expanding of the crack. In particular, by comparison of polygon ferrite, precipitations distributed on the matrix of grain boundary ferrite produced stress concentration due to the different properties between polygon ferrite and grain boundary ferrite, because of the higher value of the toughness of polygon ferrite [12,13].

Table 1. The amount of elements

	700°C to 800□		700°C to 825□	
	Mass%	Atoms%	Mass%	Atoms%
C	5.74	19.39	15.76	29.61
O	6.26	15.89	34.9	49.22
Si	0.86	1.24	1.19	0.95
Ti	0.93	0.79	0.93	0.44
Mn	2.08	1.54	1.35	0.55
Fe	84.13	61.15	44.28	19.89

In order to understand the interrelationships between the volume fraction of film ferrite and the formation of hot cracks, an analytical approach was followed utilizing SEM. Figure 9 shows some representative SEM images of the hot cracks formed after the thermo treatments. The distribution of cracks was mainly located on the film ferrite matrix of specimens and some elements of precipitations such as Ti, Nb were found on the film ferrite matrix by EDS analysis, as is shown in Table 1. In combination with the microstructure evolution, it can be concluded that the distribution of cracks was mainly located on the film ferrite matrix of specimens, and the microscopic cracks were formed while the volume fraction of film ferrite was 9.5% of the area. Factors for the formation of micro cracks in the two-phase region appear to converge on a common conclusion: coarse precipitation exiting on the film-like ferrite along austenite grain boundary weakens the continuity of the base and allows cavitations to occur, then gives an easy path for crack propagation, meanwhile, the inhomogeneous microstructure results in poor ductility. Therefore, the cracks formation of specimens charged at 700°C and reheated to 800°C is caused by a combination of precipitation at the film ferrite at along the prior austenite grain boundaries and the inhomogeneous structure, due to the transformation strain concentration within soft film-like ferrite, resulting in the formation of hot cracks [14,15]. However, it is suggested that the amount and the location distribution of precipitates for the formation of hot cracks during reheating process for Nb containing steel need further investigation.

Conclusion

The following conclusions can be obtained from the current study:
(1) Hot cracks mainly formed on the matrix of the film-like ferrite of water-quenched specimens and furnace cooling specimens, when the specimen was charged at 700°C and reheated to 800°C~1150°C.
(2) For the specimens charged at 700°C within the range of $\gamma + \alpha$ region and reheated to 800°C, the microscopic cracks formed on the matrix of grain boundary ferrite, while the volume fraction of film ferrite was 3.6% of all and 9.5% of the ferrite. The inhomogeneous microstructure results in poor ductility. On the other hand, precipitation induced by the transformation of $\gamma \rightarrow \alpha \rightarrow \gamma$ during reheating process also plays an important role on deteriorating the ductility of grain boundary.
(3) In combination of the inhomogeneous microstructure and precipitates at the film ferrite along the prior austenite grain boundaries during reheating process, cracks formed as specimen charged from 700°C to 800°C.

References

1. Y. Kamada and T. Hashimoto: ISIJ International, 30(1990), 241–247.

2. Y. Maehare and K. Nakai: Transactions ISIJ, 28(1988), 1021–1027.

3. A. Ghosh, S. Das, S. Chatterjee and B. Mishra: Materials Science and Engineering, 348 (2003), 299-308.

4. R.Z. Wang, C.I. Garcia and M. Hua: ISIJ International, 46(2006), 1345–1353.

5. G.H. Zhu, Q. Zhao, X.Y. Ma and K.B. Xu: Proc. 4th Int.Conf. on 'Continuous casting of steel in developing countries', Beijing, China, November 2008, The Chinese Society for Metals, 694.

6. J.M. Gregg, K.D.H. Bhadeshiah: Acta Materialia, 45(1997), 739-748.

7. Yongjian Lu, Qian Wang, and Jiang Li, et al.: Steel research international, 83(2012), 671.

8. J. Reiter, C. Bernhard and H. Presslinger: Materials Characterization, 59(2008), 737-746.

9. B.Mintz, S.Yue and J.J. Jones: International Materials Reviews, 36(1991), No.5, 187.

10. U.H. Lee, T.E. Park, K.S. Son, et al.: ISIJ Int., 50(2010), No.4, 540.

11. E.J. Palmiere, C.I. Garcia, A.J. DeArdo: Metall. Mater. Trans, 25A(1994), 277-286.

12. D.N. Crowther, Z. Mohamed and B. Mintz: Trans. ISIJ, 27(1987), No.5, 366.

13. Qingbo Yu, Ying Sun, Hailiang Yu, and Xianghua Liu: Journal of Mechanical engineering, (2011), 44-49.

14. M. Cabibbo, A. Fabrizi, M.Merlin, G.L. Garagnani: Mater. Sci., 43(2008), 6857-6865.

15. K.M. Banks, A.Tuling, C. Klinkenberg and B. Mintz: Materials Science and Technology, 27(2011), No.2, 537.

Materials Processing Fundamentals
Edited by: Lifeng Zhang, Antoine Allanore, Cong Wang, James A. Yurko, and Justin Crapps
TMS (The Minerals, Metals & Materials Society), 2013

A NEW METHOD FOR ULTRASONIC TREATMENT ON THE MELT OF STEEL

GANG NIE, JINWU KANG, YISEN HU, RUI YOU, JIYU MA, YONGYI HU and TIANYOU HUANG

Key Laboratory for Advanced Materials Processing Technology, Ministry of Education, Department of Mechanical Engineering, Tsinghua University, Beijing 100084, China

Keywords: Ultrasound, Amplitude Transformer, unit piece method (UPM), Induction, Steel Melt, Microstructure

Abstract

Ultrasonic treatment of the melt metals is a hot research topic; however, it is hard to introduce ultrasound into liquid steel because of the requirement of resistance to high temperature and erosion in the melt. In this paper, a new method, unit piece method (UPM), is proposed to introduce ultrasound into liquid steel. The metal to be treated is cast together with amplitude transformer as free end. During the experiment the transformer is placed upward and its free end is melt by induction coils. Then the amplitude transformer powered by ultrasound acts on the melt pool. The effect of ultrasound on the melt of 304 stainless steel is studied. The microstructure of the treated area is significantly modified and equi-axed grains are obtained. The effect of power of ultrasound is discussed. This method is proved to be effective.

Introduction

Since the beginning of the last century, ultrasonic treatment has been applied in many fields. And for foundry, ultrasonic can refine the microstructure and reduce segregation of the metal. The refinement of the microstructure can improve the mechanical properties of the metal, which means it can meet the high performance requirements that traditional casting can't achieve. Currently, the application of ultrasonic treatment on the solidification processing is gradually expanding and the research has achieved certain results at the refinement of the microstructure with ultrasonic treatment. Osawa[1] pretreated an AZ91 alloy by ultrasound, fine and uniform microstructure was obtained and the tensile strength and elongation were improved. However, for the erosion of amplitude transformer in the melt of high melting point metals, there are relatively less research about the effect of ultrasound on steel, while many researches focus on low melting metal[2-5]. Liu[6] introduced ultrasonic vibration of different power into melt through contacted steel, which is an indirect method, and studied the difference of the microstructure, segregation and mechanical properties of 1Cr18Ni9Ti stainless steel. Although this method avoided the erosion of amplitude transformer, it still had problems of power attenuation because of no direct contact between amplitude transformer and the melt. Li[7] studied different materials of amplitude transformer and found Mo-Al_2O_3-ZrO_2 ceramic metal tube which owned good high-temperature erosion resistance and vibration resistance was fit for amplitude transformer. But for the difficulty in the processing of metal ceramic tube, the application is limited.

In this paper, the author presents a new method for the introduction of ultrasound into melt. Metal to be processed are made a unit piece with amplitude transformer. The amplitude transformer is erected and its free end is melt by induction coils, which avoids the erosion and power attenuation. Using this method, the author studies the effect of ultrasound on the microstructure of a kind of austenite stainless steel.

Unit piece method (UPM) of amplitude transformer and the metal to be treated

Unit piece method, simplified as UPM, is proposed to deal with the ultrasonic treatment of metals of high melting point. Metal to be processed is made a unit piece with the amplitude transformer, locating at the free end of the transformer. During experiment the amplitude transformer is erected and the metal to be treated is melt by induction coils. The transformer is water cooled. A quartz tube is used to cover the metal to be treated and the adjacent part of the transformer, which will prevent the dripping of the melt. The schematic diagram of this method is shown in Figure . In contrast to the traditional direct method, dipping the transformer into the melt and the indirect method, transformer connecting to the contacted steel, this method avoids the erosion of the amplitude transformer and power attenuation of the ultrasound.

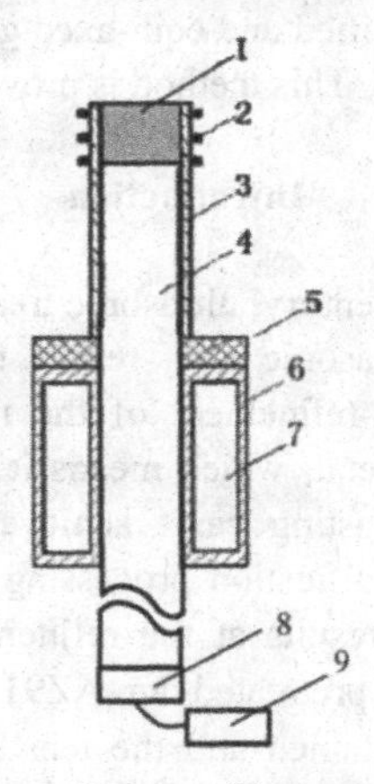

Figure 1 Schematic diagram of the unit piece method

1-metal to be treated, 2-induction coil, 3-quartz tube, 4- amplitude transformer, 5-insulation cotton, 6-water-cooled set, 7-cooling water, 8-ultrasonic transducer, 9-ultrasonic generator

Determination of the length of amplitude transformer

The amplitude transformer is one of the most important parts of the vibration system in ultrasound equipment. It can enlarge the displacement or velocity of the vibration part and focus energy on smaller area. So the property and the working condition of amplitude transformer have a great influence on the efficiency of ultrasonic treatment. Only when the length of amplitude transformer meets the resonance requirement, the efficiency of ultrasound is the best.

In general, the resonance requirement is that the length of amplitude transformer must be an integral multiple of half wave length of ultrasound. There are researches that show ultrasound velocity in steel will reduce when the temperature of steel rises[8]. So the wave length of ultrasound is the function of the temperature of amplitude transformer in the experiment, which can be calculated by:

$$c(T)=\begin{cases} 5816-0.8T & ,T<600°C \\ 6416-1.8T & ,600°C<T<800°C \\ 5516-0.6T & ,800°C<T<1400°C \\ 34915.6-21.6T & ,1400°C<T<1466°C \end{cases} \tag{1}$$

Where, c is the ultrasound velocity of ultrasound, T is the temperature of amplitude transformer.

Establish coordinate system in amplitude transformer that is shown in **Error! Reference source not found.**, and build relation between T and x which can be measured by thermocouple. According to the above, the wave length can be calculated by:

$$\lambda(x)=\frac{c(x)}{f_P} \tag{2}$$

Where, λ is the wave length of ultrasound, f_P is the frequency of ultrasound.

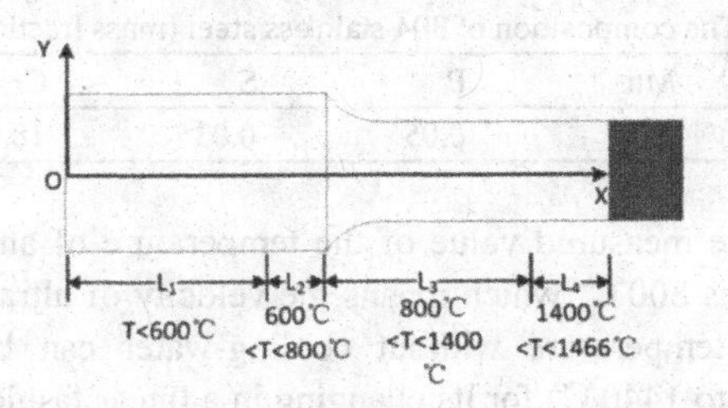

Figure 1 Coordinate diagram of amplitude transformer

Combining equations (1) and (2), the resonance requirement should be:

$$\frac{l}{\lambda}=\int_0^{l}\frac{f_P}{c(x)}dx=\int_0^{L_1}\frac{f_P}{5816-0.8T(x)}dx+\int_{L_1}^{L_1+L_2}\frac{f_P}{6416-1.8T(x)}dx+$$

$$\int_{L_1+L_2}^{L_1+L_2+L_3}\frac{f_P}{5516-0.6T(x)}dx+\int_{L_1+L_2+L_3}^{L_1+L_2+L_3+L_4}\frac{f_P}{34915.6-21.6T(x)}dx=\frac{1}{2} \quad (3)$$

Where, l is the length of amplitude transformer.

So, in order to calculate the length of amplitude transformer, the temperature in different positions of amplitude transformer is needed.

Experimental study of the effect of ultrasound on 304 stainless steel

304 stainless steel, a type of austenitic stainless steel with single as-cast microstructure, is commonly used, therefore, it is selected for the study. Its chemical composition is listed in

Table 1. And in the experiment, the amplitude transformer is also made of the same steel; the part to be treated is cast together with transformer. The length of the metal to be disposed is 36mm, and that of the amplitude transformer is 90mm. It is of the same diameter of the top diameter of the amplitude transformer 26mm. The parameters of the induction coils are frequency 15kHz, voltage 190V and current 50A.

During the experiment, the top of amplitude transformer is heated by induction coils to make sure the height of molten metal is 36mm. As the steel is melt, hold it for 5 minutes, and then introduce ultrasound into the molten metal until all the metal solidifies. The experimental apparatus is shown in Figure .In this experiment the effect of the ultrasound power on the microstructure of the melt is studied, at levels 0W, 150W, 200W and 300W. Then samples were taken from the middle section of the treated part, and the samples were ground, polished and etched for the observation of the metallographic structure using optical microscope.

Table 1 The composition of 304 stainless steel (mass fraction, %)

C	Si	Mn	P	S	Cr	Ni
0.08	1.00	2.00	0.05	0.03	18.00	9.00

In this experiment, the measured value of the temperature of amplitude transformer with cooling-water was 800℃, which means the velocity of ultrasound in this part is 4976m/s, while the temperature without cooling-water can be seen as 1120℃ (average from 800℃ to 1440℃) for its changing in a linear fashion, which means the velocity of ultrasound in this part is 4844m/s。 As a result, the length of amplitude transformer should be 123mm which is similar with the experiment parameter of 120.5mm.

For the traditional methods, amplitude transformer is permanent. When the temperature of amplitude transformer rise, the length of amplitude transformer can't meet the resonance requirement and the efficiency of ultrasound will drop. In contrast, in this experiment, the height of molten metal can be controlled by the position and the power of induction coils which can achieve the purpose of controlling the length of amplitude transformer. So with this method, the resonance requirement can be met to reduce the loss of the ultrasonic power.

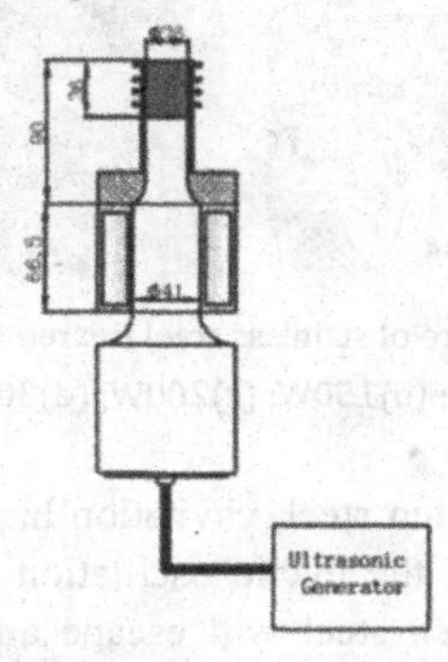

Figure 3 Schematic of ultrasonic treatment apparatus

Experimental results

The microstructure of 304 stainless steel with different powers of ultrasonic treatment is shown in Figure . The figure shows that without ultrasonic treatment, the microstructure is coarse dendrite, while the microstructure is equi-axed structure with ultrasonic treatment. The length of dendrite without ultrasonic treatment is about 1mm, and the average grain size of equi-axed structure is 160 μ m, 120 μ m and 100 μ m, respectively when the power of ultrasound is 150W, 200W and 300W. So, ultrasonic treatment can modify the microstructure of stainless steel and refine the grain size. And the effect of the refinement on the microstructure is more and more obvious when the power of ultrasound increases from 150W to 300W.

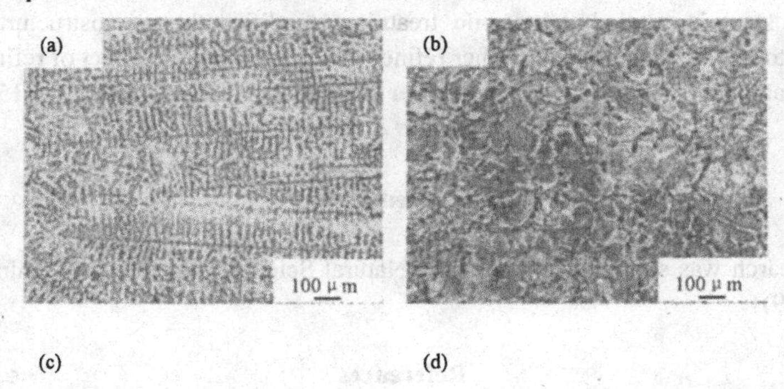

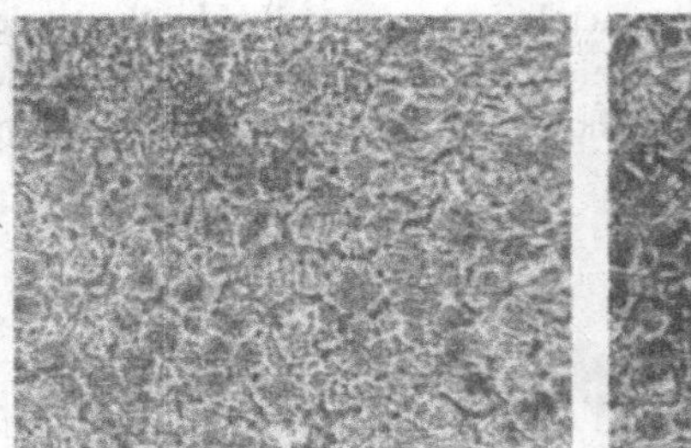
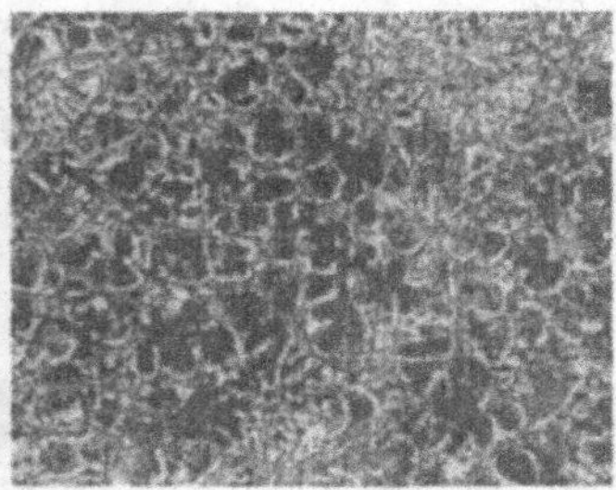

Figure 4 Solidification microstructure of stainless steel treated with different ultrasound powers: (a)0W; (b)150W; (c)200W; (d)300W.

When ultrasound travels in molten steel, cavitation happens. The local pressure will change periodically because of the cyclic oscillation of ultrasound. With negative local pressure, the gas in molten steel will escape and form small bubbles. These bubbles will oscillate and finally collapse which causes high temperature, high pressure and shock wave on the local scale. So, with ultrasound treatment, the dendrite will be broken into equi-axed grain by cavitation.

Conclusions

A new method called unit piece method (UPM) to introduce ultrasound into high temperature molten metal is proposed. This method avoids the erosion of amplitude transformer in high temperature molten metal and offers the way to control the length of amplitude transformer which is important to meet the resonance condition and improve efficiency of ultrasound. According to theoretical analysis and experimental comparison, the determination of the length of amplitude transformer is given.
By the proposed method, the effect of ultrasound on the microstructure of 304 stainless steel is studied. Ultrasonic treatment modifies the microstructure from dendrite to equi-axed grain and further refines the grains. And the effect of refinement is more and more obvious with the power of ultrasound increasing from 150W to 300W.

Acknowledgement

This research was supported by National Natural Science Foundation of China (No. 51075299)

References

[1] Y. Osawa et al, "Effect of ultrasonic vibration pretreatment on microstructural evolution and mechanical properties of extruded AZ91 alloy," *Materials Transactions*, 49(2008), 972-975.
[2] H.B. Xu, Q.Y. Han, and T.T. Meek, "Effects of ultrasonic vibration on degassing of aluminum alloys," *Materials Science and Engineering A-Structural Materials Properties Microstructure and*

Processing, 473(2008), 96-104.

[3] S.L. Zhang et al, "High-energy ultrasonic field effects on the microstructure and mechanical behaviors of A356 alloy," *Journal of Alloys and Compounds*, 470(2009), 168-172.

[4] T.V.Atamanenko, D.G.Eskin, and L. Katgerman, "Temperature effects in aluminium melts due to cavitation induced by high power ultrasound," *International Journal of Cast Metals Research*, 22(2009), 26-29.

[5] H. Puga et al, "The influence of processing parameters on the ultrasonic degassing of molten AlSi9Cu3 aluminium alloy," *Materials Letters*, 63(2009), 806-808.

[6] Q.M. Liu et al, "Influence of ultrasonic vibration on mechanical properties and microstructure of 1Cr18Ni9Ti stainless steel," *Materials & Design*, 28(2007), 1949-1952.

[7] J. Li, W.Q. Chen, and B.X. He, "Study of probe material for ultrasonic treatment of molten steel," *Journal of University of Science and Technology Beijing*, 29(2007), 1246-1249.

[8] G.L. Lü, M.J. Chu, and B,X. Wang, "Ultrasonic wave velocity and hysteresis in hot steel," *Acta Acustica*, 6(1992), 446-450.

MATERIALS PROCESSING FUNDAMENTALS

Metallurgy of Non-Ferrous Metals

Session Chair

Antoine Allanore

Materials Processing Fundamentals
Edited by: Lifeng Zhang, Antoine Allanore, Cong Wang, James A. Yurko, and Justin Crapps
TMS (The Minerals, Metals & Materials Society), 2013

CORROSION RESISTANCE OF Zn-Sn ALLOYS HORIZONTALLY DIRECTIONALLY SOLIDIFIED

Miriam B. Parra[1,2], Claudia M. Méndez[1,2], Carlos E.Schvezov[1,3,4], Alicia E. Ares[1,2,3]

[1] Materials Institute of Misiones (IMAM), Faculty of Sciences, University of Misiones, 1552 Azara Street, 3300 Posadas, Argentina.
[2] Materials Laboratory, Faculty of Sciences, University of Misiones, 1552 Azara Street, 3300 Posadas, Argentina.
[3] Member of CIC of the CONICET, Argentina. 1917 Rivadavia Street, 1033, Buenos Aires, Argentina. E-mail: aares@fceqyn.unam.edu.ar .
[4] Chairman of the Board of Development and Technological Innovation (CEDIT), 1890 Azara Street, 5th Floor, 3300 Posadas, Argentina.

Keywords: Zn-Sn alloys, horizontal directional solidification, corrosion behavior.

Abstract

The aim of this study is to evaluate the effects of the microstructural arrangement of Zn-Sn alloys horizontally directionally solidified on its resultant corrosion behavior. In this context, a water-cooled horizontal unidirectional solidification system was used to obtain alloy samples. Electrochemical impedance spectroscopy and potentiodynamic polarization curves were used to analyze the corrosion resistance in a 3% NaCl solution at 25 °C. Microscopic observation of the samples denote a higher susceptibility to pitting corrosion by the Sn, with lots of deep and localized pitting in the order of 10 microns. On the other hand, samples of Zn-Sn alloys and pure Zn showed a more generalized corrosion, with largest number of pinholes and with a depth of 4-5 microns approximately. Also, it was found that the presence of Sn in the analyzed samples affects the corrosion potential, becoming the alloys nobler respect to pure Zn.

Introduction

The results presented in this report are the continuation of previous investigation on the columnar to equiaxed transition in others alloy systems [1-3] which are now expanded to other alloys of technological interest like the Zn-Sn alloys.

In previous works, we correlated the effect of several parameters, like thermal and metallurgical ones, with electrochemical parameters on the CET macrostructure in Zn-Al alloys [4]. We were able to observe the susceptibility to corrosion of the alloys with columnar structure by analyzing the values of charge-transfer resistance (R_{ct}) obtained using the Electrochemical Impedance Spectroscopy (EIS) technique.

Another recent research [5] shows that what actually affects the response to corrosion is the way in which aluminum is distributed in the alloy, i.e., which phases are present in the solidified microstructure and how they are distributed, and not the amount of aluminum present in the alloy.

The above results show the strong relation between the solidification process parameters, the resulting structure and the mechanical and corrosion properties of the directionally solidified alloys.

The aim of this study is to evaluate the effects of the microstructural arrangement of Zn-Sn alloys horizontally directionally solidified on its resultant corrosion behavior.

Materials and Methods

The Zn, Sn and Zn-1wt.%Sn, Zn-2wt.%Sn and Zn-3wt.%Sn alloy samples were horizontally solidified. Initially the melt was allowed to reach the selected temperature and then, the furnace power was turned off and the melt was allowed to solidify. Ceramic molds of 50 mm in diameter were used for horizontal solidification experiments cooled from both ends. Eight made K-type thermocouples were used in the experimental setup. For the horizontal setup, thermocouples were fabricated with thin chromel-alumel wires of 0.5 mm diameter that were inserted into bifilar ceramics of ≅ 4.0 mm external diameter and ≅ 1.0 mm hollow diameter and introduced inside Pyrex® glass rods of 7.0 mm external diameter and ≅ 5.3 mm internal diameter.
Adjacent thermocouples were located at a distance of ≅ 20 mm. Temperatures were measured at regular intervals of 10 seconds.

Directional Solidification
A schematic drawing of the horizontal experimental device is shown in Figure 1. Small 140 mm long hemicylindrical probes of Zn-Sn alloy were solidified in the horizontal setup. The heat flux toward the ends of the sample was obtained by two cooling systems located at the ends of the ceramic crucible. In this setup, temperatures at eight different positions were measured using a TC 7003C acquisition system and recorded using SensorWatch® software every 1 minute in a compatible PC from the early beginning until the end of the solidification. Alloys were prepared with high purity metals (electrolytic Zinc and commercial grade Tin). For the horizontal setup, a set of five specimens of each alloy concentration were prepared. The alloy was first melted and mixed in a graphite crucible using a conventional furnace and then poured into a previously heated ceramic crucible. The crucible with the alloy was located into the horizontal furnace and heated up above the melting point of the alloy. The solidification of the sample was obtained by cooling down the alloy using the cooling system which extracts the heat toward both ends (Figure 1).

Figure 1. Horizontal solidification experimental setup.

Figure 2. Zn-Sn sample after directional solidification experiment.

After solidification the samples were cut in the longitudinal direction, polished with emery paper and etched to reveal the structure. The reagent used was a solution of HCl acid (70%) during 120 seconds [6-7]. Typical resulting macrographs can be seen in Figure 3 for Zn-1wt.%Sn, Zn-2wt.%Sn and Zn-3wt.%Sn alloys. To reveal the microstructure a solution containing 5 g CrO_3, 0.5 g Na_2SO_4 and 100 ml H_2O (Palmerston´s reagent) was used. The etching time varied from 15 to 20 s, depending on the alloy solute content. After etching, the samples were rinsed in a solution of 20 g CrO_3 and 100 ml H_2O before optical microscopy examination using SEM and an optical microscope in order to measure the average grain size. The grain size was measured in cross sections of the half of the sample. Each section was mounted, polished and etched and the average equiaxed grain size was determined using the ASTM E112 standard norm [8], including the width of the columnar grains.

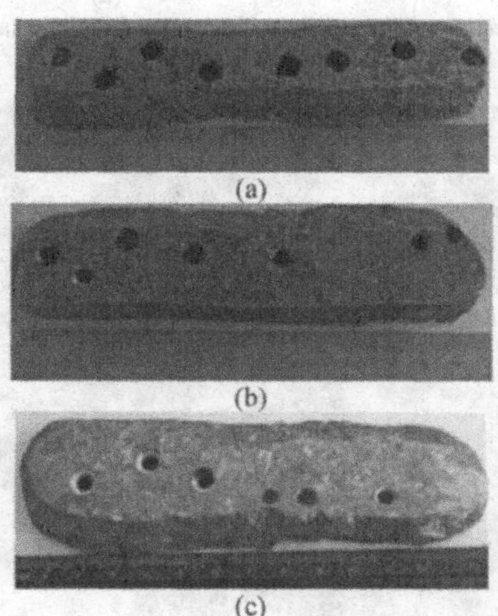

Figure 3. (a) Zn-1wt%Sn, (b) Zn-2wt.%Sn, (c) Zn-3wt.%Sn.

Electrochemical Tests

For the electrochemical tests (samples were prepared, which are used as working electrodes, approximately 2 cm long, each of the three zones (columnar, equiaxed and CET) and for each concentration from sections specimens cut longitudinally sanded to # 1200 SiC particle size, washed in distilled water and dried by natural air flow. All the electrochemical tests were conducted in a 300 ml of a 3% NaCl solution at room temperature using an IM6d ZAHNER® electrik potentiostat coupled to a frequency analyzer system, a glass corrosion cell kit with a platinum counter electrode and a sutured calomel reference electrode (SCE). Polarization curves were obtained using a scanning rate in the range of 0,002 V/s ≤ v ≤ - 0,250 V/s from open circuit potential until to 0,250 V.

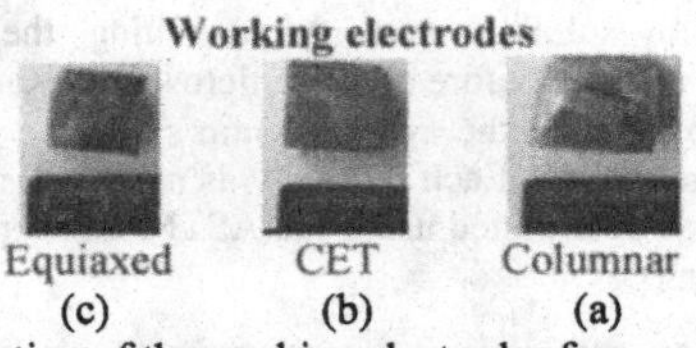

Figure 4. Detail of the selection of the working electrodes from one sample. (a) Columnar, (b) CET and (c) Equiaxed.

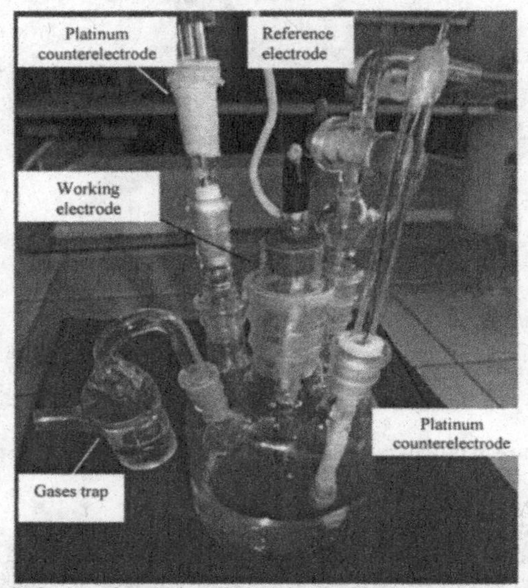

Figure 5. Conventional kit cell showing inside the working electrode.

Results and Discussion

With the data obtained we proceeded to performing the test with an electric potential sweep in anodic direction, from 100 mV below the open circuit potential at a speed of 0.16 mV/s until a current of 3.25 mA/cm^2. Thus, cyclic dynamic potentiometric curves were obtained. After completation of the tests, the cell was disassembled and proceeded to microscopic analyzes of the samples in order to evaluate the damages on the surface of the working electrodes. It was concluded that there was an increased susceptibility to a pitting corrosion in the Tin sample, with many localized and deeper pits, in the order of 10 microns.

While samples with pure Zinc and Zn-Sn alloys showed a more general corrosion, with less quantity of pits, in the order of 4-5 microns.

The polarization curves were analyzed for each samples and the corresponding corrosion and intersection potential were determined, as well as the open circuit potential, as shown in Table I.

(a) (b) (c) (d) (e) (f) (g) (h)

Figure 6. Micrographs of different samples (400 x) after the corrosion tests: (a) $Sn_{Equiaxial}$. (b) $Sn_{Columnar}$. (c) Zn-2wt.%Sn $_{Equiaxial}$. (d) Zn-2wt.%$Sn_{Columnar}$. (e) Zn-3wt.%Sn $_{Equiaxial}$. (f) Zn-3wt.%Sn $_{Columnar}$. (g) $Zn_{Equiaxial}$. (h) $Zn_{Columnar}$.

Table I. Corrosion, cutting potential and the open circuit potential for different samples.

ZnCol-Long	-1132	-1074
ZnCol-Trans	-1108	-1120
ZnEq-Long	-1146	-1052
ZnEq-Trans	-1099	-1120
ZnTCE-Long	-1141	-1096
ZnTCE-Trans	-1081	-1113
SnEq-Trans	-660	-537
SnCol-Trans	-615	-529
SnTCE-Trans	-535	-520
SnCol-Long	-560	-543
SnEq-Long	-550	-555
SnTCE-Long	-490	-519
Zn1SnCol-Long	-1150	-1083
Zn1SnCol-Trans	-1131	-1070
Zn1SnEq-Long	-1153	-1100
Zn1SnEq-Trans	-1151	-1067
Zn1SnTCE-Long	-1138	-1081
Zn1SnTCE-Trans	-1140	-1115
Zn2SnEq-Trans	-1096	-1115
Zn2SnCol-Trans	-1170	-1079
Zn2SnTCE-Long	-1132	
Zn2SnTCE-Trans	-1112	-1099
Zn2SnCol-Long	-1140	-1083
Zn2SnEq-Long	-1090	-1115
Zn3SnCol-Long	-1056	-1126
Zn3SnCol-Trans	-1155	-1072
Zn3SnEq-Long	-1151	-1086
Zn3SnEq-Trans	-1112	-1126
Zn3SnTCE-Trans	-1140	-1075
Zn3SnTCE-Long	-1130	-1083

The cutting potential is the potential at which the curves in anodic and cathodic direction are intersected. In materials which exhibit a defined passive region this potential is the one from which the material start the repassivation. The metals and alloys tested do not exhibit the phenomenon of passivation, but they corroded.
It is expected that from the analysis of the resulting polarization curves an estimation of the corrosion resistance of the samples tested is obtained. Thus, was compared for every pure metal

and alloy the curves corresponding to columnar, CET and equiaxed zones in which no marked generalized trends were observed for susceptibility to corrosion, see Figure 7.

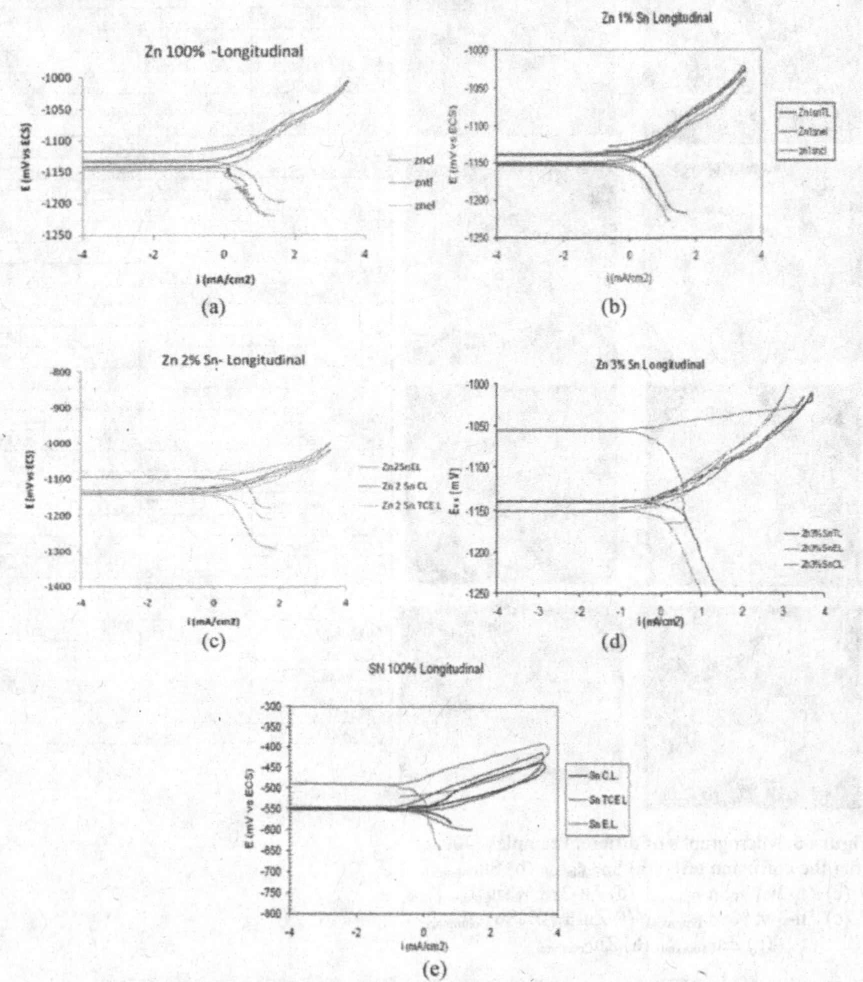

Figure 7. Potentiodynamic curves for different metals and alloys.

Where compared in turn then all curves each of which, the most representative is shown in Figure 8. As expected, was observed that the curve corresponding to Sn lies above the other curves, showing corrosion potential of -550 mV. Similarly, the Zinc presented a corrosion potential of -1130 mV.

The curves of alloys showed a trend presenting corrosion potential of: -1055 mV, -1140 mV and -1150mV for Zn-3wt.%Sn, Zn-2wt.%Sn and Zn 1wt.%Sn, respectively.

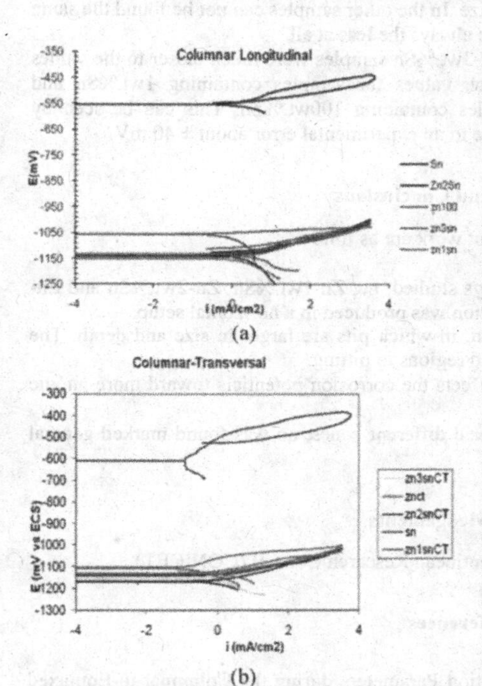

Figure 8. Cyclic potentiodynamic curves superimposed. (a) Columnar longitudinal. (b) Columnar transversal.

Table II. Current density values.

Sample	i corr	
	Long.	Transv.
Zn 3% Sn Columnar Longitudinal	0,70	
Zn 3% Sn Columnar Transversal		0,87
Zn 3% Sn Equiaxed Longitudinal	0,86	
Zn 3%Sn Equiaxed Transversal		0,90
Zn 3% Sn CET Longitudinal	1,20	
Zn 3% Sn CET Transversal		1,35
Zn 2%Sn Columnar Longitudinal	1,12	
Zn 2% Sn Columnar Transversal		0,86
Zn2%Sn Equiaxed Longitudinal	1,54	
Zn 2% Sn Equiaxed Transversal		1,20
Zn2%Sn CET Longitudinal	2,25	
Zn 2% Sn CET transversal		1,00
Zn 1% Sn Columnar Longitudinal	1,11	
Zn 1% Sn Columnar Transversal		1,13
Zn 1% Sn Equiaxed Longitudinal	1,09	
Zn 1%Sn Equiaxed Transversal		1,21
Zn 1% Sn CET Longitudinal	1,65	
Zn 1% Sn CET Transversal		0,55
Zn 100% Sn Columnar Longitudinal	1,12	
Zn 100% Sn Columnar Transversal		0,45
Zn 100% Sn Equiaxed Longitudinal	1,11	
Zn 100%Sn Equiaxed Transversal		1,49
Zn 100% Sn CET Longitudinal	1,16	
Zn 100% Sn CET Transversal		0,55
Sn 100% Columnar Longitudinal	0,73	
Sn 100% Columnar Transversal		0,37
Sn 100% Equiaxed Longitudinal	0,52	
Sn 100% Equiaxed Transversal		0,37
Sn 100% CET Longitudinal	0,61	
Sn 100% CET Transversal		0,50

Analyzing the corrosion rates using the method of the Tafel slopes, it was observed the following tendency for the columnar longitudinal sections: Sn100% < Zn3%Sn < Zn1%Sn < Zn100% < Zn2%Sn, although it is not possible to generalize. In the other samples can not be found the same tendency, but the corrosion rate of pure Sn was always the less at all.
Further, it was observed that the i_{corr} of the Zn-3wt.%Sn samples were much closer to the values of i_{corr} of the 100wt.%Sn samples, and such values for samples containing 1wt.%Sn and 2wt.%Sn were similar to those of the samples containing 100wt.%Zn. This can be seen by observing the values in Table 2 and may be due to an experimental error about ± 40 mV.

Summary and Conclusions

The main conclusions obtained from the present work are as follow:

1) For the pure metals and three type of alloys studied, i.e, Zn-1wt.%Sn, Zn-2wt.%Sn and Zn-3wt.%Sn, the columnar-to-equiaxed transition was produced in a horizontal setup.
2) Pure Sn samples suffered pitting corrosion, in which pits are larger in size and depth. The edges of the samples were the most affected regions to pitting.
3) The presence of Sn in the Zn-Sn alloys affects the corrosion potentials toward more anodic values.
4) In reference to columnar, CET and equiaxed different zones, no was found marked general trends for susceptibility to corrosion.

Acknowledgements

This work was partially supported by the Argentinean Research Council (CONICET).

References

1. A.E. Ares, and C.E. Schvezov, "Solidification Parameters during the Columnar-to-Equiaxed Transition in Lead-Tin Alloys", *Metall. Trans.*, 31A (2000), 1611-1625.
2. A.E. Ares, and C.E. Schvezov, "Influence of Solidification Thermal Parameters on the Columnar-to-Equiaxed Transition of Al-Zn and Zn-Al Alloys", *Metall. Trans.*, 38 A (2007), 1485-1499.
3. A. E. Ares, S. F. Gueijman, C. E. Schvezov, "An Experimental Investigation of the Columnar-to-Equiaxed Grain Transition in Aluminum-Copper Hypoeutectic and Eutectic Alloys", *Journal of Crystal Growth*, 312 (2010), 2154–1170.
4. A.E. Ares, L.M. Gassa, S.F. Gueijman, C.E. Schvezov, "Correlation Between Thermal Parameters, Structures, Dendritic Spacing and Corrosion Behavior of Zn–Al Alloys with Columnar to Equiaxed Transition", *J. of Crystal Growth*, 310 (2008), 1355-1361.
5. A.E. Ares, L.M. Gassa, "Corrosion Susceptibility of Zn–Al Alloys with Different Grains and Dendritic Microstructures in NaCl Solutions", *Corros. Sci.*, 59 (2012), 290-306.
6. W.J. Moffatt, *Handbook of Binary Phase Diagrams*, (New York, NY: General Electric Company Corporate Research and Development Technology Marketing Operation, 1984), 259, 419, 437, 391.
7. G. Kehl, *Fundamentos de la Práctica Metalográfica*, (Madrid: Editorial Aguilar, 1963), 112.
8. H. E. Boyer and T. L.Gall, *Metals Handbook*, Desk Edition, (Ohio, OH: American Society for Metals, 1990) 35-18, 35-19.

Materials Processing Fundamentals
Edited by: Lifeng Zhang, Antoine Allanore, Cong Wang, James A. Yurko, and Justin Crapps
TMS (The Minerals, Metals & Materials Society), 2013

High speed twin roll casting of Al-33 wt. % Cu strips with layered structure -Inspired by mathematical modeling

Seshadev Sahoo*, and Sudipto Ghosh
Department of Metallurgical and Materials Engineering, IIT Kharagpur-721302, India

Keywords: Twin roll; strip casting; solidification; Microstructure

Abstract

Mathematical modeling of casting of Al-33 wt. % Cu strips in a twin roll caster at speeds varying between 10 to 200 rpm (0.079 m/s-1.59 m/s) has been developed. The mathematical model involves solution of coupled fluid flow and energy equations, which incorporated phase change using enthalpy-porosity method. The simulation results suggests that at high casting speed strips will have layered structure, due to direct contact of the strand with Cu rolls resulting in outer solid layer having fine lamellar structure and slow cooling/solidification of the remaining liquid. Al-33 wt. % Cu alloy could be cast into strips at speed of 100 rpm (0.79 m/s) using vertical twin roll caster. As per the prediction of mathematical model, distinct layered structure forms when casting was carried out at high speed. Thus, layered structure could be directly cast using a twin roll caster.

Introduction

Materials with layered structure are of interest because they potentially offer novel mechanical and physical properties [1]. Methods for the production of layered structure material includes sputtering, vapour deposition, plasma spraying, hot rolling etc. [2, 3]. These methods are time-temperature-pressure dependent, require sophisticated processing equipment and are limited in the size and shape of the component which is produced. In order to overcome these difficulties in-situ fabrication of layered structure materials by high speed twin roll casting has been developed in the present study. The advantages of high speed twin roll caster are single step processing method, low equipment cost, low running cost, energy saving and high production rate as compared to other multistep fabrication method [4].

Previous studies done on twin roll casting at low casting speed [5-7] has found that due to low casting speed several defects like buckling, sticking, surface bleeds, micro-defects etc. are observed in the twin roll casting alloys. A mathematical model developed by Wang et al. [8] to analyze fluid flow, heat transfer and solidification in the mushy zone using finite difference technique and studied the effect of heat transfer coefficient on fluid flow and solidification in a vertical twin roll caster having roll speed 1 m/sec. The CFD model developed by Zeng et al. [9] focused on a better understanding of the melt's flow characteristics and thermal exchanges during the rapid solidification of the Mg during the twin-roll casting. They also highlighted the effect of casting speed and the gauge (twin-roll gap opening) on the melt flow and solidification. Zhao et al. [10] developed a twin roll caster to cast aluminum alloy strips and studied the effect of process parameters on microstructure and mechanical properties of the cast strip. Shan et al. [11] developed a new technology called vertical twin roll strip casting to produce Mg alloy strips and studied the influence of process parameters like casting speed (0.08-0.5 m/sec) on microstructure and mechanical properties of the as cast strips. The effect of rolling speed on the microstructure and mechanical properties of Al-Mg-Si and Al-Mg alloy studied by Das et al. [12, 13], but the limitation of the study is low casting speed. Haga and co-workers [14] used vertical twin roll caster to produce aluminium alloy strips at roll speeds up to 3 m/s and investigated the effect of casting speed on microstructure and mechanical properties of the cast strips. However all the research paid attention on the stability of the twin roll casting process and very limited research has been carried out on high speed twin roll casting process and in particular on the resulting structure of the cast strips.

The present study describes the formation of layered structure during high speed twin roll casting of Al-33 wt. % Cu alloy, as predicted by simulation study and validated experimentally.

Simulation of solidification of Al-33wt. % Cu at high speed casting

During high speed twin roll strip casting, the solidification behavior of Al-33wt. % Cu alloy has been simulated on FLUENT 6.3.16 platform. Figure 1(a) schematically shows a vertical twin roll caster. The simulations are carried out for casting of Al-33 wt. % Cu strips having thickness of 2 mm in a twin roll caster having roll diameter of 0.1524 m. The basic assumptions used for simulation are as follows (i) Process is considered to have attained steady state and symmetric around the center line of roll gap, (ii) The process is assumed to be two dimensional, (iii) The top surface of melt pool is considered flat and maintained at a fixed level, (iv) The molten metal is Newtonian fluid; Flow of the molten metal is turbulent and incompressible,(v) No slip condition exists between liquid metal and roll/solidified strip, (vi)The value of heat transfer coefficient is constant along the strip/roll interface.

Figure 1(b) shows the computational domain along with the boundary conditions. Assuming symmetry, only half of the real domain is considered for simulation. For the reliable results the grid used in the computation is having 1584 no. of cells, 3258 no. of faces and 1675 no. of nodes. The thermo-physical properties of Al-33wt. % Cu alloy has been described by Sahoo et al. [15].

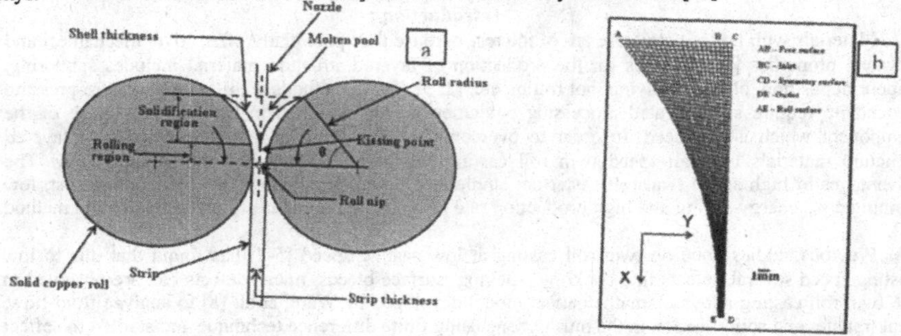

Figure 1. (a) Schematic diagram of vertical twin roll caster, (b) Computational domain used for simulation.

Experimental

Experimental conditions for the strip casting operation are given in Table 1. Strip casting of Al-33 wt. % Cu alloy is tried for the present study. The master alloy is prepared from a 99.96% pure Al and a 99.99% pure Cu by induction melting. The alloy is melted using an induction furnace at a temperature of 851 K. After melting, the molten alloy is transferred into a tundish, poured into the gap between the two counter rotating rolls through a nozzle and the casting is carried out. The solidified strips are prepared for metallographic examination using standard metallographic procedure and etched with Keller's reagent. The microstructures are studied using a JEOL JSM-6480LV scanning electron microscope (SEM).The phase presented are identified using X-ray diffractometer (XRD).

Table 1: Experimental and simulation conditions of casting operation

Casting parameters	Value
Roll diameter (m)	0.1524
Roll width (m)	0.0254
Casting speed (m/s)	0.079-1.59
Roll gap (m)	0.002
Contact angel (°)	40
Initial melt temp. (K)	851
Inlet diameter (m)	0.004
Heat Transfer coefficient ($W.m^{-2}K^{-1}$)	14938

Results and Discussion

The simulated results of solidification profile of Al-33 wt.% Cu having different roll speeds 0.079 m/s-1.59 m/s (10,50,100,200 rpm) are shown in Figure 2, which shows the variation of liquid fraction of Al-33 wt.% Cu from 0 to 1.

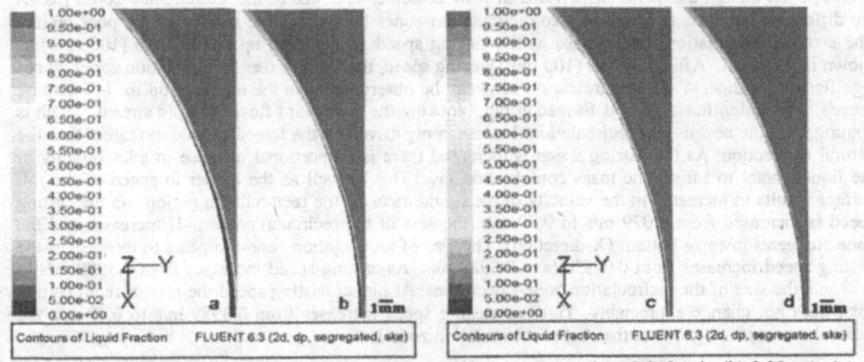

Figure 2. Solidification profile of Al-33 wt% Cu at different roll speed (a) 0.079 m/s, (b) 0.39 m/s, (c) 0.79 m/s, (d) 1.59 m/s.

Figure 2 (a) shows that, at 0.079 m/s (10 rpm) roll speed the liquid metal is completely solidified at the roll nip. At 0.079 m/s (10 rpm) roll speed the liquid metal gets sufficient time for heat transfer from the liquid metal to the roll. At 0.39 m/s (50 rpm) roll speed, some part of liquid metal is solidified at the roll nip and some part is in liquid state, as shown in Figure 2 (b).As the roll speed is increased the fraction of liquid at the roll nip increases as shown in Figure 3. The cooling of the inner liquid, which would require dissipation of heat through the solidified shell, will be significantly slower than that of the outer layer which was in direct contact with roll surface. The wide difference in the cooling rate/solidification front speed is expected to give rise to distinct structure in the outer layer and inner portion of the cast Al-33 wt. % Cu.

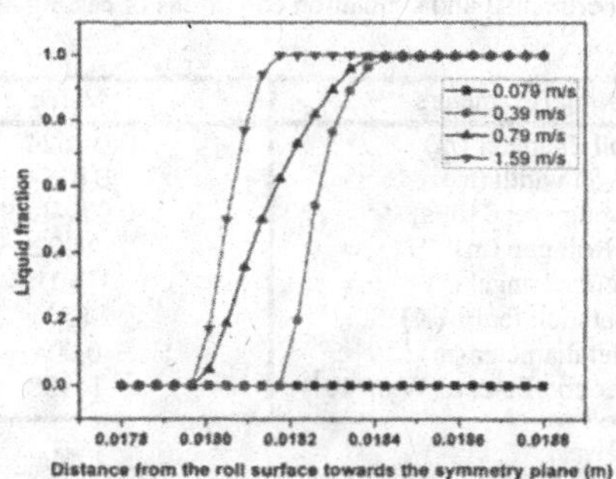

Figure 3: Liquid fraction of Al-33 wt.% Cu at the outlet section

Figure 4 shows the flow fields of liquid Al-33Cu in the molten pool region at different casting speeds, i.e., 0.079 m/s-1.59 m/s (10 rpm to 200 rpm). The flow profiles at different casting speeds are similar, even though the quantitative value of flow velocities and size of the recirculation zones (RCZ) are different. The flow field shows two recirculation zones below the free surface of the pool region. The size of recirculation zones varies as the casting speed is increased up to 0.79 m/s (100 rpm) as shown in Figure 4. After 0.79 m/s (100 rpm) casting speed, the sizes of these recirculation zones do not significantly change. A third recirculation zone can be observed above the nip position for low casting speeds, where significant solid is formed at nip, blocking the downward flow of liquid stream, which is coming from the nozzle. The recirculation flow is mainly driven by the forced convection rather than the natural convection. As the casting speed is increased there is proportional increase in inlet velocity of the liquid metal to satisfy the mass conservation law. This as well as the higher in speed of the roll surface results in increases in the velocity of the liquid metal in the recirculation region. As the casting speed is increased from 0.079 m/s to 0.79 m/s, the size of the recirculation zone-II increases and the zone elongates towards bottom (X-direction). The size of recirculation zone-I appears to increase as the casting speed increases from 0.079 m/s to 0.3989 m/s. As casting speed increases from 0.3989 m/s to 0.79 m/s the size of the recirculation zone-I decreases. At higher casting speed the size of recirculation zone does not change appreciably. Thus, as casting speed increases from 0.3989 m/s to 0.79 m/s the recirculation zone-II grows at the cost of recirculation zone-I.

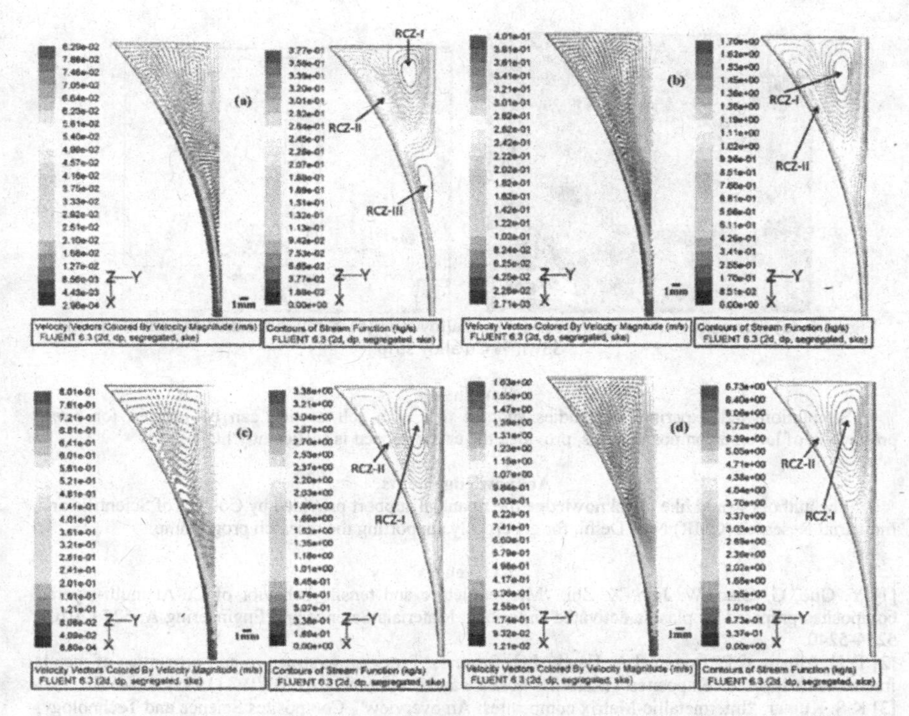

Figure 4: Velocity profile and the corresponding stream function contour of Al-33 wt.% Cu at different casting speeds (a) 0.07979 m/s, (b) 0.3989 m/s, (c) 0.79 m/s, (d) 1.59 m/s.

Al-33 wt. % Cu alloy could be cast in to strips at speed of 0.79 m/s (100 rpm). Figure 5 (a) shows the microstructure of the cast strip along thickness direction. The microstructure of the solidified Al-33 wt. % Cu strip is not uniform and two distinct zones are observed. This is the effect of the difference in the speeds of solidification front. In one zone (outer layer) the structure is lamellar and in the inner region it is wavy. It is also clear from the microstructure that Al-33 wt. % Cu strip is a layered composite structure which consists of partly lamellar and partly wavy structure. The lamellar microstructure consists of alternating layers of well bonded α-Al phase and θ-Al_2Cu phase and the phases are confirmed from the X-ray diffraction analysis as shown in Figure 5(b). Thus rapid solidification results in a layered eutectic structure. For the Cu-Al system, Neumann et al. [16] has shown that α-Al and θ-Al_2Cu phases form in Al-Cu eutectic alloy ribbons produced by melt spinning.

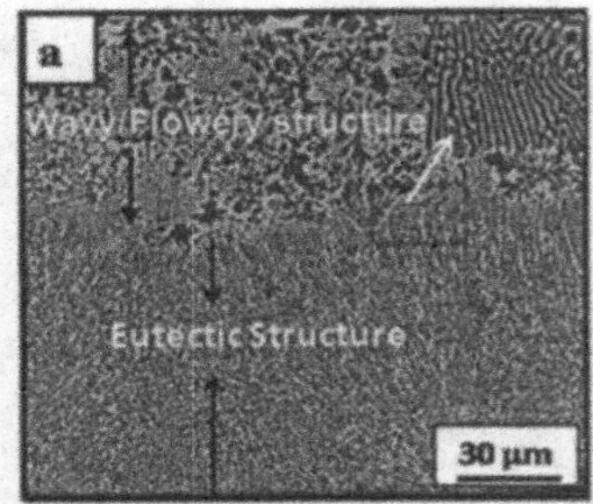

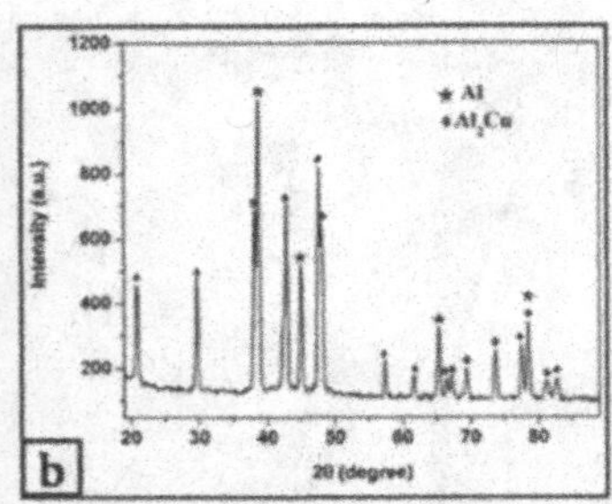

Figure 5. (a) Microstructure of the Al-33wt. % Cu alloy strip (b) X-ray diffraction pattern of the Al-33wt. % Cu alloy strip.

Conclusion

Simulation and experimental studies suggest that twin roll casting can be utilized for direct production of layered composite strips, provided the casting speed is sufficiently high.

Acknowledgements

The authors would like to acknowledge the financial support provided by Council of Scientific and Industrial Research (CSIR) New Delhi, for generously supporting the research programme.

References

[1] Y. Guo, G. Qiao, W. Jian, X. Zhi, "Microstructure and tensile behavior of Cu-Al multi-layered composites prepared by plasma activated sintering", Materials Science and Engineering A, 527 (2010), 5234-5240.

[2] D.E. Alman, C.P. Dogan, J.A. Hawk, J.C. Rawers, "Processing, structure and properties of metal-intermetallic layered composites", Materials Science and Engineering A, 192/193 (1995), 624-632.

[3] K.S. Kumar, "Intermetallic-Matrix composites: An overview", Composites Science and Technology, 52 (1994), 127-150.

[4] Y. Wang, S.B. Kang, J. Chob, "Microstructure and mechanical properties of Mg-Al-Mn-Ca alloy sheet produced by twin roll casting and sequential warm rolling", Journal of Alloys and Compounds,509 (2011), 704-711.

[5] M. Yun, S. Lokyer, J. D. Hunt, "Twin roll casting of aluminium alloys", Materials Science and Engineering A, 280 (2000), 116-123.

[6] M. Ha, J. Choi, S. Jeong, H. Moon, S. Lee, T. Kang, "Analysis and Prevention of Microcracking Phenomenon Occurring during Strip Casting of an AISI 304 Stainless Steel", Metallurgical and Materials Transactions A,33 (2002), 1487-1497.

[7] C. Gras, M. Meredith, J.D. Hunt, "Microdefects formation during the twin-roll casting of Al–Mg–Mn aluminium alloys", Journal of Materials Processing Technology, 167 (2005) 62–72.

[8] B. Wang, J. Y. Zhang, J. F. Fan, S. L. Zhao, S.B. Ren, K. C. Chou, "Modelling of Melt Flow and Solidification in the Twin-Roll Strip Casting Process. Steel Research International, 80 (3) (2009), 218-222.

[9] J. Zeng, R. Koitzsch, H. Pfeifer, B. Friedrich, "Numerical simulation of the twin-roll casting process of magnesium alloy strip", Journal of Materials Processing Technology, 209 (2009) 2321–2328.

[10] H.-Y. Zhao, D.-Y. Ju, T. Asakami, L. Hu, "An Experimental study of the strip casting process of aluminum alloy by the twin roll method", Materials and Manufacturing Processes, 16 (5) (2001), 643-654.
[11] W. Guang-shan, D. Hong-shuang, H. Feng, "Preparation of AZ31 magnesium alloy strips using vertical twin-roll caster", Transactions of Non Ferrous Metals Society of China, 20 (2010), 973-979.
[12] S. Das, N.S. Lim, J.B. Seol, H.W. Kim, C.G. Park,"Effect of the rolling speed on microstructural and mechanical properties of aluminum–magnesium alloys prepared by twin roll casting", Materials and Design, 31 (2010), 1633-1638.
[13] S. Das, N.S .Lim, H.W. Kim, C.G. Park, "Effect of rolling speed on microstructure and age-hardening behaviour of Al–Mg–Si alloy produced by twin roll casting process", Materials and Design, 32 (2011),4603-4607.
[14] T. Haga, K.Takahashi, M.Ikawa, H.Watari, "A vertical-type twin roll caster for aluminum alloy strips", Materials Processing Technology, 140 (2003), 610-615.
[15] S. Sahoo, A. Kumar, B.K. Dhindaw, S. Ghosh, "Modeling and Experimental Validation of Rapid Cooling and Solidification during High Speed Twin Roll Strip Casting of Al-33 wt. % Cu", Metallurgical and Materials Transaction B, 43 (2012), 915-924.
[16] W. Neumann, M. Leonhardt, W. Loser, R.Sellger, M. Jurisch,"Effect of ribbon-wheel contact on the microstructure of melt-spun Al-Cu eutectic alloy ribbons",Journal of Materials Science, 20 (1985), 3141-3149.

Materials Processing Fundamentals
Edited by: Lifeng Zhang, Antoine Allanore, Cong Wang, James A. Yurko, and Justin Crapps
TMS (The Minerals, Metals & Materials Society), 2013

FINITE ELEMENT MODELING OF MATERIAL REMOVAL RATE IN POWDER MIXED ELECTRIC DISCHARGE MACHINING OF AL-SIC METAL MATRIX COMPOSITES

U. K. Vishwakarma[1,*], A. Dvivedi[2], P. Kumar

Mechanical and Industrial Engineering Department, IIT Roorkee, UK-247667
*Corresponding Author Email: uvishw@gmail.com

Keywords: EDM, PMEDM, Finite Element Analysis, Al-SiC MMC.

Abstract

Electric discharge machining (EDM) is a well-established machining process which has its applications in diverse fields of medical, defense, automobile and aerospace engineering. However, EDM being a thermal process provides poor surface quality limits its area of applications. Powder mixed electric discharge machining (PMEDM) is a process variant of EDM, which is obtained by adding powder into dielectric fluid. In the present investigation, a finite element model has been developed for the single spark of PMEDM process. The temperature distribution in Al-SiC metal matrix composite (MMC) was evaluated. This model has further been utilized to calculate the material removal rate (MRR) for multi-discharge by accounting the number of pulses. For validation, further experiments were carried out on Al-SiC MMC under same process parametric conditions as in the finite element model. The result shows a good agreement between the simulation results and experimental data.

Introduction

The present scenario of technological advancement demands materials of high strength like composites, ceramics. These materials possess some specific material properties like high strength to weight ratio and have better mechanical properties than alloys. Materials that contain reinforcement supported by a binder (matrix) material are known as composites [1]. These advanced composite materials are considered to be an excellent candidate to sustain its rigidity at high temperature [2].
SiC reinforcing particles have the strongest effect on improving the strength of the composite as compared to other materials [3]. The elevated mechanical properties of composite materials create a big challenge of machine g by conventional methods. Nonconventional machining methods offer a better alternative to conventional machining processes in terms of tool wear and surface finish [4, 5].
A comprehensive review of mathematical models for electric discharge machining (EDM) has been given by Erden et al. [6]. In the EDM process dielectric fluid serves many purposes like (i) it acts as an insulator between tool and work-piece electrodes, (ii) it keeps the expanding plasma channel confined to a small diameter so that the intensity of the heat flux is very high over the discharge area, (iii) it flushes away the molten particles, known as debris, from the discharge crater, (iv) it also acts as a coolant. All the heating and cooling cycles generate thermal and residual stresses in the work-piece [7]. Yadav et al. [8] developed a finite element model to estimate the temperature field and thermal stresses for EDM process. The accuracy of an FEA model mainly depends upon the process parameters considered during the modelling of the process. Whereas, temperature dependency of thermal conductivity is most crucial parameters

affecting the simulation results [9]. Kansal et al. [10] developed an axisymmetric model for powder mixed EDM (PMEDM) using finite element method.

Literature reports extensive studies on various aspects of the EDM process, but very less attention has been given to FEA modelling of PMEDM process for metal matrix composite (MMC). There are certain difficulties involved in the modelling of MMC like, a macroscopic combination of two or more distinct materials having a recognizable interface between them, non-uniform distribution due to improper mixing etc.

In the present investigation an FEA model has been developed, for a single discharge of PMEDM. Different aspects of machining have been considered like pulse-on time, pulse-off time, number of pulses, current, voltage etc. Further, the single discharge model has been utilized to predict the material removal rate (MRR). Later, the predicted results obtained from the model have been verified with the experimental results and good correlation has been found between predicted and experimental results.

Thermal model of PMEDM

In PMEDM of Al-SiC composite, at higher discharge energy, crater forms are larger. This is due to de-bonding of SiC reinforcement that causes by the early melting of matrix material. It leaves the voids on the machined surface [11, 12]. The PMEDM process involves many complex phenomena, to simplify problem few assumptions have been made to make problem mathematically feasible:

Assumptions

- The work-piece domain is considered to be axisymmetric.
- The composition of the work-piece material is quasi-homogeneous.
- The mode of heat transfer to the work-piece is conduction.
- The initial temperature was set to room temperature.
- Analysis is done by considering 100% flushing efficiency.
- The thermal properties of work-piece material are considered as a function of temperature. It is assumed that due to thermal expansion, density and element shape are not affected.
- The heat source is assumed to have a Gaussian distribution of heat flux on the surface of the work-piece.

Governing Equation

The differential governing equation for heat transfer without internal heat generation written in a cylindrical coordinates of an axially symmetric thermal model for calculating the heat flux is given by Eq. (1) [13].

$$\rho Cp\left[\frac{\partial T}{\partial t}\right]=\left[\frac{1}{r}\frac{\partial}{\partial r}\left(K_r\frac{\partial T}{\partial r}\right)+\frac{\partial}{\partial z}\left(K\frac{\partial T}{\partial z}\right)\right] \quad (1)$$

Where ρ is density, C_p is specific heat, K is the thermal conductivity of the work-piece, T is temperature, t is the time and r & z are the coordinates of the work-piece.

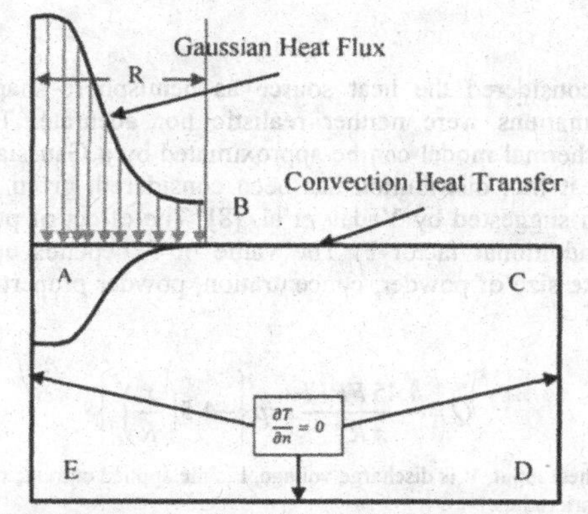

Figure 1. An axisymmetric model for the PMEDM process simulation.

Heat distribution

During the pulse-on time continuous flow of electrons from cathode to anode creates a plasma channel. Many authors have assumed the plasma channel as a uniform disk source [14-17], while Gaussian heat distribution has proven more realistic and accurate [8, 18-21]. Figure 1 shows the thermal model with boundary conditions applied to PMEDM process.

Boundary condition

The work-piece domain is considered to be axisymmetric about z axis. Work-piece domain considered for analysis is shown in Figure 1.
On the top surface of the work-piece, upto spark radius, the heat transferred is given by the Gaussian heat flux distribution, beyond spark radius convection takes place due to dielectric fluids. As boundary CD and DE are far away from the discharge location, no heat transfer conditions have been assumed. For boundary AE, as it is the axis of symmetry, the heat transfer is zero.
In mathematical terms, the applied boundary conditions are given as follows:

$$K\frac{\partial T}{\partial Z} = Q, \text{ when } R < r \text{ for boundary AC}$$

$$K\frac{\partial T}{\partial Z} = h_f(T - T_0), \text{ when } R \geq r \text{ for boundary AC}$$

$$\frac{\partial T}{\partial n} = 0, \text{ at boundary } CD, DE, AE$$

where h_f is the heat transfer coefficient of dielectric fluid, Q is the heat flux due to the spark and T_0 is the initial temperature.

Material Properties

The Al-SiC MMC was used as a workpiece, prepared by using SiC as reinforcement with mean particle size of 30μm by melt stir–squeeze –quench casting sequence [22].

Heat flux

Most researchers have considered the heat source as hemisphere shape or a disk source. However, these approximations were neither realistic nor accurate. The isotherms curves obtained from the EDM thermal model can be approximated by a Gaussian distribution. In this analysis, the Gaussian heat flux distribution has been considered, given in Eq. (2). The same distribution has also been suggested by Yadav et al. [8]. The effect of powder additional has been taken care by an additional factor F. The value of F depends upon different powder dependent parameters, like size of powder, concentration, powder properties, shape of particles etc.

$$Q = \frac{4.45\,FPVI}{\pi R^2} exp\left\{-4.5\left(\frac{x}{R}\right)^2\right\} \tag{2}$$

Where, P is the percentage of heat input, V is discharge voltage, I is the applied current, x is the radial coordinate of the work-piece and R is the spark radius.

Determination of MRR

MRR prediction depends upon the crater morphology. The morphology of the crater is assumed to be spherical dome shape, as shown in Figure 2 [23].

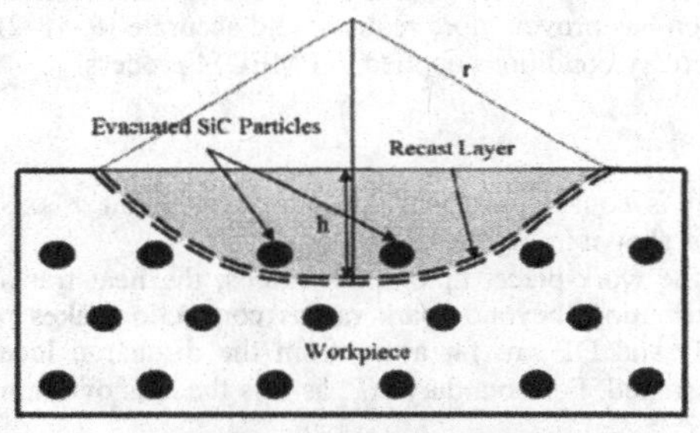

Figure 2. Assumed bowl-shaped cavity

where r is the radius of spherical dome and h is depth of the crater. The crater volume can be calculated by using Eq. (3). given in Eq. (3).

$$C_v = \frac{1}{6}\pi h(3r^2 + h^2) \tag{3}$$

The NOP can be calculated by dividing the total machining time to pulse duration as given in Eq. (4).

$$NOP = \frac{T_{mach}}{T_{on} + T_{off}} \tag{4}$$

where T_{mach} is the machining time, T_{on} is pulse-on time and T_{off} is pulse-off time.
By using Eq. (3) and Eq. (4), the MRR can be calculated for multi-discharges, as given in Eq. (5).

$$MRR = \frac{C_v \times NOP}{T_{mach}} \tag{5}$$

Table 3 shows the process parameters and their range used for the FEA modeling.
During the electric discharge sudden increase in temperature causes some materials to melt and evaporate from the work-piece surface. The melting temperature of aluminum matrix is lower than the SiC reinforced particles. Hence, aluminum melts before the SiC particles. In the absence of matrix materials the reinforced particle evacuates from the discharge crater, as shown in Figure 2.

Results and discussion

To obtain the results from developed FEA model for PMEDM, Al-SiC was chosen as the work-piece. The model was developed using the process parameters given in Table 3 for a single discharge. Later, the model was extended to predict the MRR in multi-discharge machining and results were validated by comparing the predicted results with the experimental data.
Figure 3 shows the meshed model used for the modeling. The number of reinforced particles was calculated by "Rule of Mixture". Figure 4 shows the temperature profile in the work-piece obtained at the end of the spark. It is clear from Figure 4, that the maximum temperature occurs at the center of the discharge channel and descends along the radius of the discharge channel. Figure 5 shows the two dimensional expanded model of the crater formed after material removal. The temperature distribution in the work-piece after the material removal is shown in Figure 6. It can be seen from the Figure 6 that SiC particle has the maximum temperature of 1224K, which is very less to melt the reinforced particles.

Table I. Chemical Composition and Thermal Properties of Al6063 at Room Temperature

Chemical Composition									
Element	Al	Cr	Cu	Mg	Fe	Mn	Si	Ti	Zn
Content (%)	97.5	0.1	0.1	0.45-0.9	0.35	0.1	0.4	0.1	0.1
Thermal Properties									
Thermal Conductivity						210 W/mK			
Specific Heat						900J/kgK			
Density						2700 Kg/m^3			
Melting Temperature						952K			

Table II. Chemical Composition and Thermal Properties of SiC at Room Temperature

Chemical Composition		
Element	SiC (Pure)	Others
Content (%)	98	2
Thermal Properties		
Thermal Conductivity		120 W/mK
Specific Heat		750J/kgK
Density		3210 Kg/m^3
Melting Temperature		3003K

Table III. Process Parameters and Their Values

Properties	**Value(s)**
Fraction of heat input (P)	0.08
Voltage (V)	50 V
Current (I)	8, 10, 12 A
Initial temperature (T_0)	300 K
Coefficient of heat transfer of dielectric fluid (h_f)	10000 W/m^{2}K
Pulse-on time (T_{on})	65, 120, 190 μs
Pulse-off time (T_{off})	28, 40, 47 μs
Spark radius (R)	120 μm
Powder Concentration (%)	2, 4
F	2

Table IV. Process Parameters Conditions For Experimental Run

I	Volt	Powder Conc.	Duty factor	T_{on}	T_{off}	Experimental MRR
8	50	2	0.8	190	47	48.57
10	50	2	0.75	120	40	65.14
12	50	2	0.7	65	28	49.56
8	50	2	0.7	65	28	45.15
10	50	4	0.8	190	47	53.46
12	50	4	0.75	120	40	65.44

Model validation

The validation of proposed model has been done by conducting some random experiments, as given in Table 4, under the same machining conditions at which simulation results were obtained. The experiments were performed on EMS 5030 EDM machine with additional attachment (built in-house) provided to perform PMEDM. The machining was done on Al-SiC MMC with a copper tool of 6mm diameter. Figure 7 shows the comparison of predicted MRR and experimental MRR. A good correlation of 91.94% was found between the experimental and predicted MRR.

Conclusions

In this paper, an FEA was developed to predict the MRR for PMEDM process. This model finds its applicability to predict the MRR with a good degree of correlation. Simulations have been performed to predict the temperature distribution in Al-SiC MMC for single discharge. The single discharge model was utilized to predict the MRR for multi-discharge. It was observed that the maximum temperature gained by the SiC particle was about 1200K in single discharge, which was insufficient to melt the reinforced particle, although, a considerable amount of thermal and residual stresses might have been developed in the reinforced particles. The FEA model can further be extended to predict residual and thermal stresses in PMEDM process.

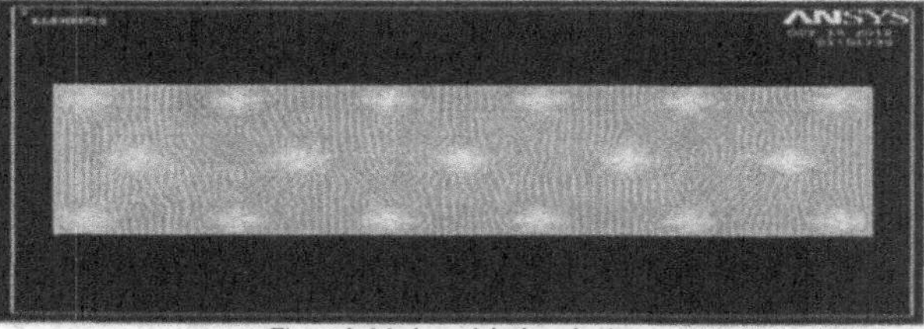

Figure 3. Mesh model of work-piece

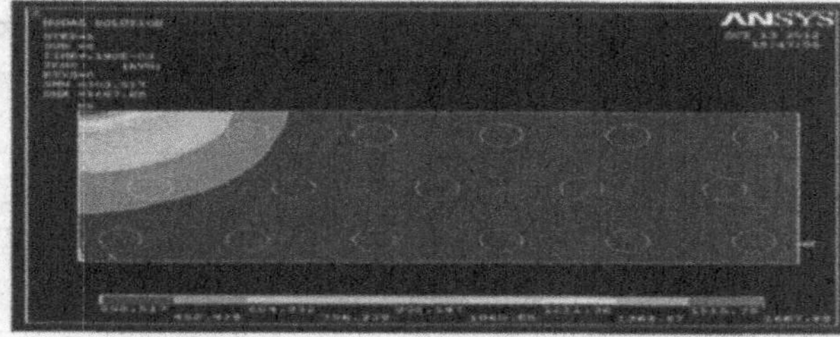

Figure 4. Temperature Distribution at the end of the spark at P = 0.08, V = 50V, I = 8A and F = 2.

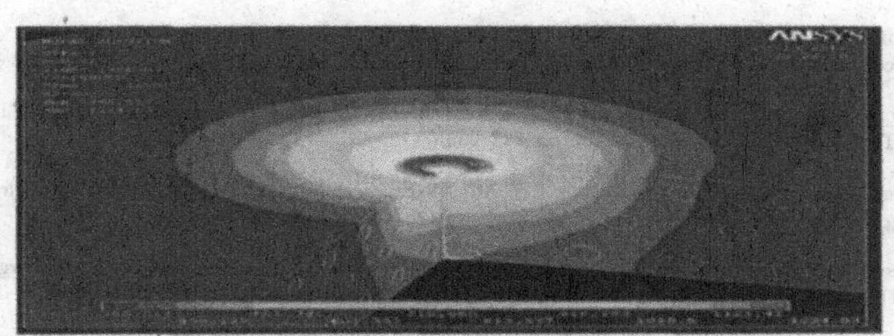

Figure 5. Expanded model of the crater formed after material removal

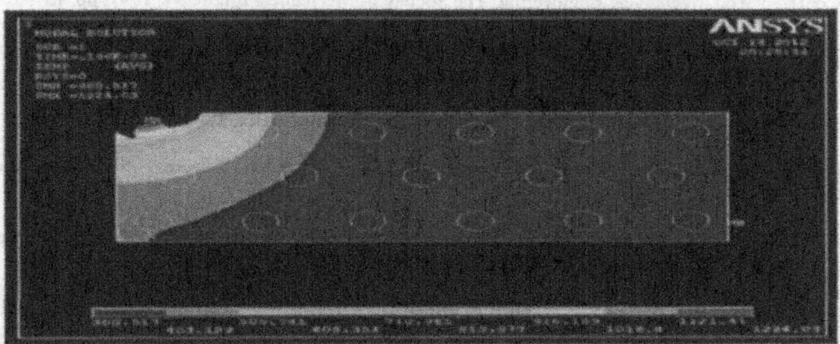

Figure 6. 2D representation of work-piece after material removal.

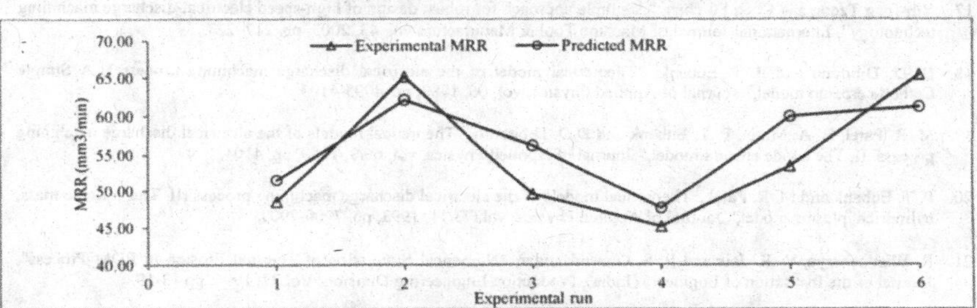

Figure 7. Comparison of experimental MRR and Predicted MRR

References

1. W. König, "Machining of new materials", Annals CIRP vol. 39 (2), 1990, pp. 673–681.

2. L. Cronjäger, "Machining of fibre and particle-reinforced aluminium", Annals CIRP vol. 41 (1), 1992, pp. 63–66.

3. J. Li, B.Y. Zong, Y.M. Wang and W.B. Zhuang, "Experiment and modeling of mechanical properties on iron matrix composites reinforced by different types of ceramic particles", Materials Science and Engineering vol. A 527, 2010, pp. 7545–7551.

4. P. Narender Singh, "Electric discharge machining of Al–10%SiCP as-cast metal matrix composites", Journal of Materials Processing Technology, vol. 155–156, 2004, pp. 1653–1657.

5. R. Komanduri, "Machining fiber-reinforced composites", Mech. Eng. , April 1993, pp. 58–64.

6. A. Erden, F. Arinc, M. Kogmen, Comparison of mathematical models for electric discharge machining, Journal of Material Processing and Manufacturing Science 4 (1995) 163–176.

7. Shuvra Das, Mathias Klotz and F. Klocke, "EDM simulation: finite element-based calculation of deformation, microstructure and residual stresses", Journal of Materials Processing Technology, vol. 142, 2003, pp. 434–451.

8. V. Yadav, V. Jain and P. Dixit, "Thermal stresses due to electrical discharge machining", International Journal of Machine Tools Manufacturing, vol. 42, 2002, pp. 877–888.

9. Nizar Ben Salah, Farhat Ghanem and Kaïs Ben Atig, "Numerical study of thermal aspects of electric discharge machining process", International Journal of Machine Tools & Manufacture, vol. 46, 2006, pp. 908–911.

10. H.K. Kansal, Sehijpal Singh and Pradeep Kumar, "Numerical simulation of powder mixed electric discharge machining (PMEDM) using finite element method", Mathematical and Computer Modelling, vol. 47, 2008, pp. 1217–1237.

11. B.H. Yan, H.C. Tsai and Y.C. Lin, "Study of EDM characteristics of cemented carbides", in: Proceedings of the 14th National Conference on Mechanical Engineering, The Society of Mechanical Engineers, 1997, pp. 157–164.

12. C.C. Wang and B.H. Yan, "Blind hole drilling of Al_2O_3/6061 Al composites using rotary EDM", J. Mater. Process. Technol, 102 (2000), pp. 90–102.

13. D.S. Kumar, Heat and Mass Transfer, 10th edition, S.K. Kataria & Sons, Delhi, 2000.

14. M. Kunieda, K. Yanatori, "Study on debris movement in EDM gap, International Journal of Electrical Machining", vol. 2, 1997, pp. 43–49.

15. Y.F. Luo, "The dependence of interspace discharge transitivity upon the gap debris in precision electro-discharge machining", Journal of Materials Processing Technology, vol. 68, 1997, pp. 127–131.

16. K Furutani, A. Saneto, H. Takezawa, N. Mohri, H. Miyake, "Accertation of titanium carbide by electrical discharge machining with powder suspended in working fluid", Precision Engineering vol. 25, 2001, pp. 138–144.

17. Yih-fong Tzeng and Chen Fu-chen, "A simple approach for robust design of high-speed electrical-discharge machining technology", International Journal of Machine Tool & Manufacture vol. 43, 2003, pp. 217–227.

18. D. D. Dibitono and P. T. Eubank, "Theoretical model of the electrical discharge machining process I. A Simple Cathode erosion model," Journal of Applied Physics, vol. 66, 1989, pp. 4095-4103.

19. M. R. Patel, B. A. Maria, P. T. Eubank and D.D. Dibitonto, "Theoretical models of the electrical discharge machining process. II. The anode erosion model," Journal of Applied Physics, vol. 66/9, 1989, pp. 4104.

20. P. T. Eubank and M. R. Patel, "Theoretical models of the electrical discharge machining process.III. The variable mass, cylindrical plasma model," Journal of Applied Physics, vol. 73/11, 1993, pp. 7900-7909.

21. R. Bhattacharya, V. K. Jain and P. S. Ghoshdastidar, "Numerical Simulation of Thermal Erosion in EDM Process", Journal of the Institution of Engineers (India), Production Engineering Division, Vol.77, 1996, pp.13-19.

22. Akshay Dvivedi, Pradeep Kumar and Inderdeep Singh, "Experimental investigation and optimisation in EDM of Al 6063 SiCp metal matrix composite", vol. 3 (3), 2008, pp. 293-308.

23. U. K. Vishwakarma , A. Dvivedi and P. Kumar, "FEA Modeling of Material Removal Rate in Electrical Discharge Machining of Al6063/SiC Composites", International Journal of Mechanical and Aerospace Engineering, vol. 6, 2012, pp. 398-403.

Materials Processing Fundamentals
Edited by: Lifeng Zhang, Antoine Allanore, Cong Wang, James A. Yurko, and Justin Cropps
TMS (The Minerals, Metals & Materials Society), 2013

Multicriteria optimization of rotary tool electric discharge machining on metal matrix composite

Manjot Singh Cheema, Akshay Dvivedi, Apurbba K Sharma, Sudip Biswas
Indian Institute of Technology Roorkee, Roorkee, Uttarakhand, 247667, India

Keywords: REDM, User Preference Rating, Multicriteria optimization

Abstract

Rotary tool electric discharge machining (REDM) is a process variant of electric discharge machining (EDM). The advantage with REDM is in ease of debris removal due to effective flushing. Present work focusses on multicriteria optimization of REDM on Metal matrix composite (MMC). The output characteristics are Material removal rate, tool wear rate and surface roughness. Different researchers have different opinions regarding output results with multicriteria optimization problems. So in a number of multicriteria optimization techniques, the weights given to outputs are decided on intuition basis. To solve this problem a systematic and unbiased approach for calculation of weights known as "User preference rating" has been used in this research work. The calculated weights have been used further for multicriteria optimization by utility technique. The optimum combination has been validated through experimental results.

Introduction

Rotary tool electric discharge machining (REDM) is one of the variants of electric discharge machining (EDM) in which tool electrode is rotated with simultaneous electric discharges to the workpiece. The advantage of using a rotating tool electrode is the facilitation of debris removal in an efficient manner. Experimental results have confirmed that a high material removal rate (MRR) and nearly same surface finish has been observed with REDM in comparison with EDM [1]. In EDM debris accumulation on the candidate surface leads to inactive pulses which hinder material removal, moreover these pulses cause damage to the surface integrity of machined specimen [2]. The Rotary tool electric discharge machining has proved to be an effective solution to provide proper flushing of the debris. While machining several output responses like MRR, Tool wear rate (TWR), surface roughness (SR), hole oversize, wear ratio etc. are to be taken into consideration for optimization of parameters. Ample investigations regarding effect of various input parameters like current, gap, duty cycle, tool rotation speed etc. have been carried out by several researchers [1, 3-6]. In a manufacturing industry, optimization of the process is must to yield the best pragmatic results. In order to optimize the multiple output responses a priori articulation of preferences method has been generally used [7]. In this method the user has to specify the degree of relative importance between the output responses i.e. weights have to be assigned. The weighted summation technique is easy to apply but the major problem faced was the calculation of weights. Though there are many techniques for calculation of weights like AHP and ANP, but their use is complex. Thus, in order to simplify things, constant numbers have been selected as weights. Further problems arise when two or more researchers have different viewpoints regarding assignment of weights.

Considering the limitations discussed above, a new method for calculation of weights has been discussed in this work. This method has been further incorporated in multicriteria optimization of process parameters in REDM by utility method [8]. This technique is an addendum to existing decision making technique known as "Customer preference rating" [9]. Another advantage of this technique is that incomplete information can also be used for calculation of weights. The voice of researcher is converted into a mathematical form; the corresponding concept is mathematically represented by graph theory technique known as preference graphs [10]. The technique described in this paper has been validated using experimental data obtained while machining a popular MMC material.

In the present investigation current, gap, duty cycle, tool rotation and flushing pressure were used as the input parameters machining of Al/SiC MMC. On the other hand SR, MRR and TWR are the output responses considered for optimization of the process parameters.

REDM machining details

The experiments were conducted on a die sinking EDM machine (EMS 5030) developed by Electronica M/c tools. It was energized with 35 A current pulse generator and the controller produces rectangular shaped current pulses. The existing EDM was modified to provide rotation to the tool. The rotation speed of motor was controlled by a step variator. The various process parameters identified and their levels have been shown in Table 1. The ranges were decided by pilot experimentation on the basis of one factor at a time approach. Taguchi's L_{27} orthogonal array was used for design of experiments.

Table 1. REDM process parameters and their levels

S. No.	Process Parameters	Process parameter designation	Level 1	Level 2	Level 3
1	Current (Amp)	A	10	15	20
2	Gap control	B	5	6	7
3	Duty cycle	C	0.55	0.63	0.7
4	Tool rotation (rpm)	D	500	600	700
5	Flashing pressure (kg/cm^2)	E	0.6	0.7	0.8

Calculation of weights

The calculation of weights has been carried out by considering the viewpoint of three different researchers with different preferences. The first researcher preferred the SR to MRR followed by TWR. The second researcher preferred MRR to SR followed by TWR. The third researcher preferred SR as the most important parameter in comparison to MRR and TWR. However no relationship was defined between MRR and TWR by the third researcher. The preference graphs (PG's) for all the researchers have been shown in Figure 1 a, b and c.

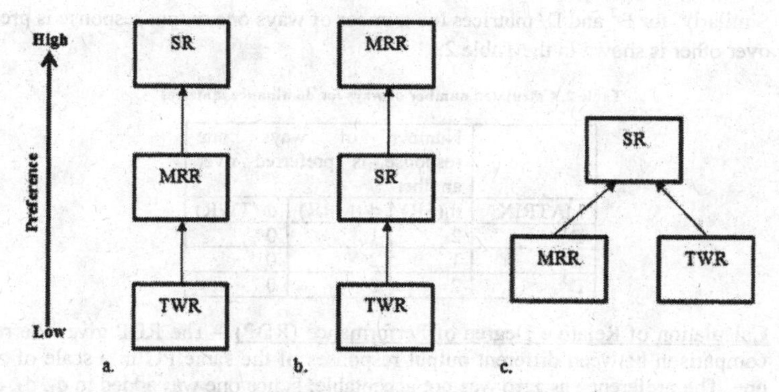

Figure 1. Preference graphs for (a) first researcher (b) second researcher and (c) third researcher

The steps involved in calculation of weights are [9]

1. Creation of the Adjacency Matrix – Adjacency matrix represents relationship of a researcher's PG in the form of a matrix. Numeral 'one' represents a relationship between the output responses whereas numeral 'zero' represents no relationship. The adjacency matrix for different PG's (as in Figure 1) have been constructed as

$$\begin{array}{c} \text{SR MRR TWR} \\ \uparrow \quad \uparrow \quad \uparrow \end{array}$$

$$PG_1 = \begin{matrix} 0 & 1 & 0 \\ 0 & 0 & 1 \\ 0 & 0 & 0 \end{matrix} \begin{matrix} \rightarrow SR \\ \rightarrow MRR \\ \rightarrow TWR \end{matrix}$$

Similarly, $PG_2 = \begin{matrix} 0 & 0 & 1 \\ 1 & 0 & 0 \\ 0 & 0 & 0 \end{matrix}$ and $PG_3 = \begin{matrix} 0 & 1 & 1 \\ 0 & 0 & 0 \\ 0 & 0 & 0 \end{matrix}$

2. Calculation of Dominance Matrix – Dominance matrix (D) for a particular PG tells about the number of ways one output response is preferred more over other responses. In this case the dominance matrix for first PG can be calculated as

$$D^1 = [PG_1] + [PG_1]^2. \qquad (1)$$

Thus, the Dominance matrices can be calculated as

$$D^1 = \begin{matrix} 0 & 1 & 1 \\ 0 & 0 & 1 \\ 0 & 0 & 0 \end{matrix} \begin{matrix} \rightarrow 2 \\ \rightarrow 1 \\ \rightarrow 0 \end{matrix}$$

Similarly $D^2 = \begin{matrix} 0 & 0 & 1 \\ 1 & 0 & 1 \\ 0 & 0 & 0 \end{matrix}$ and $D^3 = \begin{matrix} 0 & 1 & 1 \\ 0 & 0 & 0 \\ 0 & 0 & 0 \end{matrix}$

The number of ways (d) for Dominance matrix D^1 are given as $d_1{}^1 = 2$, $d_2{}^1 = 1$ and $d_3{}^1 = 0$. It can be then said that SR is preferred in 0+1+1 = 2 ways, MRR is preferred in 0+0+1 = 1 way and TWR is preferred in 0+0+0 = 0 way.

Similarly, for D^2 and D^3 matrices the number of ways one output response is preferred over other is shown in the Table 2.

Table 2. Calculated number of ways for dominance matrices

	Number of ways one response is preferred over another		
MATRIX	d_1(SR)	d_2(MRR)	d_3(TWR)
D^1	2	1	0
D^2	1	2	0
D^3	2	0	0

3. Calculation of Relative Degree of Performance (RDP) – The RDP gives the relative comparison between different output responses of the same PG in a scale of zero to one. The preference as zero was not acceptable; hence one was added to d_1, d_2, and d_3. In every PG case, the relative preference (denoted by rdp) for first researcher can be given by the following formula

$$rdp^1 = (1+d_1)/max(1+d_m) \quad (2)$$

where m is number of way (d) with the max value
here $max(1+d_m) = 1+2 = 3$
therefore $rdp^1 = (1+2)/3 = 1$
similarly $rdp^2 = (1+1)/3 = 0.66$ and $rdp^3 = (1+0)/3 = 0.33$
In vector form, the different RDP values for the three Dominance matrices D^1, D^2 and D^3 are

$RDP_1 = (1, 0.66, 0.33)$

$RDP_2 = (0.66, 1, 0.33)$

$RDP_3 = (1, 0.33, 0.33)$

4. Calculation of Relative Importance Rating (RIR) - The RDP values obtained in step 3 were obtained for three different researchers. In order to combine the viewpoints of all the three researchers into a single value RIR has been calculated. The rir represents the combined rating by the three researchers for a single output response. It is given by

$$rir = (rdp_1{}^1 + rdp_2{}^1 + rdp_3{}^1)/\ max\,(\sum rdp) \quad (3)$$

where $rdp_1{}^1$, $rdp_2{}^1$ and $rdp_3{}^1$ are the relative preference of SR by the first, second and third researcher.
$max\,(\sum rdp)$ is the maximum sum of relative preference for a single output response by the three different researchers.
In this case $max\,(\sum rdp) = (1+0.66+1) = 2.66$
The total RIR's for SR, MRR and TWR in vector form can be given by
RIR = (2.66/2.66, 1.99/2.66, .99/3) = (1, 0.748, 0.33)

5. Calculation of Weights – In order to satisfy the condition $\sum_{i=1}^{n} Wi = 1$, normalization was done to get the final weights. The values of RIR in the step 4 were normalized and following values were obtained
$W_m = (1/2.078, 0.748/2.078, 0.33/2.078) = (0.481, 0.359, 0.16)$

Calculation of Utility values

Utility method has been one of the simplest methods for multi criteria optimization of process parameters [8]. In the utility method a composite index is calculated which represents overall utility (usefulness) of a product. In this case the overall utility of machined component is the sum of each output response (SR, MRR and TWR). If X_i is the measure of effectiveness of the attribute i and there are n attributes in the outer space, then the joint utility $U_i(X_i)$ for i^{th} attribute is represented as

$$U(X_1, X_2, \ldots, X_n) = f(U_1(X_1), U_2(X_2), \ldots, U_n(X_n))$$

The attributes are assumed to be independent. The overall utility function is given by sum of individual utilities.

$$U(X_1, X_2, \ldots, X_n) = \sum_{i=1}^{n} U_i(X_i)$$

Weights are assigned according to the requirements such that $\sum_{i=1}^{n} W_i = 1$ and then the overall utility function can be explained as

$$U(X_1, X_2, \ldots, X_n) = \sum_{i=1}^{n} W_i U_i(X_i) \quad (4)$$

Preference Scale construction

Preference scale for each quality response was constructed to obtain the utility values. In this scale nine was chosen as the preference number for maximum acceptable level of each output response. The constant A was derived using optimum values obtained by Taguchi's analysis. Logarithmic scale was used in this study which is given by

$$P_i = K \log (X_i / X_i') \quad (5)$$

where, X_i = Experimental values obtained for response i

X_i' = Just acceptable value of output response i

K = Constant

The value of K can be calculated by substituting $X_i = X^*$ in above equation. X^* is the optimum value which has been derived from Taguchi's analysis. The calculated preference values are shown in Table 3.

Table 3. Preference scale formulae for SR, MRR and TWR

	Response		
	SR	**MRR**	**TWR**
Optimum values	0.74 (μm)	47.3 (mm^3/min)	0.15 (mm^3/min)
Just acceptable values	2.6	10	5.9
Preference scale P	-14.67*Log(X_i/2.6)	13.35*Log(X_i/15)	-5.64*Log(X_i/5.9)

The utility values were finally calculated using the following expression

$$U = P_{SR} \times W_{SR} + P_{MRR} \times W_{MRR} + P_{TWR} \times W_{TWR} \quad (6)$$

W_{SR}, W_{MRR} and W_{TWR} are the weights calculated in step 5. The weights selected are 0.481, 0.359 and 0.16 respectively. The preference scale values P_{SR}, P_{MRR} and P_{TWR} were derived from the relation 5. The corresponding utility values calculated have been shown in Table 4.

Table 4. Calculated preference values and utility values

Trial no.	P_{SR}	P_{MRR}	P_{TWR}	Overall Utility, U	Logarithmic value
1	5.36	2.49	3.86	4.09	12.24
2	4.74	5.49	5.41	5.12	14.18
3	3.16	4.21	3.51	3.59	11.11
4	5.54	5.72	4.37	5.42	14.67
5	7.68	7.07	6.36	7.25	17.21
6	7.37	7.09	5.02	6.89	16.77
7	4.33	1.31	6.66	3.62	11.17
8	3.20	5.02	6.20	4.33	12.74
9	6.16	7.33	5.95	6.55	16.32
10	6.63	4.08	3.49	5.21	14.34
11	2.31	7.56	5.15	4.65	13.35
12	0.26	6.53	3.46	3.02	9.60
13	3.29	5.51	4.05	4.21	12.48
14	2.82	7.78	5.15	4.97	13.94
15	4.23	8.50	4.90	5.87	15.37
16	7.29	4.55	6.48	6.18	15.81
17	2.66	7.41	5.41	4.81	13.64
18	2.39	7.97	4.72	4.76	13.56
19	2.19	3.77	4.75	3.17	10.02
20	4.43	3.15	3.60	3.84	11.68
21	0.55	6.21	3.47	3.05	9.68
22	3.03	4.87	4.40	3.91	11.85
23	3.74	4.35	4.50	4.08	12.22
24	3.84	8.41	4.64	5.61	14.98
25	2.58	3.82	5.82	3.55	10.99
26	3.98	5.54	4.03	4.55	13.16
27	1.86	7.23	4.18	4.16	12.39

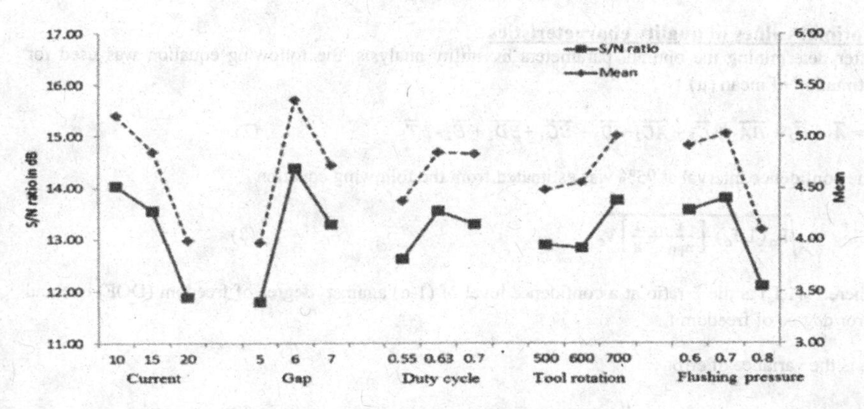

Figure 2. Effect of individual process parameters used in the REDM process

Analysis of Utility Data for optimal settings

The main effect plots for means and S/N ratio are shown in figure 2. Higher the better type condition was chosen for S/N ratio of utility value. The ANOVA results for mean have been shown in the Table 5. The Table 5 and Figure 2 shows that the second level of Gap (B_2), first level of Current (A_1) and second level of flushing pressure (E_1) have the maximum contribution on the utility values as compared to interaction effects. Though the interaction effects were not so significant, but there values were also considered while calculating mean. Tool rotation was the least significant factor.

Table 5. ANOVA for Means

	SS	DOF	Variance	SS'	F ratio	Percentage contribution (%)
A	7.05	2	3.52	6.89	45.61	20.0
B	8.66	2	4.33	8.51	56.05	24.6
B*A	1.76	2	0.88	1.60	11.36	4.6
C	1.31	2	0.65	1.15	8.46	3.3
A*C	1.92	2	0.96	1.76	12.40	5.1
D	1.47	2	0.73	1.31	9.49	3.8
B*C	1.55	2	0.77	1.40	10.03	4.0
D*E			pooled			
B*D	1.37	2	0.68	1.21	8.85	3.5
E	4.68	2	2.34	4.52	30.26	13.1
A*E			pooled			
Error	4.79	62	0.08	6.18	0.18	17.9
Total	34.54	80	0.4318			100.0

SS= Sum of Squares, DOF = degree of freedom, SS' = Pure sum of squares, F= Fisher test factor

Optimal values of quality characteristics

After determining the optimal parameters by utility analysis, the following equation was used for estimation of mean (μ)

$$\mu = \overline{A}_1 + \overline{B}_2 + \overline{BA}_2 + \overline{C}_3 + \overline{AC}_2 + \overline{D}_3 + \overline{BC}_1 + \overline{BD}_2 + \overline{E}_2 - 7\overline{T} \quad (7)$$

The confidence interval at 95% was estimated from the following equation

$$CI_{CE} = \sqrt{F_\alpha(1, f_e)\left[\frac{1}{n_{eff}} + \frac{1}{R}\right]V_e} \quad (8)$$

where $F_\alpha(1,f_e)$ is the F ratio at a confidence level of (1-α) against degree of freedom (DOF) one and error degree of freedom f_e,

V_e is the variance of error.

$$n_{eff} = \frac{N}{1 + \text{D.O.F of all factors used in the estimate of mean}} \quad (9)$$

N= Total number of trials and R = number of repetitions

a) *Surface finish*

The predicted mean of the SR given by eq. (7) is

$\mu = \overline{A}_1 + \overline{B}_2 + \overline{BA}_2 + \overline{C}_3 + \overline{AC}_2 + \overline{D}_3 + \overline{BC}_1 + \overline{BD}_2 + \overline{E}_2 - 8\overline{T} = 0.67$

$\overline{A}_1 = 1.28$, $\overline{B}_2 = 1.41$, $\overline{BA}_2 = 1.49$, $\overline{C}_3 = 1.72$, $\overline{AC}_2 = 1.5$, $\overline{D}_3 = 1.31$, $\overline{BC}_1 = 1.45$, $\overline{BD}_2 = 1.54$, $\overline{E}_2 = 1.42$ and $\overline{T} = 1.56$. The 95% confidence interval of confirmation experiments (CI_{CE}) was calculated by putting the following values in eq. (8).

N = Total number of trials = 81, R = number of repetitions = 3, $n_{eff} = 3.521$, $F_{0.05}(1, 62) = 4$, $V_e = 0.01$.

Therefore, $CI_{CE} = \pm 0.17$ (μm)

Consequently the predicted optimal range is $0.52 < SR < 0.85$ (μm)

b) *Material removal rate*

The predicted mean of the MRR given by eq. (7) is

$\mu = \overline{A}_1 + \overline{B}_2 + \overline{BA}_2 + \overline{C}_3 + \overline{AC}_2 + \overline{D}_3 + \overline{BC}_1 + \overline{BD}_2 + \overline{E}_2 - 8\overline{T} = 37.42$

$\overline{A}_1 = 25.34$, $\overline{B}_2 = 32.11$, $\overline{BA}_2 = 28.61$, $\overline{C}_3 = 34.48$, $\overline{AC}_2 = 28.66$, $\overline{D}_3 = 24.7$, $\overline{BC}_1 = 28.92$, $\overline{BD}_2 = 29.64$, $\overline{E}_2 = 28.17$ and $\overline{T} = 27.9$. The 95% confidence interval of confirmation experiments (CI_{CE}) was calculated by putting the following values in eq. (8).

N = Total number of trials = 81, R = number of repetitions = 3, $n_{eff} = 3.521$, $F_{0.05}(1, 62) = 4$, $V_e = 9.078$

Therefore, $CI_{CE} = \pm 4.74$

Consequently the predicted optimal range is $32.68 < MRR < 42.16$ (mm^3/min)

c) *Tool Wear Rate*

The predicted mean of the TWR given by eq. (7) is

$\mu = \overline{A}_1 + \overline{B}_2 + \overline{BA}_2 + \overline{C}_3 + \overline{AC}_2 + \overline{D}_3 + \overline{BC}_1 + \overline{BD}_2 + \overline{E}_2 - 8\overline{T} = 0.69$

$\overline{A}_1 = 0.76$, $\overline{B}_2 = 0.85$, $\overline{BA}_2 = 0.86$, $\overline{C}_3 = 1.02$, $\overline{AC}_2 = 0.88$, $\overline{D}_3 = 1.01$, $\overline{BC}_1 = 0.91$, $\overline{BD}_2 = .846$ $\overline{E}_2 = 0.73$ and $\overline{T} = 0.89$. The 95% confidence interval of confirmation experiments (CI_{CE}) was calculated by putting the following values in eq. (8).

N = Total number of trials = 81, R = number of repetitions = 3, n_{eff} = 3.521, $F_{0.05}(1, 62)$ = 4, V_e = 9.078.
Therefore, $CI_{CE} = \pm 0.07$

Consequently the predicted optimal range is 0.721 < TWR < 0.86 (mm^3/min).

The values of the process responses obtained in the confirmatory experiments and the predicted ranges of the responses are presented in Table 6. The experimentally obtained values are found to be well within the predicted range indicating the validity of the technique presented.

Table 6. Confirmatory experiments and predicted ranges of the process responses.

S.No.	SR (µm)			MRR (mm^3/min)			TWR(mm^3/min)		
Trial	R1	R2	R3	R1	R2	R3	R1	R2	R3
Confirmatory experiment	0.81	0.83	0.89	34.931	31.49	35.621	0.763	0.872	0.839
Predicted range	0.52 < SR < 0.85			32.68 < WR < 42.16			0.721 < WR < 0.86		

Conclusions

The following major conclusions can be drawn from present study.

1. The developed technique provides a logical and a simple way of calculation of the weights in multicriteria optimization.
2. The weights can be calculated even with incomplete information.
3. This technique converts the opinions of different researchers into a mathematical form.
4. In the present REDM machining case, gap control contributes maximum in the utility value followed by current. The interaction effects are less significant compared to main effects.
5. This technique has the potential to be coupled with other multicriteria optimization techniques to derive the weights and get the optimum results.

References

[1] J.S. Soni and G. Chakraverti, "Machining characteristics of titanium with rotary electro-discharge machining," Wear 171(1994), 51-58.

[2] H.T. Lee, F.C. Hsu, and T.Y. Tai, "Study of surface integrity using the small area EDM process with a copper–tungsten electrode," Materials Science and Engineering A 364 (2004), 346-356.

[3] B. Mohan, A. Rajadurai and K.G. Satayanarayana "Effect of SiC and rotation of electrode discharge machining of Al-SiC composite," Journals of Materials Processing Technology 124(2002), 297-304.

[4] B. Mohan, A. Rajadurai and K.G. Satayanarayana, "Electric discharge machining of Al-SiC metal matrix composites using rotary tube electrode," International Journal of Advance Manufacturing Technology 38 (1-2) (2004), 74-84.

[5] B.H. Yan, C.C. Wang, W.D. Liu and F.Y. Huang, "Machining Characteristics of Al2O3/6061Al Composite using Rotary EDM with a Disklike Electrode," International Journal of Advanced Manufacturing and Technology 16(2000), 322–333.

[6] M. Ghoreishi and J. Atkinson, "A comparative experimental study of machining characteristics in vibratory, rotary and vibro-rotary electro-discharge machining," International Journal of Advanced Manufacturing Technology, 120 (1-3) (2002) 374-384.

[7] R.T. Marler and J.S. Arora, "Survey of multi-objective optimization methods for engineering," Structural and Multidisciplinary Optimization, 26(2004), 369–395.

[8] R.S. Walia, H.S. Shan, and P. Kumar, "Multi-Response Optimization of CFAAFM Process Through Taguchi Method and Utility Concept," Materials and Manufacturing Processes, 21 (8) (2006), 907-914.

[9] Y.E. Nahm, H. Ishikawa, and M. Inoue, "New rating methods for prioritizing customer requirements in QFD based on customer preference and satisfaction analysis with incomplete information," The International Journal of Advanced Manufacturing Technology, Springer, DOI 10.1007/s00170-012-4282-1. (2012).

[10] Y.E. Nahm and H. Ishikawa, "A new 3D-CAD system for set-based parametric design," International Journal of Advanced Manufacturing Technology 29 (2006), 137-150.

Materials Processing Fundamentals
Edited by: Lifeng Zhang, Antoine Allanore, Cong Wang, James A. Yurko, and Justin Crapps
TMS (The Minerals, Metals & Materials Society), 2013

CHARACTERIZATION OF PORE FORMATION IN A356 ALLOY WITH DIFFERENT OXIDE LEVELS DURING DIRECTIONAL SOLIDIFICATION

Hengcheng Liao[1*]; Wan Song [1]; Qigui Wang [2]
[1]School of Materials Science and Engineering, Southeast University, Jiangsu Key Laboratory for Advanced Metallic Materials, District of Jiangning, Campus of Southeast University, Nanjing 211189, China
[2]Materials Technology, GM Global Powertrain Engineering, 823 Joslyn Ave., Pontiac, MI 48340, USA

Keywords: Al-Si alloy, directional solidification, X-ray imaging, porosity formation, aluminum oxides

Abstract

Characterization of porosity formation in an A356 alloy with different oxide levels during directional solidification was investigated using micro-focus X-ray imaging and directional solidification technology. Stirring melt is thought to provide more active nucleation sites for pore formation, thus lead to a remarkable rise in the nucleation temperature of pores. The fast growth of those pores formed at higher temperatures further restrains the succeeding nucleation operations in local regions, and results in a considerable reduction in the pore volume density but a significant increases in pore volume fraction. Fluctuations of pore volume fraction and pore volume density along solidification length is thought to be closely related to a competition mechanism of pore nucleation with pore growth for hydrogen supplement. The increase in oxide content by stirring melt completely changes the pore size distribution and considerably increases the average size of pores formed.

Introduction

Porosity in cast aluminum alloys has long been recognized as one of the most detrimental factors affecting mechanical properties [1,2], especially fatigue resistance of the material [3-8]. Porosity forms in castings due to gas re-partitation and volumetric shrinkage. The large difference in hydrogen solubility between the solid and liquid is the main cause for hydrogen gas porosity [9]. Shrinkage porosity refers to the density difference between the solid and the liquid. Porosity formation can be divided into nucleation and growth. In cast aluminum alloys, the nucleation of porosity is generally considered to be heterogeneous. Chen and Engler [10] reported that the formation of the first gas bubble in their experiment only required a very low superstauration pressure, about 0.1kPa, indicating that the formation of pores is heterogeneously nucleated. The heterogeneous nucleation occurs at sites such as oxide inclusions with gas gaps, grooves of mould wall, and undissolved fine gas bubbles in liquid, etc. [10-12]. During melting or pouring of aluminum alloys, oxide inclusions are prone to form on which there are a great number of nano-scale gas gaps which have strong ability to absorb hydrogen. The oxide inclusions are almost not wetted by aluminum liquid and thus they are ideal nucleation sites for porosity formation. Laslaz and Laty [13] reported that stirring melt or adding crippled aluminum pieces aggravated melt oxidation, increasing the amount of pores and pore fraction in the castings. Chen

* Corresponding author, Ph. D., professor of Southeast University, School of Materials Science and Engineering, Jiangning Campus of Southeast University, Nanjing 211189, China. Tel: +86 25-52090686; Email: hengchengliao@seu.edu.cn

and Gruzleski [14] reported the pore density decreased with the reduction of oxides in melt. Mohanty et al. [11,12] intentionally added Al_2O_3 particles into A356 melt to arise the oxide level and found that addition of Al_2O_3 particles resulted in an increase in the pore density and hence the increase in volume fraction of pores. Recently, Campbell [9] has claimed that nucleation of all pores is originated from bifilms of Al_2O_3 oxides, which have a dual-layer structure and no bonding strength between the two dry layers. And the formation of porosity is thought to be that the bifilm is pulled open by all kinds of forces produced during solidification. The solidification shrinkage force exerts the bifilms to deform along the profile of solidified solid to form the irregular morphology of pores.

Previous experimental studies on porosity formation mainly rely on the metallographic analysis of solidified specimens or liquid quenching technology for directional solidification [15-18]. Recently, X-ray imaging technology has been used to investigate the pore formation in aluminum alloy [19-23]. Although significant advances in understanding of physics with respects to porosity formation have been achieved with an in-situ X-ray technique, more research is still needed on porosity formation under different oxide levels. In this paper, a micro-focus X-ray and directional solidification technology (XIDS) are utilized to study the influence of oxide level on the formation of pores in A356 alloy.

Experimental procedures

Alloy and casting

The base alloy A356 with a nominal composition of Al-7wt.%Si-0.4wt.%Mg -0.2wt.%Ti was melted at 1033K(760°C) in an electrical resistance furnace with a graphite-clay crucible of 5kg capacity. Two oxide levels were utilized. One melt (Specimen Z, referred as a low oxide level) was just melted from the ingots without any melt treatment. The other melt of high oxide content (Specimen G) was obtained by first mechanically stirring the melt for 20min., and then degassing with Ar for 10 min. to reduce the hydrogen rise by stirring to obtain a close level to the Z specimen. Prior to pouring, small amount of the liquid metal was taken out to solidify in a heavy copper Ransely mold to obtain a chilled sample for hydrogen content measurement. The hydrogen content is measured by vacuum remelting technique (LECO® Technical Services Laboratory).

After processed, the melt was then poured into a stainless steel specimen container with a cavity of 300×10×3 mm^3 and a wall thickness of 0.1mm. The specimen container was placed in the XIDS set-up [22] prior to pouring. Once the melt was poured, the specimen container moved downward at a constant velocity. Simultaneously the cooling system in the XIDS was activated to start directional solidification from the bottom to the top in the specimen container and X-ray radiographs and video were recorded automatically at a frequency of 25 Hz. The temperature gradient during solidification was measured using two "K" type thermocouples (with an accuracy of ±1.0K) placed 5 mm apart from each other inside the specimen cavity. The melting processes, hydrogen contents, and solidification variables are listed in Table 1. In order to reduce the rise of hydrogen content by stirring, the G melt is degassed by Ar bubbles after stirred, however, the hydrogen level in it is still slightly higher than that in Z melt due to the high oxide level in it.

Table 1 Melt quality and solidification conditions of studied alloys

Specimen	Z	G
Stirring the melt for 20min.	N	Y
De-gassing by Ar bubbles for 10min.	N	Y
Hydrogen content, mL/100g Al	0.20	0.23
Solidification velocity, mm/s	0.1	0.1
Temperature gradient, °C/mm	4.28	4.28

***In-situ* characterization of porosity formation**

Porosity formation was *in-situ* observed using a micro-focus X-ray and directional solidification technology (XIDS). The detailed description of the XIDS and directional solidification experimental setup has been published elsewhere [22]. During the steady state of solidification in the XIDS set-up, the temperature gradient was kept constant and the temperature field in the X-ray window of view was thus fixed. According to the position of thermocouples in the window of view, the temperature field is determined. In XIDS experiments, the positions (corresponding to temperature) of the selected pores from their first emergence to final disappearance were tracked from a series of time-dependent photos captured by *in-situ* XIDS. The equivalent circle diameter (ECD), the changes of the ECD and growth rate (dr/dt, r the equivalent circle radius, t the time) of the pores with temperature were then obtained from image analyzer. As the radiographic resolution was insufficient to resolve the nucleation event, the nucleation temperatures of pores were estimated by extrapolating the growth curve for each pore to 10 μm. The distribution of the projected pore nucleation temperatures were fitted with Gaussian curves from the nucleation temperature histogram (2 °C bin size). The average nucleation temperature (T_{avg}) was then determined from the Gaussin distribution.

Statistic characteristics of porosity in solidified specimen

After solidification is completed, the containers on the XIDS specimens were taken off to improve the resolution and then they were further evaluated by the XIDS detection. The X-ray detection image of each specimen was divided into five sections: Section 1 representing the beginning of solidification and Section 5 the end of solidification. Photographic enhancement was adopted to improve the contrast between pores and the matrix for image analysis. The porosity was quantified in terms of ECD (Equivalent circle diameter) of each pore, Pxp (volume fraction of pores in one section) and Pxs (volume fraction of pores in one specimen), Pdxp (density of pores in one section, the counts of pores per unit mm^3) and Pdxs (density of pores in one specimen), and number of large pores (defined as pores with a configuration size over 0.5mm). The pore size distribution in each specimen was made with a bin of 0.02 mm.

Results and discussion

From the time-dependent X-ray imagings obtained by XIDS, twenty pores in each specimen are tracked from their first emergence to final solidification in the window of view of X-ray. The ECDs of these pores in both specimens are plotted as functions of temperature as shown in Figure 1. The fluctuation of those curves reveals the growth complexity of pores, consistent with the real time observation. The concurrent of nucleation and growth of pores results in the fluctuation of pore growth. In both specimens, pores grow fast at the early stage, corresponding to the high temperature region, and then slow down at the late stage corresponding to the low temperature region. The increase of oxide content in the melt (specimen G), leads to a significant increase in temperature of porosity first emergence. The final sizes of the pores tracked are also larger with a much wider size distribution too in comparison with low oxide content specimen (Z).

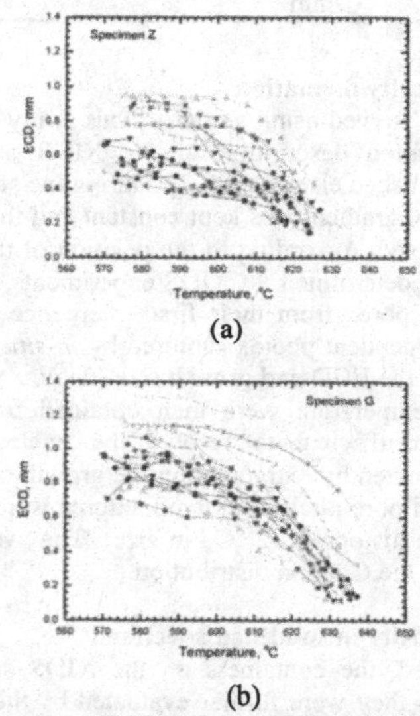

Figure 1. Equivalent circle diameters (ECD) of the pores tracked v.s. temperature; (a) Specimen Z and (b) Specimen G.

The nucleation kinetics of pores during directional solidification mainly depends on the super-saturation of hydrogen solute and the nucleation power of the potential nucleation sites of pores in the liquid. The solute super-saturation is a combination of original hydrogen content in the melt and the solute amount rejected by solid that is dependent on the solidification velocity. The potential nucleation sites for gas pores have been thought to be oxide inclusions. As listed in Table 1, the original hydrogen contents in both Specimens Z and G are approximately the same and the solidification velocities applied in the experiments are the same too. However, the melt of Specimen G was stirred for 20min., and it is believed that the oxide level in the specimen was high, especially the number of active sites for pore nucleation.

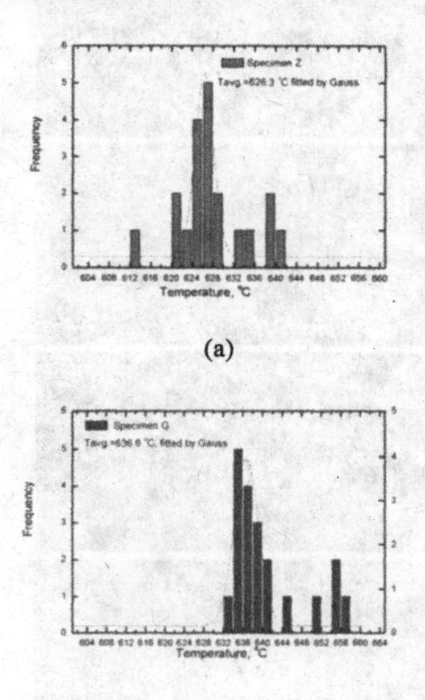

(b)

Figure 2. Histogram and fitted Gaussian distribution of the projected nucleation temperature of the tracked pores (2°C bin size); (a) Specimen Z and (b) Specimen G

As the radiographic resolution was insufficient to resolve the nucleation event, the nucleation temperatures of pores were estimated by extrapolating the growth curve for each pore to 10 μm. The histogram distribution of the nucleation temperature of pores is shown in Figure 2. For a low oxide level in Specimen Z, the derived nucleation temperatures are between 612 °C and 642°C. Whereas for a high level of oxides in Specimen G the nucleation temperatures range from 632 °C to 658°C. The average nucleation temperature, *T*avg. fitted by Gaussian in Specimen Z is about 626.3°C, much lower than that in Specimen G (636.6 °C). It is thought that stirring the melt (in specimen G) produced many high active nucleation sites for pore formation. During solidification, pores can easily nucleate from these sites with a low supersaturation of hydrogen required, and grow at a higher temperature region. As a result, the nucleation operation of new pores at the low temperature region is suppressed by the growth of existing pores formed in the high temperature region.

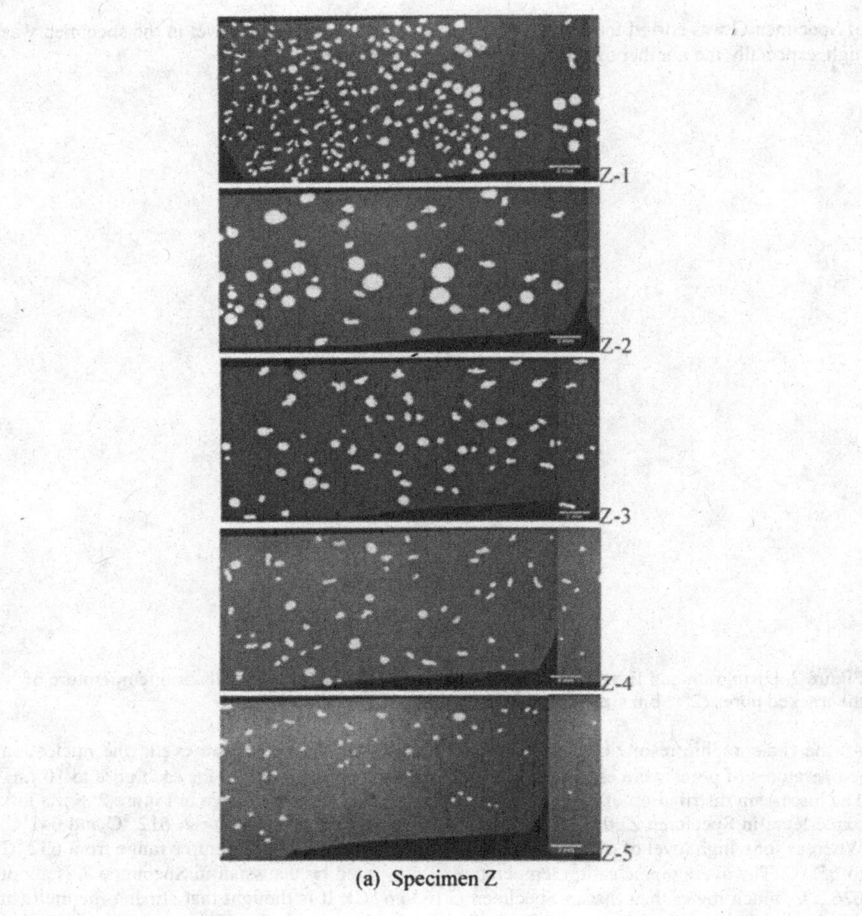

(a) Specimen Z

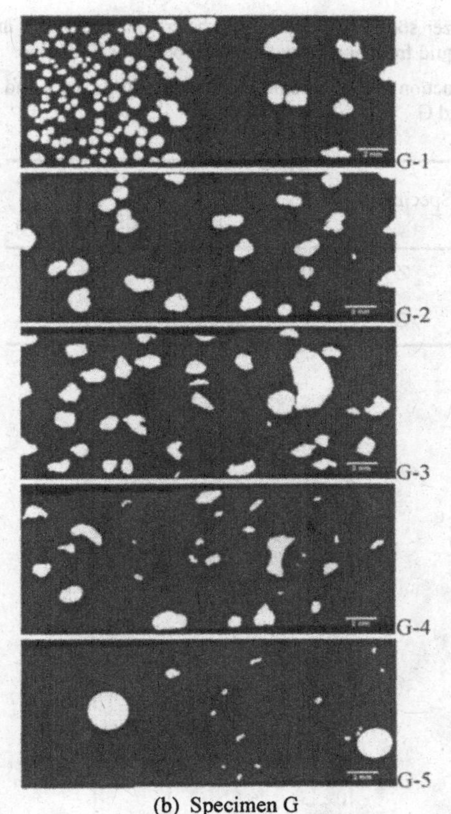

(b) Specimen G

Figure 3 X-ray detection radiographs showing porosity distribution in Specimens Z (a) and G (b).

Figure 1 shows X-ray detection radiographs, revealing the formed pores in the solidified XIDS specimens. The statistical results of porosity quantification with a X-CT detection are listed in Table 2 and shown in Figures 4 and 5. In general, the pore sizes in Specimen G with high oxide level are much larger than those in Specimen Z. The majority of pores in both specimens are not really in spherical shape. The irregularity of pore morphology indicates that in case of such hydrogen content studied in this work, solidification shrinkage exerts an important influence on the growth and especially the morphology evolution. However, in the alloy with high hydrogen content solidification shrinkage has little influence on pore morphology[24]. In addition, there are many more small pores formed in Section 1 (early solidified portion) of both Specimens Z and G compared with other regions, as shown in Figure 3. This is because of the extra chilling effect from the freezer in the XIDS. When liquid metal is poured into the XIDS container, the melt contacting the freezer solidifies faster than other regions resulting in high hydrogen super saturation in the solid/liquid front and more nuclei for porosity formation.

Table 2 Pore volume fraction (Pxs), volume number density (Pdxs), and total number of large pores in Specimens Z and G

Specimen	Pxs vol.%	Pdxs mm^{-3}	Total number of large pores
Z	0.77	0.102	63
G	3.13	0.051	137

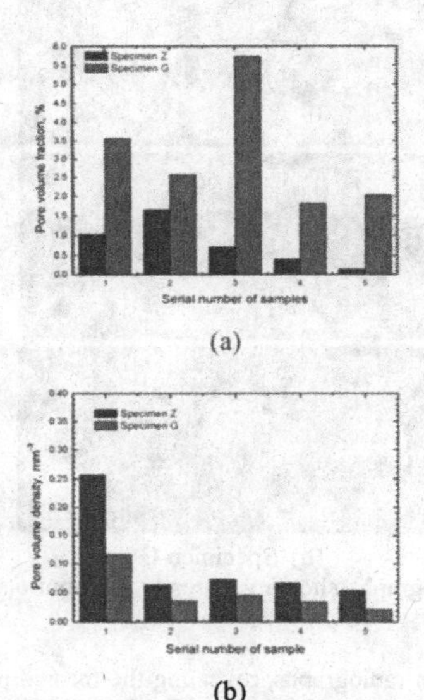

(a)

(b)

Figure 4 Pore volume fraction (a) and volume number density (b) along the solidification length of Specimens Z and G

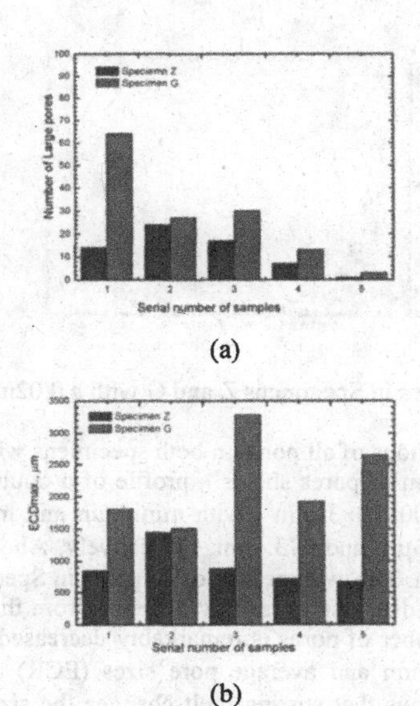

Figure 5 Number of large pores (a) and ECDmax. (b) along the solidification length of Specimens Z and G

Porosity quantification results in Table 2 indicate that high oxide level results in a considerable rise in porosity volume fraction, Pxs, and the total number of large pores, but remarkably decreases the overall pore volume number density, Pdxs. Stirring the melt do not increases the amount of pores formed, however, Laslaz and Laty [13] reported that stirring melt or adding crippled aluminum pieces to aggravate oxidening increased the amount of pores and pore fraction in the castings with the same hydrogen content in the melt. Figure 4 shows the porosity volume fraction and volume number density along the solidification length of Specimens Z and G. It is seen that the volume fraction of porosity (Pxp) in all sections of Speciem G is higher than that of Speciem Z, but the results of Pdxp are just inversed. Along the solidification length, there are considerable fluctuations of the volume fraction of porosity in both specimens (as shown in Figure 2a), particularly in Specimen G. The highest volume fraction of porosity is in Section 2 for Specimen Z and in Section 3 for Specimen G. However, the fluctuation of pore volume number density in both specimens is not obvious (as shown in Figure 2b), and the density in Section 2 is only slightly less than that in Section 3. A general trend is that the pore density is decreased along the solidification length with the highest pore density in Section 1. For the total number of large pores and ECDmax., as shown in Figure 5, there are also considerable fluctuations along the solidification length. This may be contributed to the competition of pore nucleation with pore growth for hydrogen supplement.

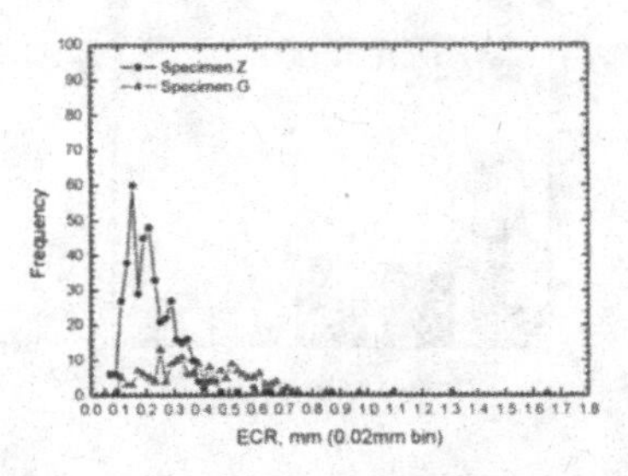

Figure 6 Size distribution of pores in Specimens Z and G with a 0.02mm bin

Figure 6 shows the size distributions of all pores in both specimens with a bin size of 0.02mm. In Specimen Z, the size distribution of pores shows a profile of a double-humps. Total 440 pores are formed in the specimen ($300\times10\times3$ mm^3) with minimum and maximum pore sizes (ECR, equivalent circle radius) of 65.3μm and 773.4μm, respectively. About 80% of the pores are in the size range of 100μm -300μm. The average size of all pores in Specimen Z is about 222μm. In Specimen G, however, the size distribution is fully different from that in Specimen Z. There is no obvious peak. The total number of pores is remarkably decreased. There are total 190 pores formed with minimum, maximum and average pore sizes (ECR) of 52.1μm, 1820.9μm and 410μm, respectively. It is obvious that stirring melt changes the size distribution of pores and considerably increases the average size of pores formed.

As described above, there is no considerable difference in hydrogen content between two specimens. The remarkable difference in porosity volume fraction and especially the total number of large pores indicates that oxides play a very important role in both nucleation and growth. This is generally in agreement with the findings reported in the literature. Mohanty *et al.* [11,12] intentionally added Al_2O_3 particles into A356 melt to increase the oxide level and found addition of Al_2O_3 particles resulted in an increase in the pore density and hence the increase in volume fraction of porosity, however, a reduction in the pore size. Report from Laslaz and Laty [13] indicated that with the same amount of hydrogen in the melt, stirring melt or adding crippled aluminum pieces to aggravate oxidation increased the number density of pores and the volume fraction of porosity in the castings. Chen and Gruzleski [14] reported the pore density decreased with the reduction of oxides in the melt. Because their data of porosity properties were obtained from 2D metallographic analysis, one irregular shrinkage porosity formed during conventional solidification may be sectioned into a few small porosities, which will lead to an increase in number density and a reduction in size of porosities. X-ray detection reveals the actual configuration of porosities. In our study, the reduction of pore density with increasing oxide content may be explained by the competition mechanism of pore nucleation and growth for hydrogen supplement. During solidification, the nucleation behavior of pores is not explosive, but continual. Pores should first nucleate and grow from those more active nucleation sites which require less supersaturation of hydrogen. This has been demonstrated in a study of directional solidification of eutectic Al-Si alloy [22]. The pores were found to form far away from solidification front, where the supersaturation of hydrogen is lower than that near the solidification front. The growth of those previously formed pores needs much lower superaturation of hydrogen than that for nucleation of new pores. As a result, the continuous growth of those existing pores further restrains local nucleation operations of new pores. In our study, stirring melt is believed to increase the oxide level in the melt. More importantly, it also increases the amount of more active nucleation sites for nucleation of pores, which has been demonstrated by the remarkable rise in nucleation temperature of pores. In Specimen G with a high level of oxide, a great amount of pores formed far above the liquidus were observed in real time by XIDS. Growth of those early formed pores restrains succeeding pore nucleation operations in local regions, thus the pore density is considerably decreased in specimen with stirring. Furthermore, those pores formed at higher temperature have more time to grow to larger sizes. As a result, both volume fraction of porosity and the total number of large pores are remarkably increased in the sample solidified from the stirred melt.

Conclusions

(1) Stirring melt is thought to provide more active nucleation sites for pore formation, thus leads to a remarkable rise in the nucleation temperature of pores. The fast growth of those pores formed from the oxides at higher temperatures further restrains the succeeding nucleation operations in local regions and results in a considerable reduction in the pore density. As a result, stirring melt significantly increases the volume fraction of porosity and the total number of large pores.

(2) Fluctuations of volume fraction of porosity, volume pore density, and the number of large pores during directional solidification is caused by the competition mechanism of pore nucleation with pore growth for hydrogen supplement. Stirring melt completely changes the pore size distribution and considerably increases the average size of pores formed.

Acknowledgement

The authors would like to thank Prof. Ye Pan at Southeast University for his valuable discussions. This work was funded by GM Research Foundation under contract No. GM-RP-07-211.

References

1 R. C. Atwood, S. Sridhar and P. D. Lee, "Equations for nucleation of hydrogen gas pores during solidification of aluminium seven weight percent silicon alloy," *Scripta Mater.*, 41(12) (1999), 1255-1259.

2 C. D. Lee, "Tensile properties of high-pressure die-cast AM60 and AZ91 magnesium alloys on microporosity variation," *J. Mater. Sci.*, 42(24) (2007), 10032-10039.

3 Q.G. Wang, D. Apelian and D. A. Lados, "Fatigue behavior of A356-T6 aluminum cast alloys. Part I. Effect of casting defects," *Journal of Light Metals*, 1 (2001): 73-84.

4 J. Y. Buffière et al., "Experimental study of porosity and its relation to fatigue mechanisms of model Al-Si7-Mg0.3 cast Al alloys," *Mater. Sci. Engn.* A, 316(1-2) (2001), 115-126.

5 J. Z. Yi et al., "Statistical modeling of microstructure and defect population effects on the fatigue performance of cast A356-T6 automotive components," *Mater. Sci. Engn.* A, 432 (1-2) (2006), 59-68.

6 L. Dietrich and J. Radziejewska, "The fatigue damage development in a cast Al–Si–Cu alloy," *Materials and Design*, 32 (2011), 322–329.

7 J.B. Jordon et al., "Microstructural Inclusion Influence on Fatigue of a Cast A356 Aluminum Alloy," *Metall. Mater. Trans.* A, 41 (2010), 356-363.

8 De-Feng Mo et al., "Effect of microstructural features on fatigue behavior in A319-T6 aluminum alloy," *Mater. Sci. Engn.* A, 527 (2010), 3420–3426.

9 J. Campbell, "Castings," Butterworth-Heinemann, 2003.

10 X. G. Chen and S. Engler, "Formation of Gas Porosity in Aluminum Alloys," *AFS Trans.*, 102 (1994), 673-682.

11 P. S. Mohanty, F. H. Samuel and J. E. Gruzleski, "Mechanisum of heterogeneous nucleation of pores in metals and alloys," *Metall. Mater. Trans.* A, 24 (1993), 1845-1856.

12 P. S. Mohanty, F. H. Samuel and J. E. Gruzleski, "Experimental study of pore nucleation by inclusions in alumiaum casings," *AFS Trans.*, 103 (1995), 555-564.

13 G. Laslaz and P. Laty, "Gas Porosity and Metal Cleanliness in Aluminum Casting Alloys," *AFS Trans.*, 99 (1991), 83-90.

14 X. G. Chen and J. E. Gruzleski, „Influence of Melt Cleanliness on Pore Formation in Aluminum-Silicon Alloys," *Cast Metals*, 9 (1996), 17-26.

15 R. Fuoco, E. R. Correa and M. de Andrade Bastos, "Microporosity morphology in A356 aluminum alloy in unmodified and in Sr modified conditions," *AFS Trans.*, 108 (2001), 659-768.

16 K. Tynelius, J. F. Major and D. Apelian, "A Parametric Study of Microporosity in the A356 Casting Alloy System," *AFS Trans.*, 101 (1993), 401-13.

17 O. Savas and R. Kayikci, "Application of Taguchi's methods to investigate some factors affecting microporosity formation in A360 aluminum alloy casting," *Mater. & Design*, 28(7) (2007), 2224-2228.

18 J. R. Kim and R. Abbaschian, "Influence of processing variables on microporosity formation in Al-4.5% Cu alloy," TMS Annual Meeting, *Frontiers in Solidification Science* - Proceedings of Symposium, held during the 2007 TMS Annual Meeting, 2007, p 35-45.

19 L. Omid et al., "X-ray microtomographic characterization of porosity in aluminum alloy A356," *Metall. Mater. Trans.* A, 40(4) (2009), 991-999.

20 P. D. Lee and J. D. Hunt, "Hydrogen porosity in directional solidification aluminum-copper alloys: in situ observation," *Acta Mater.*, 45(10) (1997), 4155-4169.

21 R.C. Atwood et al., "Diffusion-controlled growth of hydrogen pores in aluminum-silicon castings: in situ observation and modelling," *Acta Mater.* 48 (2000), 405-417.

22 L. Zhao et al., "In-situ observation of porosity formation during directional solidification of Al-Si Casting Alloys," China Foundry, 8(1) (2011), 14-18.

23 L. Zhao et al., "Abnormal segregation induced by gas pores during solidification of Al-Sn alloy," *Scripta Mater.*, 65 (2011), 795-798.

24 Hengcheng Liao et al., "Effect of solidification velocity and hydrogen content on pore formation in A356 alloy based X-ray detection," TMS Light Metals, March 11- 15, 2012, Orlando, FL, United states, 2012, p 345-348

MATERIALS PROCESSING FUNDAMENTALS

Process Metallurgy of Non-Ferrous Metals

Session Chair

James Yurko

Materials Processing Fundamentals
Edited by: Lifeng Zhang, Antoine Allanore, Cong Wang, James A. Yurko, and Justin Crapps
TMS (The Minerals, Metals & Materials Society), 2013

TMAH wet etching of silicon micro- and nano-fins for selective sidewall epitaxy of III-Nitride semiconductors

L. Liu[1], D. Myasishchev[1], V. V. Kuryatkov[2], S. A. Nikishin[2], H. R. Harris[3], and M. Holtz[1]

[1] Department of Physics and Nano Tech Center, Texas Tech University, Lubbock, Texas 79409
[2] Department of Electrical Engineering and Nano Tech Center, Texas Tech University, Lubbock, Texas 79409
[3] Department of Electrical Engineering, Texas A&M University, College Station, Texas 77843

Keywords: TMAH, Wet Etching, Nano-fins

Abstract

We describe wet etch experiments to form silicon micro- and nano-fins, with (111)-plane sidewall facets, to be used in selective epitaxy of III-Nitride semiconductors. Starting material was (110)-oriented silicon wafers. Silicon dioxide is used for producing a hard mask and patterned using photo- and electron-beam lithography for micro- and nano-fins, respectively. Wet etching to produce silicon fins was carried out using tetramethyl ammonium hydroxide (TMAH) diluted with isopropyl alcohol (IPA). We experimented with silicon doped TMAH/IPA solution, using a sacrificial wafer, and with surfactant (Triton X-100). Scanning electron microscopy (SEM) and atomic force microscopy (AFM) were used to determine the morphology including surface roughness of the area between fins and etching rate of silicon. By controlling the etching time, temperature, percentage of IPA and concentration of Triton X-100 we obtain good surface morphology on both (111) and (110) planes. Nanofins as small as 30 nm in width and ~ 250 nm in height are prepared, with corresponding aspect ratio of ~ 8:1.

Introduction

Nano scale materials are of great interest due, primarily, to the differences in electronic structures from their bulk counterparts [1, 2]. Growth of these structures has been based on both top-down and bottom-up approaches. The growth substrate is a key factor in material quality for the latter case. Silicon remains an excellent substrate material for design of novel electronic nanostructures due to the wide-scale availability, low cost, and breadth of technologies which are based on this material. However, growth of high-quality group III-V and III-nitride materials is challenging on silicon due to mismatches in epitaxial relationship.

Fin architectures similar silicon field effect transistors (FinFETs and Multi-FinFETs) are of current interest for achieving high device density and much faster devices [3-5]. The high mobility and high breakdown field of III-nitride semiconductors, along with a wide range of alloy compositions for band structure engineering, makes devices founded on vertical III-nitride fins interesting for a range of applications. To explore these possibilities, experiments are needed to obtain silicon fins with vertical (111) side walls on Si (110) substrates for the III-nitride selective epitaxy [6].

In this paper we utilize directional wet etching to produce silicon fins from bulk substrates. The fins range from the micro- to nanoscale. Directional wet etching of silicon is well developed for applications in microelectronics. Notable etchants include potassium hydroxide (KOH) [7], ethylene diaminepyrocatechol (EDP) [8], and tetramethyl ammonium hydroxide (TMAH) typically mixed with isopropyl alcohol (IPA) [9]. A desirable process for etching nano fins must provide control over width and height, along with smooth sidewalls and smooth planes between the fins. In addition, the corners where the sidewalls and planar regions meet should have controlled morphology without over- or underetching, respectively yielding the formation of the undesirable trenches or shoulders at the base of the fins.

Previous investigators have carried out IPA/TMAH wet etching of micro-structures [10]. The desired etch temperature was in the 70 to 90 °C range [11, 12]. The effect of IPA and TMAH concentrations have also been investigated [13]. The IPA concentration was varied and the etch results investigated by scanning electron microscope. The primary result of this prior work was to show that IPA was effective at reducing the etching rate on select crystallographic planes, including (100) and (311). Furthermore, the effect of surfactants for producing smooth etch surfaces has been investigated using various additives. To produce structures needed for micro-electro-mechanical systems (MEMS), Gosalvez, et al.[14], experimented with the addition of surfactant Triton X-100 in order to improve etch anisotropy and root-mean squared (RMS) surface roughness.

In this paper we report etching with a solution of isopropyl alcohol (IPA) and TMAH for achieving < 100 nm wide silicon fins with variable aspect ratio, based on etch time. The IPA/TMAH solution etches (110) surfaces considerably faster than (111) planes [15, 16], thus making it a good choice for producing nanofins with (111)-oriented sidewalls on (110)-oriented silicon wafers. We also report the result of etching with commercial surfactant (Triton X-100) and doping of the etchant using blank silicon wafers.

Experimental Details

Our starting material was silicon wafers with (110) - oriention. A ~ 50-nm thick SiO_2 layer is first deposited as the hard mask material using plasma enhanced chemical vapor deposition. The mask is patterned using photolithography or electron-beam lithography (EBL) for the micro- and nanofins, respectively. The wafer was next cleaned to remove organic contaminants, rinsed in DI water and baked for 2 min at 115 °C. For the photolithography, adhesion promoter hexamethyldisilazane (HMDS, 99%) was applied by spin coating, followed by a soft bake (115 °C). The S1813 resist was applied and a pre-bake completed (115 °C). The lithography masks were aligned using the wafer flat so that fins would be oriented along the (110) direction. Stripe widths ranged from 1 to 4 μm, and were spaced 50 μm apart to minimize micro loading effects. These areas also provide representative regions for the AFM images. The sample was exposed and developed using MF319. Following a post-bake, the SiO_2 is etched in buffered HF.

Scanning electron microscope (SEM) based EBL was used to pattern the nanoscale structures. The EBL followed a similar procedure, with polymethylmethacrylate (950 PMMA A4) serving as the resist. Following spin coating, the PMMA is soft baked (170 °C). Line widths in the 100 to 300 nm range were written, with lengths of 50 to 100 μm also oriented along the crystalline (110) direction. The nano fins were developed using MIBK: IPA (1:3). The post bake was carried out at 170 °C. All other steps were described above for the micro fins.

The etching process was carried out in a bath placed on a temperature-controlled hot plate, as depicted in Figure 1. Bath temperature was measured directly using a digital thermometer. The etchant consisted of IPA (>99.5%) and TMAH (25% in water) solutions measured by volume and specified according to these volume amounts. The 20/100 notation for IPA/TMAH corresponds to 20 parts IPA (98%) in 100 parts TMAH (25%). IPA is highly volatile in the temperature range investigated here. Therefore, the TMAH and IPA containers were separately heated to the processing temperature and mixed at the time of etching. This allows us to specify these concentrations prior to substantial evaporation of IPA. In addition, we also "doped" the TMAH solution by pre-immersion of a sacrificial silicon wafer in the TMAH for several minutes, followed by etching the patterned silicon wafer in the doped TMAH solution. For the studies utilizing surfactant (Triton X-100), we added this to the pre-heated IPA immediately prior to mixing with the TMAH.

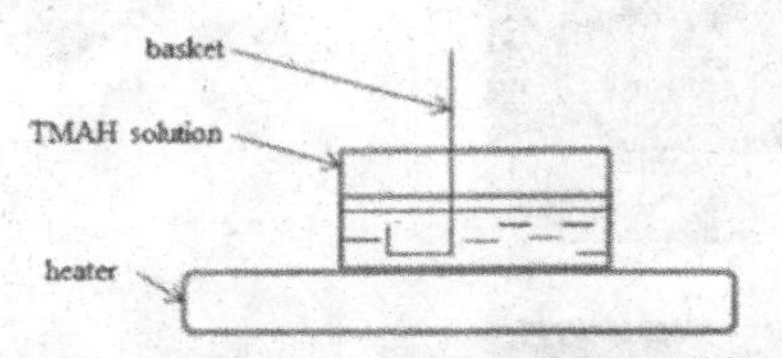

Fig.1. Apparatus for TMAH wet etching

Temperature and IPA Concentration Dependence

We first carried out studies of the temperature dependence of wet etching using micro fins. The etchant used 15, 20, or 25 parts IPA in 100 parts of TMAH solution, and at temperatures ranging from 60 to 78 °C. SEM images show that vertical sidewalls were readily obtained due to anisotropic etching of silicon. The fins are narrower than the mask patterns due to the undercutting by lithography bias and etch bias [17]. For the 20/100 IPA/TMAH concentration, smooth sidewalls were obtained after etching at all temperatures investigated. The SEM cross-section is shown in Figure 2(a). Corners where the sidewalls meet the (110) planes are sharp. Upon investigating the 25/100 IPA/TMAH concentration, we found that shoulders developed at these corners. For each sample, we used AFM to determine surface roughness. The measurement was carried out using tapping mode with a silicon tip at scan sizes of 1 μm × 1 μm and 5 μm × 5 μm. The root mean square roughness (RMS) is the square root of the arithmetic mean of the square of the vertical deviations from the reference line:

$$RMS^2 = \frac{1}{L^2}\int_0^L \int_0^L [z(x,y) - \bar{z}]\, dxdy \quad (1)$$

where L is the scan length (1 and 5 μm here), z is the surface height at point (x,y), and $\bar{z}$ is the average height in the image area. The AFM images of the etched areas between the fins, hereafter denoted planar regions, reveal RMS roughness below 5 nm ($5\ \mu m \times 5\ \mu m$), with improvement at higher etching temperatures where it drops below 2 nm. This is shown in Figure 3 (b) and (c). The $1\ \mu m \times 1\ \mu m$ AFM images show that at lower temperatures, the etched surface is rough. Higher temperature decreases the size and density of bumps, which nearly disappear at etch temperature 78 °C. This is consistent with the results achieved by Chandrasekaran et al. [18]. For the 15/100 IPA/TMAH concentration, we observed that the (110) silicon surfaces were considerably rougher. Therefore, from a morphological point of view, the 20/100 IPA/TMAH mixture was found to be the best.

Both the etching rate and smoothness of silicon (110) plane increase as etchant temperature increases. Figure 4 shows the etching rates obtained from the three different etchant concentrations versus temperature. As expected, the etch rate increases with higher temperature.

Analyzing this using an Arrhenius dependence (not shown) results in activation energies in the 0.52 to 0.69 eV range. This is consistent with previously published results of Charbonnieras, et al. [19].

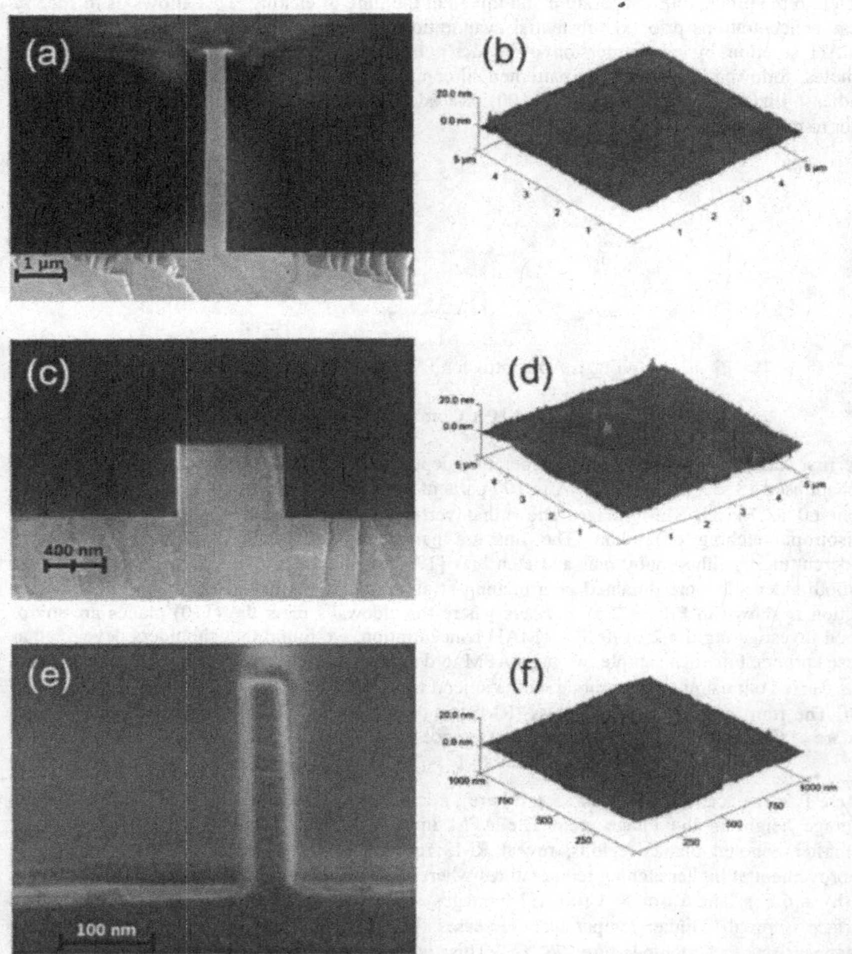

Fig. 2.Cross-section SEM image and plan-view AFM images following different etches: (a) & (b) 20/100 IPA/TMAH concentration at 78 °C for 10 min. (c) & (d) 30/150 IPA/TMAH concentration doped with silicon and 3 drops surfactant at 78 °C for 2 min. (e) & (f) 30/150 IPA/TMAH doped with silicon and 3 drops surfactant at 78 °C for 1 min with nanoscale pattern.

Timed Etching

To examine the effect of etch time on surface roughness, we processed a series of wafers with time varied from 2 to 20 min using the 20/100 IPA/TMAH concentration described above and etchant temperature of 78 °C. Figure 2(a) is the corresponding cross-section SEM image following 10 min. The fin is ~ 500 nm wide, and the etch depth is ~ 4.5 µm. Figure 2(b) is the corresponding AFM image. The etch depth depends linearly on time, establishing a rate of 418 ± 11 nm/min with no etch delay. In addition, the surface exhibits high RMS roughness (~ 8 nm after 2 min) becomes consistently smoother with etch time, reaching values < 1 nm after 5 min and remaining in this range for longer etch times. The better surface morphology may result from the increased concentration of dissolved silicon in IPA/TMAH solution. This is consistent with previously described results [20] where a silicon powder was used for a doping of IPA/TMAH solution and motivates intentional silicon doping.

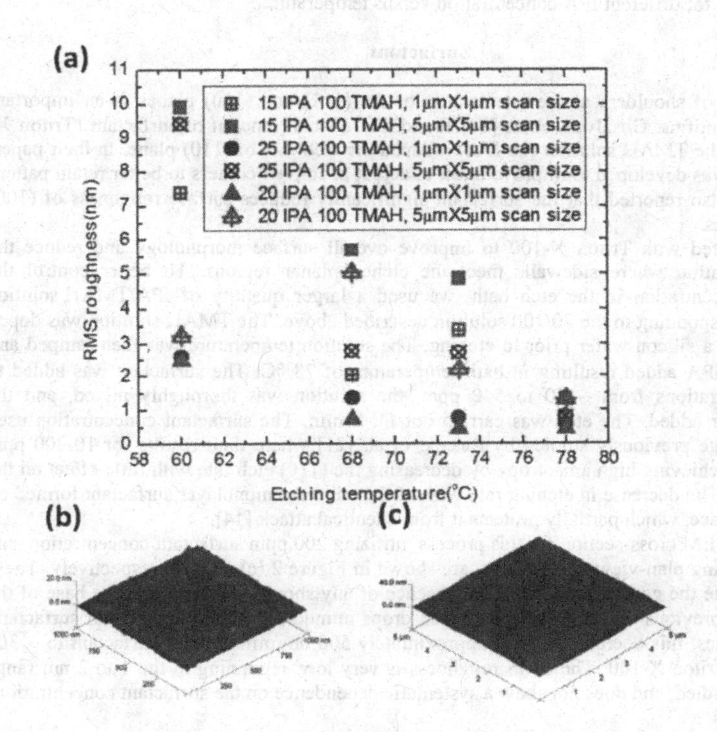

Fig.3. (a) RMS roughness based on AFM measurements following etching at different IPA concentrations and different temperature. (b) 1 µm × 1 µm, (c) 5 µm × 5 µm plan-view AFM image following 25/100 IPA/TMAH concentration at 78 °C for 10 min.

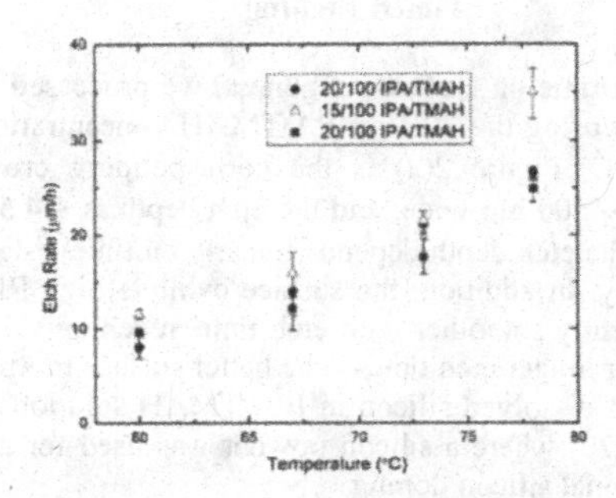

Fig.4. Etch rate for different IPA concentration versus temperature.

Surfactant

The formation of shoulders at the intersection of the (111) and (110) planes is an important problem for nanofins. Gosalvez, et al. [14] found that a small amount of surfactant (Triton X-100) added in the TMAH solution (no IPA) reduced the etch rate of (110) plane. In their paper, this approach was developed to improve mask undercut at convex corners to better retain pattern shapes. They also reported that the surfactant significantly reduces surface roughness of (100) and (110) planes.

We experimented with Triton X-100 to improve overall surface morphology and reduce the shoulder formation where sidewalls meet the etched planar regions. To better control the surfactant concentration in the etch bath, we used a larger quantity of IPA/TMAH solution (30/150), corresponding to the 20/100 solution described above. The TMAH solution was doped by submersing a silicon wafer prior to etching. The solution temperature was then ramped and the preheated IPA added resulting in bath temperature of 78 °C. The surfactant was added to obtain concentrations from ~ 60 to 540 ppm, the solution was thoroughly mixed, and the patterned wafer added. The etch was carried out for 2 min. The surfactant concentration used was in the range previously studied by Resnik, et al. [21] where their results for 10-200 ppm were best for achieving high anisotropy by decreasing the (110) etch rate with little effect on the (100) etching. The decrease in etching rate was attributed to the monolayer surfactant formed on the silicon surface, which partially protects it from chemical attack [14].

The resulting SEM cross-section for this process, utilizing 200 ppm surfactant concentration, and the corresponding plan-view AFM image, are shown in Figure 2 (c) and (d), respectively. These images illustrate the good morphology and absence of any shoulder or trench at the base of the fin. As in the previous report [21] the etch rate drops immediately upon addition of surfactant. From our studies, this decrease is from approximately 500 nm/min without surfactant to ~ 300 nm/min with Triton X-100. The RMS roughness is very low, remaining in the 1 to 2 nm range for all cases studied, and does not show a systematic dependence on the surfactant concentration.

Nanofins

Using our results, we etched nanoscale silicon fins using two approaches. In the first approach, the 20/100 IPA/TMAH solution was used without surfactant or silicon doping. Etch temperature was 78 °C and duration was up to 5 min. In each case we obtained shoulders where the (111) and (110) facets meet. A possible explanation for this difference with the micro-scale processing

experiments is the proximity between nano-fins (down to 1 μm) and resulting micro-loading of the etchant.
To reduce the shoulders in nano-fins, we added the optimized concentration of 200 ppm surfactant to the silicon-doped 30/150 IPA/TMAH solution. The etch rate of the (100) shoulder facets is not appreciably affected by the presence of surfactant. The reduced etch rate for (110) plane, with Triton X-100 added, is intended to provide ample time to etch (100) shoulder facets, thereby aiding in their removal. Etch bath temperature was 78 °C and a 1 min duration was used to achieve ~ 250 nm fin height. Shown in Figure 2(e) is a representative SEM cross-section. Straight sidewalls were achieved having smooth surfaces. The etch rate of ~ 250 nm/min is slightly lower than what we obtained from the microfin process development. This is also due to microloading near the nanofins. The AFM image in Figure 2(f) shows the areas between fins to be very smooth, with RMS roughness in this case to be ~ 0.9 nm.

Summary

We have systematically carried out a series of experiments for developing silicon nanofins with (111)-oriented sidewalls and (110)-oriented planar regions between the fins. These fins are needed for subsequent selective epitaxy of III-nitride semiconductors, which grow on the (111) facets but not appreciably on the (110) surfaces. The process is developed using microfins produced by optical lithography and SiO_2 hard mask. We employ a solution of IPA and TMAH. Smooth sidewalls are obtained by virtue of the directional etching process, although this does not guarantee smooth regions between the fins, nominally corresponding to the (110) surfaces. As in previous work for developing MEMS structures doping the etchant with silicon (a sacrificial wafer, in our case) improves the surface roughness of the planar regions. Addition of surfactant helps to reduce the presence of shoulders where the (111) and (110) crystallographic surfaces meet. Control over the fin width, down to 30 nm, and fin height (~ 250 nm) is achieved using EBL and a SiO_2 hard mask.

Acknowledgments

The TTU authors acknowledge partial support from NSF award ECCS 1028910 and the State of Texas NHARP program. Work at TAMU was partially supported by NSF award ECCS 1028791.

References

1. D. Bouvet, *et al.*, "Materials and Devices for Nanoelectronic Systems Beyond Ultimately Scaled CMOS," *Nanosystems Design and Technology,* (2009), 23-44.
2. T. Kudernac, *et al.*, "Nano-electronic switches: Light-induced switching of the conductance of molecular systems," *Journal of Materials Chemistry,* vol. 19, (2009), pp. 7168-7177.
3. V. Jovanović, *et al.*, "FinFET technology for wide-channel devices with ultra-thin silicon body," *Proceedings of the 31st International Convention MIPRO,* (2008), *79-83.*
4. E. Suzuki, *et al.*, "Emerging double-gate MOS devices technology," (2003), pp. 16-29.
5. Y. Liu, *et al.*, "Nanoscale Wet Etching of Physical-Vapor-Deposited Titanium Nitride and Its Application to Sub-30-nm-Gate-Length Fin-Type Double-Gate Metal-Oxide-Semiconductor Field-Effect Transistor Fabrication," *Japanese Journal of Applied Physics*, (2010), vol. 49.
6. V. V. Kuryatkov, *et al.*, "GaN stripes on vertical {111} fin facets of (110)-oriented Si substrates," *Applied Physics Letters,* Feb 2010, vol. 96, p. 3.
7. K. Mathwig, *et al.*, "Bias-assisted KOH etching of macroporous silicon membranes," *Journal of Micromechanics and Microengineering,* (2011), vol. 21, p. 035015.

8. D. Fang, *et al.*, "Methods for controlling the pore properties of ultra-thin nanocrystalline silicon membranes," *Journal of Physics: Condensed Matter*, (2010), vol. 22, p. 454134.
9. S. Yan, *et al.*, "A novel fabrication method of silicon nano-needles using MEMS TMAH etching techniques," *Nanotechnology*, (2011), vol. 22, p. 125301.
10. S. Chandrasekaran, *et al.*, "The effect of anisotropic wet etching on the surface roughness parameters and micro/nanoscale friction behavior of Si (100) surfaces," *Sensors and Actuators A: Physical*, vol. 121, (2005), pp. 121-130.
11. I. Zubel and M. Kramkowska, "The effect of isopropyl alcohol on etching rate and roughness of (1 0 0) Si surface etched in KOH and TMAH solutions," *Sensors and Actuators A: Physical*, (2001), vol. 93, pp. 138-147.
12. K. B. Sundaram, *et al.*, "Smooth etching of silicon using TMAH and isopropyl alcohol for MEMS applications," *Microelectronic Engineering*, (2005), vol. 77, pp. 230-241.
13. I. Zubel, *et al.*, "Silicon anisotropic etching in alkaline solutions IV:: The effect of organic and inorganic agents on silicon anisotropic etching process," *Sensors and Actuators A: Physical*, (2001), vol. 87, pp. 163-171.
14. M. Gosálvez, *et al.*, "Orientation-and concentration-dependent surfactant adsorption on silicon in aqueous alkaline solutions: explaining the changes in the etch rate, roughness and undercutting for MEMS applications," *Journal of Micromechanics and Microengineering*, (2009), vol. 19, p. 125011.
15. K. B. Sundaram, *et al.*, "Smooth etching of silicon using TMAH and isopropyl alcohol for MEMS applications," *Microelectronic Engineering*, (2005), vol. 77, pp. 230-241.
16. M. H. Jones and S. H. Jones, "Wet-Chemical Etching and Cleaning of Silicon," *Fredericksburg, Va.: Virginia Semiconductor*, (2003).
17. S. A. Campbell, *The science and engineering of microelectronic fabrication* (Oxford University Press Oxford, UK, 1996).
18. S. Chandrasekaran, *et al.*, "The effect of anisotropic wet etching on the surface roughness parameters and micro/nanoscale friction behavior of Si(1 0 0) surfaces," *Sensors and Actuators A: Physical*, (2005), vol. 121, pp. 121-130.
19. A. R. Charbonnieras and C. R. Tellier, "Characterization of the anisotropic chemical attack of {hk0} silicon plates in a T.M.A.H. solution: Determination of a database," *Sensors and Actuators A: Physical*, (1999), vol. 77, pp. 81-97.
20. G. Yan, *et al.*, "An improved TMAH Si-etching solution without attacking exposed aluminum," *Sensors and Actuators A: Physical*, (2001), vol. 89, pp. 135-141.
21. D. Resnik, *et al.*, "The role of Triton surfactant in anisotropic etching of {1 1 0} reflective planes on (1 0 0) silicon," *Journal of Micromechanics and Microengineering*, (2005), vol. 15, p. 1174.

Materials Processing Fundamentals
Edited by: Lifeng Zhang, Antoine Allanore, Cong Wang, James A. Yurko, and Justin Crapps
TMS (The Minerals, Metals & Materials Society), 2013

Purification of Indium by Vacuum Distillation

Yong Deng[1, 2, 3] Bin Yang [1,2,3*] DongSheng Li, Baoqiang Xu, Heng Xiong

(1 National Engineering Laboratory for Vacuum Metallurgy, Kunming University of Science and Technology, Kunming 650093, China;

2 Key Laboratory of Vacuum Metallurgy for Nonferros Metal of Yunnan Province, Kunming 650093, China

3 Faculty of Metallurgical and Energy Engineering, Kunming University of Science and Technology, Kunming 650093, China)

Keywords: indium; purity indium; high-purity; indium vacuum instillation

Abstract: The average content of impurity elements of crude indium has been reduced, by employing the two-step vacuum distillation method. The two-step vacuum distillation were carried out to study the influence of distillation temperature, distillation time on the impurities. At the first step the content of impurities which has low boiling point namely Cd、Zn、Tl、Pb in crude indium could be reduce to the standard of 5N about 950□, 60min and 5Pa. At the second step the content of impurities which has high boiling point such as Cu、Fe、Sn、Ni in crude indium could be reduce to the standard of 5N about 1075□, 60min and 5Pa. The average content of impurity elements of crude indium(99.7%) has been reduced, by employing the two-step vacuum distillation method, to achieve 5N(99.9993) indium under the best conditions. The impurity analysis carried out using ELEMENT GD.

Introduction

High purity indium is a atrategic electronic material , used in electrical industries for its excellent chemical, physical and mechanical performance. Indium and its compounds have numerous industrial applications, and it is extensively used in the manufacture of liquid crystal displays, semiconductors, low-temperature solders and infrared photodetectors[1]. These device applications demand high purity indium metal. Much attention has been paid to novel and efficient ways of purification for device grade materials including tellurium[2-3], cadmium[4], nickel[5], Bismuth[6] Silicon[7-8]etc, because the residual impurities even in sub-ppm concentration in semiconductor materials play a major role on the electronic properties by creating deep energy levels. The high purity indium can be prepared by several methods such as vacuum distillation[9], electrorefining[10], zone refining[11], thin layer fused salt and low halide etc. The purfication techniques involving electrorefining followed by zone refining are normally adopted for preparation of high pure indium[12-13]. Electrorefining and zone refining processes have distinct purification roles. Majority of impurities from indium can be removed by electrorefining whereas cadmium, thallium, lead, but tin is more diffcult to remove due to their proximity of electrode potential. The preparation technologies for the high-purity indium were comparaed and the method combining vacuum distillation and other purification technique was predicted to be prosperous[14-15]. In this paper the purification of crude indium by vacuum distillation was studied. In this study, indigenous indium was used as starting material and vacuum distillation process alone was employed. A two-step distillation method was adopted to reduce the content of low vapour pressure impurities. The element contents of the samples were tested by ELEMENT-GD produced by Thermo Scientific Company, which is one of the most advanced equipment for testing the high-purity metals with its accuracy up to 10-9(ppb).

2.Experimental

The experiment was carried out in a self-designed vacuum distillation furnace. It consists of an resistance heated vacuum furnace, rotary pump and temperature controlling system, that is made up of thermocouple, electrical source controller and temperatyre instrument. The vacuum distillation furnace is made up of evaporator and condenser and the schematic details is shown in Fig.1. The condenser can be moved up and down to change of condensation strength.

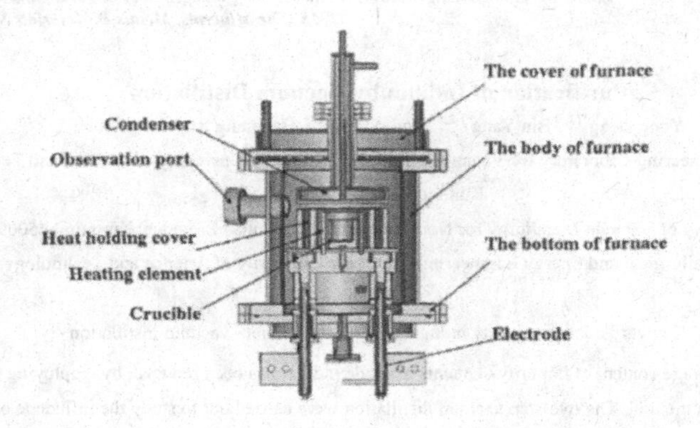

Fig.1 The schematic details of the vacuum distillation furnace

Composition of indium metal as raw material is listed in Table 1. Two parameters including the various temperatures and distillation time were determined according to thermodynamic calculations. The experiment was conducted in two steps. In the first phase, called as "low temperature distillation", the distillation experiment was carried out at a comparatively lower temperature to remove high vapour pressure impurities under a dynamic vacuum of 5 Pa, such as Zn, Cd, Tl, Pb. In this stage, high vapour pressure impurities were collected on the condenser and indium and other impurities left in the bottom of the crucible were used in the second phase. In the second phase, called as "high temperature distillation", low vapour pressure impurities can be remove at a comparatively higher temperature under a dynamic vacuum of 5 Pa. The indium were collected on the condenser and the low vapour pressure impurities, such as Sn, Fe, Cu, Ni were left over in the bottom of the crucible.

Samples of condensate were collected at time intervals of 1, 1.5, 2 and 2.5 hour in case of low fraction experiment and 2, 3, 3.5 and 4 hour in case of high fraction experiment, after the melt has attained the respective distillation temperature.

Table 1 GDMS analysis results of indium (all values in ppm)

element	Fe	Cu	Pb	Zn	Sn	Cd	Tl	Mg	Al	As	Si	S	Ag	Ni	Total impurity
content	0.067	6.548	49.350	0.155	66.627	40.253	2848.042	0.002	0.007	0.949	0.741	3.973	0.009	0.243	3016.966

*Footnote: The impurity elements in table are rooting in the standard of high indium in worldwide,such as China ,USA,Japan,England and Russia.

3.Results and Discussion

The vacuum distillation results have been studied for most common impurities in indium. In the low temperature distillation, the influence of distillation temperature and distillation time was shown in Fig.2 and Fig.3. A diagram of the saturation vapor tension of indium and the impurity element was shown in Fig.4[16-17].

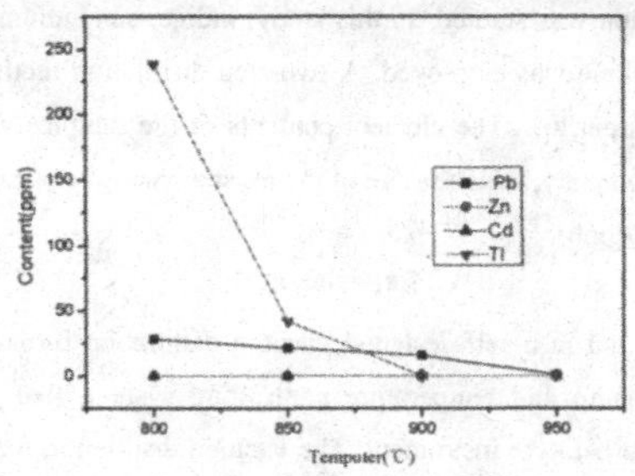

Fig.2. Effect of distillation temperature on the concentration of impurity of the residue (5Pa, 1hour)

From Fig.2 we could see that the low temperature distillation, distillation time 60min and residual gas pressure 5Pa, the indium remained in the residue with less than 0.0177ppm Cd, 0.0037ppm Zn, 0.0016ppm Tl, 0.8321ppm Pb at 950□ which was inferior to the standard of 5N of China, as may be seen from Fig.4. The vapour pressures of the Pb was very similar to that of indium. At lower temperature distillation such as 800□, the removal rate was not more than 10%. But the content of Pb in residue could be kept particularly low if the distillation temperature was rised to 950□.

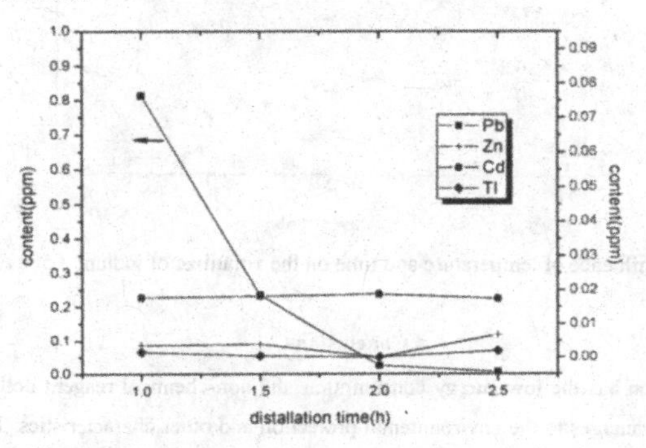

Fig.3. Effect of distillation time on the concentration of impurity during vacuum diatillation（5Pa，950□）

The Fig.3. was the effect of distillation time on the concentration of impurity during vacuum diatillation about distillation temperature 950□ and residual gas pressure 5Pa. It shown that the high vapour pressure impurities was reduced to less than 2 ppm. A first fraction, about 12% of the total quantity, was distilled off at 950□. It contained the most volatile elements, such as Zn, Cd, Tl, Pb,Mg,As.

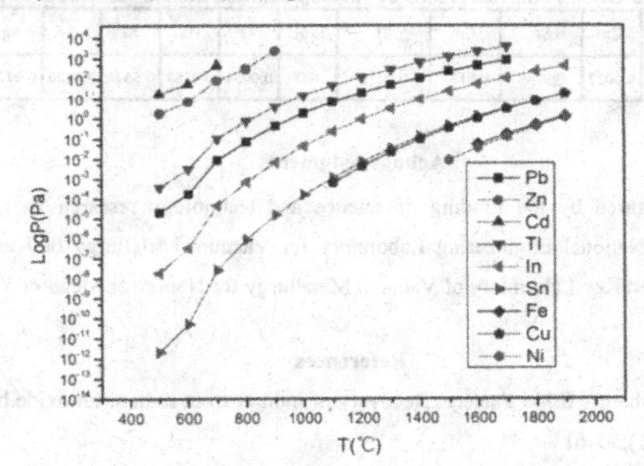

Fig.4 Saturation Vapor Tension of Indium and the Impurity Element

In the high temperature distillation, the main quantity of the indium was distilled at 975□~1100□: The inflence of temperature and time on the volatilize of indium was shown in Fig.5. In the second stage, the metal evaporation rate increased along with the distillation temperature ascension. When the temperature was 1050□ and the diatillation time was 2hour, the evaporation rate was more than 80% and even to 99%. Distillation at 1075□ and a residual gas pressure of 5Pa, the residue form the distillation—about 1% of the starting material—contained the less volatile elements, such as Cu,Fe,Sn,Ni. Considering the effect of impurity removal of impurities and the indium metal evaporation rate, the best conditions of high temperature distillation purifying indium were as follows: distillation temperature for 1075 □, distillation time for 60min, residual gas pressure 5Pa .

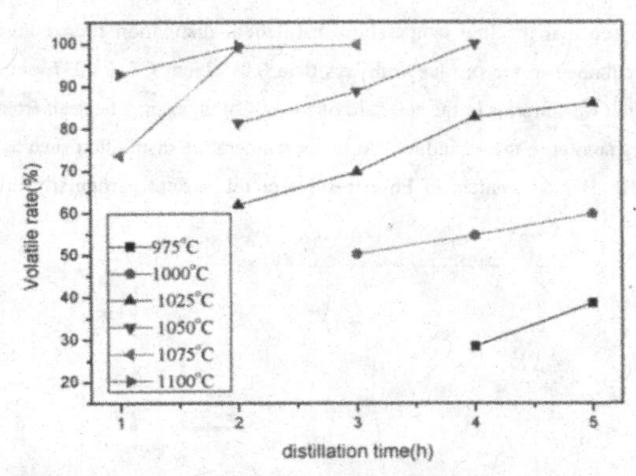

Fig.5. Influence of temperature and time on the volatilize of indium（5Pa）

4.Conclusions

The vacuum distillation has the low energy consumption, the non-chemical reagent pollution, the gas flows too easy to control and advantages to the environmental protection and other characteristics. The residues and the distillate material are easy to recycle processing by the vacuum distillation processing. The impurities content in the vacuum distilled indium are given in Table2. The average content of impurity elements of crude indium(99.7%) has been reduced, by employing the two-step vacuum distillation method, to the extent of achieving 5N(99.9993) indium under the best conditions. The indium material purified by this method of vacuum distillation can be subjected to zone refining process or electrolysis refining to attain ultra high purity.

Table 2 GDMS analysis results of vacuum distilled indium (all values in ppm)

element	Fe	Cu	Pb	Zn	Sn	Cd	Tl	Mg	Al	As	Si	S	Ag	Ni	Total impurity
content	0.385	0.1532	0.8132	0.0037	0.9286	0.0177	0.0016	ND	0.0051	0.0187	0.5574	3.8326	0.0773	0.0087	7.0288

*Footnote：ND:not detected

Acknowledgments

This work was supported by the funding of science and technology research form Yunnan provincial department of education, National Engineering Laboratory for Vacuum Metallurgy of Kunming University of Science and Technology and Key Laboratory of Vacuum Metallurgy for Nonferros Metal of Yunnan Province.

References

[1] Sami Virolainen, Don Ibana , Erkki Paatero. Recovery of indium from indium tin oxide by solvent extraction. Hydrometallurgy, 107(2011),56–61

[2] S.T. Ali, D.S. Prasad, N.R. Munirathnam, T.L. Prakash. Purification of tellurium by single-run multiple vacuum distillation technique. Separation purification technology,2005,43:263-267

[3] D.S. Prasad, N.R. Munirathnam, J.V. Rao, T.L. Prakash . Purification of tellurium up to 5N by vacuum distillation. Materials Letters, 59(2005),2035-2038

[4] C Sudheer, R. C. Reddy, T. L. Prakash. Purification of cadmium by vacuum distillation and its analysis. Materials Letters, 58(10)(2004),1638-1641

[5] D.C. Liu, B. Yang, F. Wang, Q.C. Yu, L. Wang, Y.N.Dai. Research on the Removal of impurities from Crude Nickel by Vacuum Distillation [J]. Physics procedia , 2012, 32:363-371

[6] Xiong Lizhi, He Zeqiang, Liu Wenping, Ma Chengjin, Dai Yongnian. Preparation of high-purity bismuth by sulphur deleadization in vacuum distillation. Transactions of Nonferrous Metals Society of China.

14(6)(2004),1210-1214

[7] Wei Kuixian, Ma Wenhui, Dai Yongnian, Yang Bin, Liu Dachun, Wang jingfu. Vacuum distillation refining of metallurgical Grade silicon(I)-Thermodynamics on removal of phosphorus from metallurgical grade silicon. Transactions of Nonferrous Metals Society of China. 17(A02)(2007),1022-1025

[8] Ma Wenhui, Wei Kuixian, Yang Bin, Liu Dachun, Dai Yongnian. Vacuum distillation refining of metallurgical Grade silicon(II)-Kinetics on removal of phosphorus from metallurgical grade silicon. Transactions of Nonferrous Metals Society of China. 17(A02) (2007),1026-1029

[9] Li Dongsheng, Yang Bin, Dai Yongnian, Deng Yong, Yang Xiankai. Purification of indium up to 5N by vacuum distillation. Proceedings of the 10th International Conference on Vacuum Metallurgy and Surface Engineering , Album(2011), 59-63

[10] Zeng Dongming, Zhou Zhihua, Shu Wangen, Liu Younian and Hu Aiping. Preparation of 5N high purified indium by the method of chemical purification-electrolysis. Rare Metals,2002,Jun21(2):137-141

[11] Zhou Zhihua, Mo Hongbing, Zeng Dongming. Preparation of high-purity indium by electrolytic refining and zone melting. Rare Metals,28(4) (2004),807-810

[12] Zhou Zhi-hua,Mo Hong-bing,Zheng Dong-ming. Preparation of high-purity indium by electrorefining.Trans. Nonferrous Met.Soc. China. 14(3)(2004),637-640

[13] Hidenori Okamoto,Kazuaki Takebayashi.Method for recovering indium by electrowinning appartus therefore.us patent,5,543,031.1996

[14] DU Guoshan YANG Bin DAI Yongnian. Removal of Cd, Zn, Tl and Pb from Indium by Vacuum Distilla tion. Yunnan Chemical Technology. 6(33)(2006),28-31

[15] Deng Yong, Yang Bin, Du Guoshan, Xu Baoqiang. Preparation Technology and Application of High-Purity Indium. Rare Metals, (30)(2006), 78-81

[16] Dai Yongnian, Yang Bin. Vaccum metallurgy of Nonferrous metal materials. (China,BJ:Metallurgical industry Press,(2000),80-82

[17] M.Olette. Physical Chemistry of Process Metallurgy, 8(2)(1961),1065-1088

Materials Processing Fundamentals
Edited by: Lifeng Zhang, Antoine Allanore, Cong Wang, James A. Yurko, and Justin Crapps
TMS (The Minerals, Metals & Materials Society), 2013

INVESTIGATING CURRENT EFFICIENCY OF ALUMINUM ELECTROLYSIS IN NaF-KF-AlF_3 SYSTEM

Huanhuan Ma, Jilai Xue, Jigang Li, Yanan Zhang

School of Metallurgical and Ecological Engineering,
University of Science and Technology Beijing
Xueyuan Road 30, 100083 Beijing, China

Keywords: Aluminum electrolysis, current efficiency, KF addition

Abstract

Existing indusial aluminum electrolysis is operated at about 950 °C. In this work, NaF-KF-AlF_3 based electrolytes were used at lower temperatures and the current efficiency of laboratory aluminum electrolysis was measured by mass loss method. The effects of KF addition, electrolyte compositions (cryolite ratio 1.8-2.5), and temperatures (890-960 °C) were studied in a laboratory cell with carbon electrodes. The obtained results suggest that alternative technology of lower temperature electrolysis has to overcome some problems to reach the operational goal in energy savings and production efficiency.

Introduction

Metal aluminum is currently produced using Hall-Heroult cells where cryolite-alumina melts serve as electrolyte at operating temperature around 950 °C. This high-temperature process is energy-intensive, which has stimulated great effort for decades in looking for an alternative technology, especially a lower temperature electrolyte system for retrofitting the existing H-H process. However, current efficiency is one of the most important issues to be considered in applying the low temperature electrolysis.

In recent years, KF based electrolyte or KF addition in cryolitic melts have received great attention because of its lower operation temperature and high alumina solubility in aluminum electrolysis process [1-4]. Yang studied vertical inert electrodes in potassium cryolite with CR=1.3 at 700 °C and 750 °C, showing stable cell voltage with current efficiency (CE) of 79.9 % [5-6]. Zaikov used KF based electrolyte in the traditional cell with high current densities, where the cell voltage went up rapidly after about 15 min in electrolysis, and then the operation had to stop [7]. Apisarov used Potassium Aluminum Fluoride (PAF) based electrolytes, in which the cell voltage with 17 % KF addition was higher than that with 10 % NaF addition in the traditional cell. The cell voltage was relatively stable with different current densities in the vertical cells [8]. Wang studied corrosion behavior of inert anode in the low temperature electrolyte [9]. KF-AlF_3 electrolyte was also applied in preparing Al-Sc alloy at 750 °C [10].

Current efficiency has been extensively investigated in laboratory electrolysis. Peterson showed that cryolite ratio and bath temperature were the major factors in electrolyte influencing cell current efficiency [11]. Solli and Stern designed a laboratory cell for determination of current efficiency, which could make good reproducibility with deviation 0.2 % [12-15]. Haarberg used the same set-up to measure CE, where the CE decreased with decreasing CR (from 3 to 2) [16]. Qiu measured CE with CR (1.2-2.3) that showed the current efficiency decreased with increasing CR when MgF_2 addition was 4.5 % and 6 % [17]. Thonstad demonstrated that Al_2O_3 layer at the interface was beneficial to a higher current efficiency but it could decease again with increased temperature [18]. Sterten reported a model of the current efficiency and the effects of cathode current density and local electrode overvoltage, based on laboratory testing data [13, 15].

The advantage of studying current efficiency in laboratory is a low cost approach to verify the influencing factors for scientific understanding and industrial improvement in Hall process. It is, however, found in open literature that reported current efficiency data can vary to a great extend in respect with aluminum electrolysis in various KF based or KF added cryolite melts. Such information are especially needed for re-testing or evaluating the alternative low temperature electrolyte systems for selecting or optimizing the processing parameters such as temperature, cryolite ratio, KF addition, alumina content, etc.

The aim of this work is to investigate the effects of KF addition in the cryolitic melts with varying cryolite ratio and temperature between 890-960 °C on the current efficiency of aluminum electrolysis. The experimental studies were performed in a laboratory cell with carbon anode, and the cell voltages were monitored during these electrolysis tests.

Experiments

Materials and Chemical

The melt composition was NaF (-KF) -AlF_3-5 wt % CaF_2-6 wt % Al_2O_3 with cryolite ratio (CR) of 2.5, 2.3, 2.1 and 1.8, respectively. The chemicals of NaF, KF, AlF_3 and alumina were analytically pure. Before mixing, they were pre-treated at 400 °C for 8 h to eliminate the influence of the moisture, and then were kept in a closed oven at 120 °C. The chemicals were mixed using a mortar and pestle in stoichiometric amounts.

Set-up for Aluminum Electrolysis

The laboratory cell, as shown in Figure 1, was designed specifically with the aim of giving a relative flat metal surface (cathode) during electrolysis. Anode, cathode and NaF-KF-AlF_3 based electrolytes were contained in a graphite crucible (Φ50 mm) with a cylindrical sintered alumina lining (~43 mm). The carbon anode was cylindrical with a sintered alumina tube covering its surface.

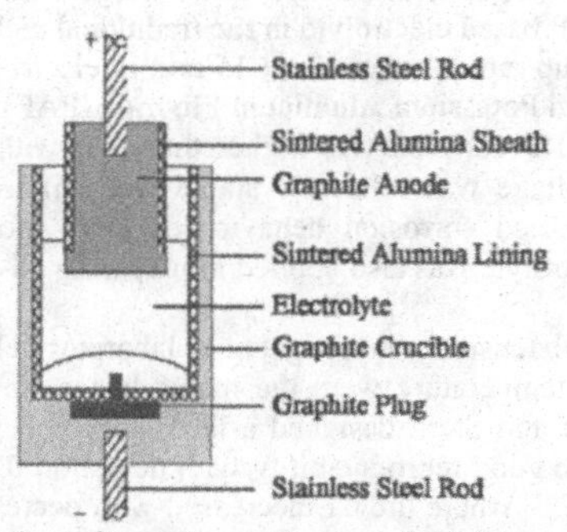

Figure 1. Schematic drawing of experimental set-up for laboratory aluminum electrolysis

The electrical contact to the aluminum metal pad was established using a small graphite plug through a hole at the lining bottom. The idea behind this design was intended to avoid Al_4C_3 formation and the leakage of the molten electrolyte during electrolysis. The cell was placed in a vertical tube furnace and positioned at the middle heating zone to avoid temperature gradients. During electrolysis, the electrical current was provided by TRADEX MPS 302 DC power supply, and the cell voltage was recorded by Tektronix DMM 4050 digital voltage meter.

Determination of Current Efficiency

In the experiment, a portion of 3 % alumina was mixed into the electrolyte (180 g in total) before heating. When the desired temperature was reached and kept for 30 min, Al metal was added into the molten electrolyte. Then, at 20 min later upon a complete melting of Al metal, another portion of 3 % alumina was poured into the cell. The anode was lowered to the surface of electrolyte. The electrolysis started at another 25 min, and lasted for 2 h. The electrical current through the cell system was 5 A that made a current density of 0.7 A / cm^2 at the anode.

When electrolysis ended, the cell was pulled down to the bottom area of the furnace where the temperature was about 300 °C. The molten electrolyte and aluminum were cooled down so quickly that the metal loss after electrolysis could be minimized. When the electrolyte and Al metal became solid state, the crucible was taken out of the furnace. At room temperature, the crucible under tests was broken into pieces for collecting Al metal.

To remove the adhered fluorides after electrolysis, the collected aluminum was re-melted in NaCl-KCl molten salt at 800 °C for 5 min. The CE values were determined by weighing the metal mass before and after the electrolysis in respect to the theoretical mass, which can be expressed as below:

$$\eta = (m_2 - m_1) / m \qquad (1)$$

where m_1 and m_2 are the mass of Al before and after electrolysis; and m, theoretical mass of aluminum calculated from:

$$m = I\,t\,M_{Al} / 3F \qquad (2)$$

where I is cell current; t, electrolysis time; M_{Al}, the molar mass of aluminum, and F, Faraday constant.

Results and Discussions

Selection of Testing Temperature

Table I shows the liquidus temperature and testing temperature of the electrolyte with varying CR. The liquidus temperature reduced with lowering CR with KF and without KF addition. Figure 2 showed that the liquidus temperature reduced with KF addition at the same CR. The operating temperature depended on the liquidus temperature of the electrolyte. The testing temperature was chose to keep appropriate superheat.

Table I. Liquidus Temperature and Testing Temperature with Varying CR and KF Addition

Electrolyte composition	Liquidus Temp. /°C	Testing Temp./ °C	Superheat / °C
*CR=2.5	953	960	7
CR=2.5, KF/[KF+NaF]=10%	940#	950	10
CR=2.3	923	930	7
CR=2.3, KF/[KF+NaF]=10%	910#	920	10
CR=2.1	911.5	920	8.5
CR=2.1, KF/[KF+NaF]=10%	900	910	10
CR=1.8	885	900	15
CR=1.8, KF/[KF+NaF]=10%	856.4	890	33.6

*CR=[KF+NaF]/AlF_3 [in mol], and KF/[KF+NaF]=10% in mass.
Value estimated.

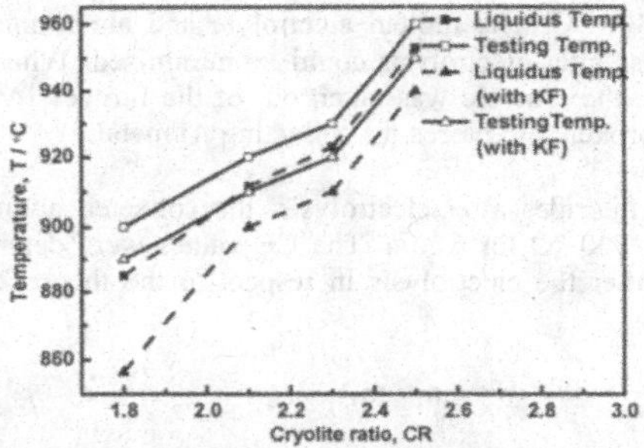

Figure 2. The liquidus temperature and testing temperature vs. cryolite ratio

Effect of Cryolite Ratio (CR)

Table II shows that aluminum loss and current efficiency obtained in laboratory aluminum electrolysis were up and down, depending upon a combination of changes in both the CR and temperature of the melts.

Table II. Aluminum Loss and Current Efficiency with Varying CR and Temperature

Cryolite ratio	Temperature / °C	Mass of aluminum / g			Current efficiency / %
		Before electrolysis	After electrolysis	Mass change	
CR=2.5	960	30.3426	32.8783	2.5357	75.56
CR=2.3	930	30.3550	33.2303	2.8753	85.68
CR=2.1	920	30.3991	33.2222	2.8231	84.12
CR=1.8	900	30.3842	32.3023	1.9181	57.15
CR=1.8	900	30.3943	32.2424	1.8481	55.07

In Figure 3, the highest value of CE is 85.68 % with CR=2.3. This is because when CR is higher than 2.3, the rising CR would increase the solubility of aluminum in the melts. A higher temperature could also increase the aluminum metal loss [19-20]. When CR is lower than 2.3, the CE value reduces with lowering CR, especially with CR = 1.8. As it was very interesting in this investigation to check out the CE value at this point, two parallel measurements were made to verify the testing results. However, it seems to be disappointed with such low current efficiency values (55.07 % and 57.15 %) by the combined effects of low CR and low temperature (900 °C). To check out the operational cause for the low current efficiency, the testing cell was broken after the experiment, as shown in Figure 4. Many small Al metal drops appeared within the electrolyte on the cell bottom after electrolysis test at 900 °C and CR=1.8. It is that the poor collecting behavior of metal in cryolitic melts with low temperature and low CR causes heavy metal loss during aluminum electrolysis process.

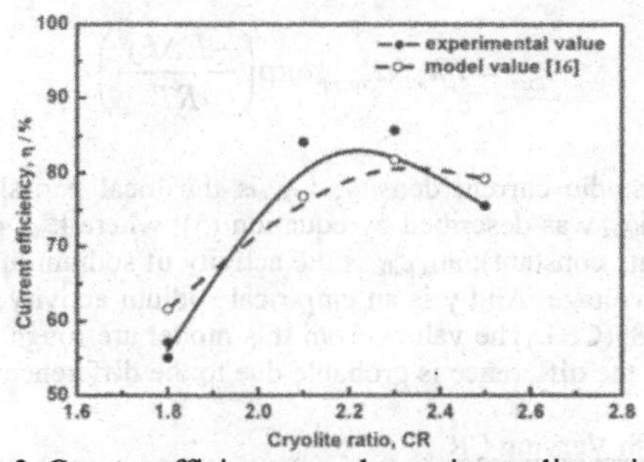

Figure 3. Current efficiency vs. change in cryolite ratio (CR)

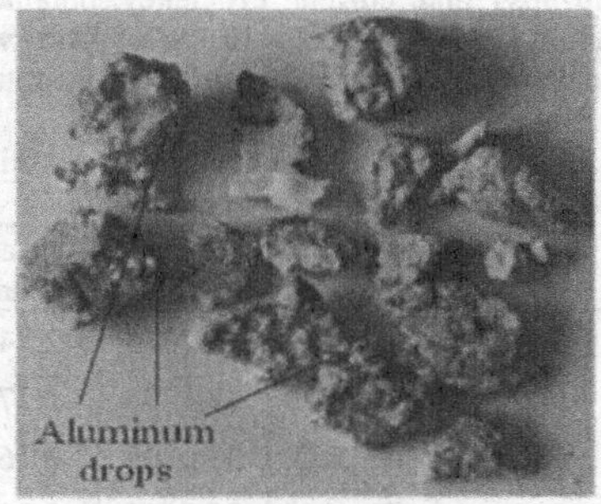

Figure 4. Aluminum drops appearing within the electrolyte on cell bottom after electrolysis at 900 °C and CR=1.8

The major reaction for metal loss is the re-oxidation of aluminum by the anode gas:

$$Al + 3/2\ CO_2 = 1/2\ Al_2O_3 + 3/2\ CO \qquad (3)$$

The reaction (3) takes place in the anode-cathode space where the rate process is diffusion-controlled. When electrolysis temperature rises, the rate of metal re-oxidation goes up and CE decreases with high CR (high liquidus temperature) and large superheat. However, when CR is as low as 1.8, the metal drops become poorly collected so the re-oxidation rate increase due to enhanced mixing action with dissolved CO_2 gas. This can offset the benefit offered by low temperature and low CR. In Dewing's work, the CE increased with decreasing CR (1-1.5) and temperature [21]. Haarberg also found that CE decreased with increasing CR (2.0-3.0) at the same temperature [16]. The mechanism behind seems more complex.

The predicted values of CE model as function of cryolite ratio and temperature is also shown in Figure 3. It was based on a laboratory cell with uniform cathode current density and assuming no short circuits or metal dispersion, and the current efficiency was defined as below [14-15]

$$\eta = \frac{i_c - i_{loss}}{i_c} \times 100 \tag{4}$$

$$i_{loss} = Fk_{mix} a_{Na,eq}^{y} \exp\left(\frac{-F\Delta Ey}{RT}\right) \tag{5}$$

where i_c is the local cathodic current density; i_{loss} is the local partial current density for all cathodic loss reactions., i_{loss} was described by equation (5); where k_{mix} is a mixed mass transfer coefficient (or a mixed rate constant); $a_{Na,bulk}$ is the activity of sodium in the bulk electrolyte; ΔE is the concentration overvoltage. And y is an empirical sodium activity exponent, which can be obtained by y=-0.87+0.485(CR). The values from this model are roughly in agreement with the experimental data, where the difference is probable due to the difference in the testing conditions.

Effect of KF Addition with Varying CR

Table III shows aluminum loss and current efficiency obtained in laboratory aluminum electrolysis with KF addition (KF/(KF+NaF)=10 wt%) and varying CR. The operating temperature varied with change in CR. It was partly different from the result without KF addition.

Table III. Aluminum Loss and Current Efficiency vs. KF Addition, CR and Temperature

Cryolite ratio	Temp. / °C	Mass of aluminum / g			Current efficiency / %
		Before electrolysis	After electrolysis	Mass change	
CR=2.5	950	30.4156	32.7613	2.3457	69.90
CR=2.3	920	30.347	33.1204	2.7734	82.64
CR=2.1	910	30.3552	33.5207	3.1655	90.55
CR=1.8	890	30.416	32.9501	2.5341	75.51
CR=1.8	890	30.3840	32.8151	2.4311	72.44

Figure 5 shows that CE rises and then reduce with lowering.CR and electrolysis temperature. With 10 % KF addition, the melting point of electrolyte became lower and the electrolysis temperature reduced to keep the suitable superheat. When CR was more than 2.1, CE reduced with increase of CR, which was the same as Haarberg's results of CR (2-3) [16]. However, when CR was lower than 2.1, CE value decreased. The highest value of CE was 90.55 % at CR= 2.1

with 10 wt % KF. Comparison with Figure 3 and Figure 5, the operating temperature reduced 10 °C with KF addition at the same CR. The current efficiency rose and reduced. At CR= 2.1 and at 920 °C with KF addition, the current efficiency was the highest (90.55 %), i. e. CE could also increase with KF addition at certain CR, but the mechanism was not known so far.

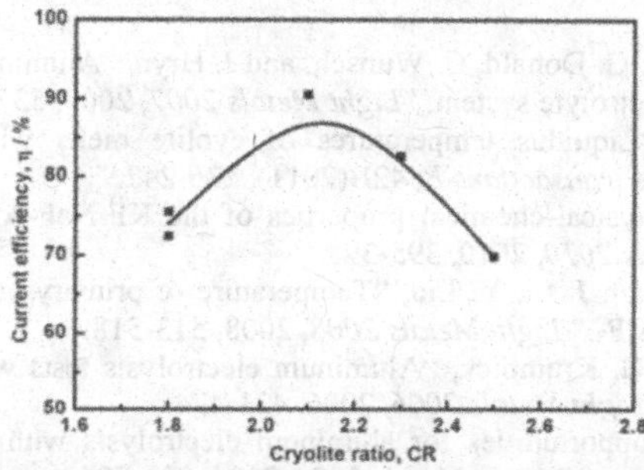

Figure 5. Current efficiency vs. varied cryolite ratio (CR) along with KF addition

Observations

Figure 5 is the photograph of a lab cell, which was cut open after solidification and cooling. There were some carbon dusting on the surface of electrolyte and sintered alumina sheath, which may increase anode consumption. The electrolyte around aluminum showed gray, while the up one was white. This was because aluminum was soluble in electrolyte, which increased the metal loss. Alumina and electrolyte of high CR may deposit, which may induce unstable electrolysis.

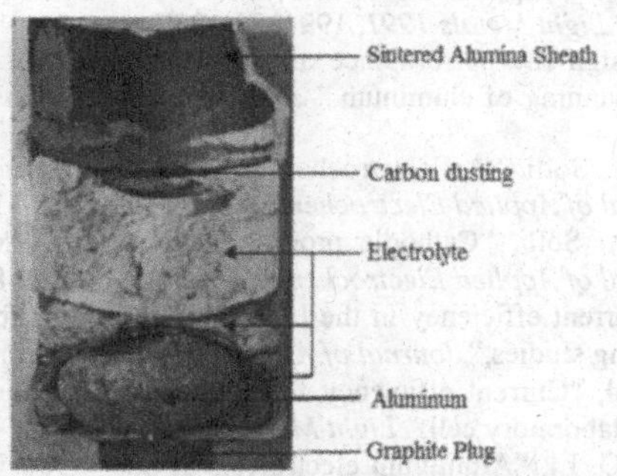

Figure 6. The cross-section of cell after electrolysis (CR=2.3)

Conclusions

1. For lowering cryolite ratio in the melt, current efficiency first increases and then decreases with the highest current efficiency at CR=2.3.

2. For 10 % addition of KF and varying cryolite ratio in the melts, current efficiency shows the similar up-down trend with the highest current efficiency at CR=2.1

3. The benefit of low temperature electrolysis may be offset by enhanced re-oxidation of Al metal due to the mixing action in melts with low cryolite ratio.

References

1. J. Yang, G. D. Graczyk, G. Donald, C. Wunsch, and J. Hryn, "Alumina solubility in $KF-AlF_3$-based low-temperature electrolyte system," *Light Metals 2007*, 2007, 537-541.
2. A. Apisarov et al, "Liquidus temperatures of cyolite melts with low cryolite ratio," *Metallugical and materials transactions B*, 42B(2011), 236-242.
3. A. Apisarov et al, "Physical-chemical properties of the $KF-NaF-AlF_3$ molten system with cryolite ratio," *Light Metals 2010*, 2010, 395-398.
4. J. Wang, Y. Lai, Z. Tian, J. Li, Y. Liu, "Temperature of primary crystallization in party of system $Na_3AlF_6-K_3AlF_6-AlF_3$," *Light Metals 2008*, 2008, 513-518.
5. J. Yang, J. Hryn, and G. Krumdick, "Aluminum electrolysis tests with inert anodes in KF-AlF3-based electrolytes," *Light Metals 2006*, 2006, 421-424.
6. J. Yang et al, "New opportunities for aluminum electrolysis with metal anodes in a low temperature electrolyte system," *Light Metals 2004*, 2004, 321-326.
7. Z. Yurii et al, "Electrolysis of aluminum in the low melting electrolytes based on potassium cryolite," *Light Metals 2008*, 2008, 505-508.
8. A. Apisarov et al, "Redution of the operating temperature of aluminum electrolysis: low temperature electrolyte," *Light Metals 2012*, 2012, 783-786.
9. J. Wang et al, "Investigation of $5Cu-(10NiO-NiFe_2O_4)$ inert anode corrosion during low-temperature aluminum electrolysis," *Light Metals 2007*, 2007, 525-530.
10. Q. Liu, J. Xue, J. Zhu, and C. Guan, "Preparing aluminum-scandium inter-alloys during reduction process in $KF-AlF_3-Sc_2O_3$ melts," *Light Metals 2012*, 2012, 685-689.
11. R. D. Peterson, and X. Wang, "The influence of dissolved metals in cryolitic melts on Hall cell current inefficiency," *Light Metals 1991*, 1991, 331-337.
12. P.A. Solli et al, "Design and performance of a laboratory cell for determination of current efficiency in the electrowinning of aluminum," *Journal of Applied Electrochemistry*, 26(1996), 1019-1025.
13. A. Sterten and P. A. Solli, "An electrochemical current efficiency model for aluminum electrolysis cells," *Journal of Applied Electrochemistry*, 26(1996), 187-193.
14. A. Sterten and P. A. Solli, "Cathodic process and cyclic redox reactions in aluminum electrolysis cells," *Journal of Applied Electrochemistry*, 25(1995), 809-816.
15. P. A. Solli et al, "Current efficiency in the Hall-Heroult process for aluminum electrolysis: experimental and modeling studies," *Journal of Applied Electrochemistry*, 27(1997), 939-946.
16. G. M. Haarberg et al, "Current efficiency for aluminum deposition from molten cryolite-alumina electrolytes in a laboratory cell," *Light Metals 2011*, 2011, 461-463.
17. Z. Qiu, M. He, and Q. Li, "Aluminum electrolysis at 800 ~ 900 °C — a new approach to energy savings," Transactions of Nonferrous Metals Society of China, 3(4) (1993), 11-18.
18. J. Thonstad and Y. Liu, "The effect of alumina layer at the electrolyte / alumina interface," *Light Metals 1981*, 1981, 303-312.
19. R. Ødegård, Å. Sterten and J. Thonstad, "Solubility of aluminum in cryolitic melts," *Light Metals 1987*, 1987, 389-398.
20. X. Wang, R. Peterson, and N. Richards, "Dissolved metals in cryolitic melts," *Light Metals 1991*, 1991, 323-330.
21. E. W. Dewing, "Loss of current efficiency in aluminum electrolysis cells," *Metallurgical Transaction B*, 22B(4) (1991), 177-182.

MATERIALS PROCESSING FUNDAMENTALS

Recirculation of Materials and Environments

Session Chair

Justin Mandel Crapps

Materials Processing Fundamentals
Edited by: Lifeng Zhang, Antoine Allanore, Cong Wang, James A. Yurko, and Justin Crapps
TMS (The Minerals, Metals & Materials Society), 2013

Effects of Ca and Na Containing Additives on WO_3 Content in Sheelite Concentrates after Floatation Separation Process

Jun Zhu, Jilai Xue, Kang Liu

School of Metallurgical and Ecological Engineering, University of Science and Technology Beijing; Xueyuan Road 30; Beijing, 100083, China

Keywords: Floatation separation, WO_3 concentrate, Tungsten processing

Abstract

Separation of scheelite from fluorite and calcite has technical difficulty in the mineral processing for metallurgical industry. In this paper, floatation separation of scheelite is researched, which aims to make full use of mineral resources and evaluate floatability of the tungsten ore resources. The raw ore selected in research containing WO_3 0.50-0.77 *wt%* is low and medium grade ore. According to the nature of scheelite, it prepares to choose following process flow to process the scheelite: scraping the floating black materials - room temperature desulfurization – one time ambient temperature roughing floatation – twice ambient temperature cleaning process – one time heating cleaning process. The concentrate WO_3 grade is more than 65% and seems to be quite suitable for the production of tungsten metal and tungsten based products.

Introduction

Among 15 tungsten-containing minerals in nature, only scheelite ($CaWO_4$) and wolframite [(Fe, Mn)WO_4] have industrial values [1]. Recently, the extensive exploitation of wolframite resources is parallel developed in tungsten industry with increasing usage of scheelite resources. Because sheelite ores are low-grade tungsten-containing minerals with complex mineralogy of sulphide, quartz, fluoride, and other calcareous minerals as major gangues, the major separation processes are gravity separation, magnetic separation and froth floatation.

Floatation is a surface-chemistry based process for separation of fine solids through difference in surfaces wettability of the solid particle, and the useful part is concentrated in froth or in solution [2]. This becomes an important separation technology in mineral processing, industrial wastewater treatment [3] and hydrometallurgy [4], especially in producing scheelite concentrates for metallurgical industry [5, 6]. It is usually not difficult to remove quartz (SiO_2) and sulphide (Me_xS) gangue during floatation process, but problems may appear in separation of scheelite and fluorite (CaF_2) or apatite ($Ca_{10}(PO_4)_6F_2$) minerals due to their similar surface properties and high surface reactivity with collectors [7-9]. The major influencing factors include collector adsorption, minerals dissolution, the interference of released species, and the interaction of adsorbed ions with mineral surface [10-14]. Slime coating around the mineral particles could also give adverse effect on scheelite separation [15]. Therefore, the technical challenge in scheelite separation is to depress selectively of the other calcium-containing minerals from the scheelite, where the technical solution have to vary according to the raw ores available.

This work is focused to improve the selectivity of scheelite separation during floatation separation process through suitable reagents combination of depressant, modifier, and collector. WO_3 content of raw ores used is 0.5-0.7 *wt.%*, and is to be upgraded above 65 *wt.%* in the concentrate for metallurgical application. The effects of Ca and Na containing additives on WO_3 content in scheelites concentrates have been investigated to improve the separation efficiency.

Experimental

Characterization of Sheelite Raw Ores

Figure 1 shows XRD spectrum of the sheelite raw ores, in which the major gangue minerals were silicon-, calcium- and fluorine-contained gangue, such as quartz, fluorite, fluorapatite, grossular, Na-rich anorthite and vesuvianite. And the weak peaks of the scheelite indicate the low WO_3 grade of the raw ore.

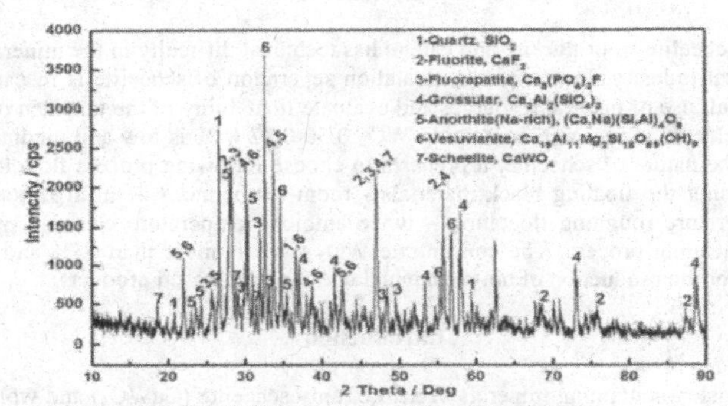

Figure 1. XRD spectrum of Sheelite raw ores.

The chemical compositions of raw ores were tested by ICP-AES and XRF, as shown in Table I. The small-scope fluctuations of components content were due to the samples mined from various areas in the same mine. The raw ores containing WO_3 0.50-0.77% is low grade, which is agreed with the phase analysis result by XRD.

Table I. Chemical Composition of Raw Ore Samples (wt.%)

Elements	Content / %	Elements	Content / %	Elements	Content / %
WO_3	0.50-0.77	MgO	0.96-2.25	Cr_2O_3	0.04-0.19
SiO_2	30.55-43.57	Na_2O	0.82-1.71	F	0.02-5.81
CaO	25.65-43.67	TiO_2	0.23-0.41	S	0.02-0.10
Al_2O_3	10.61-15.22	K_2O	0.19-1.14	LOI	3.66
Fe_2O_3	2.78-3.98	SnO_2	0.07-0.09		
P_2O_5	1.84-5.38	MnO	0.07-0.10		

Floatation Separation Process

For the raw ores containing fluorite and fluorapatite without calcite, the effective floatation separation process is lime method [14]. At present, however, there was no universal process flow sheet and reagent system for the selective floatation of scheelite [5, 14, 16, 17]. In this work, a proprietary rough floatation flow sheet is applied, as shown in Figure 2, to treat the scheelite raw ores. The process was divided into two steps: 1) Step 1 is desulphurizing floatation process to remove sulfides in scheelite ores firstly, because sulfides are harmful to the final WO_3 concentrate grade. 2) Step 2 is scheelite rough floatation process for its crucial effects on the separation efficiency. Lime (emulsion) was used as modifier and depressant, sodium carbonate was modifier, and sodium silicate played as depressant. Collectors were a mixture of oleic acid, naphthenic acid and paraffin oil. Hereinto, paraffin oil was an assistant nonpolarity collector with oleic acid and naphthenic acid, and it was used to improve the properties of froth, strengthen the hydrophobic interaction, and facilitate the reunite of hydrophobic functional groups which was benefit to upgrade WO_3 content in the concentrate products and increase tungsten recovery.

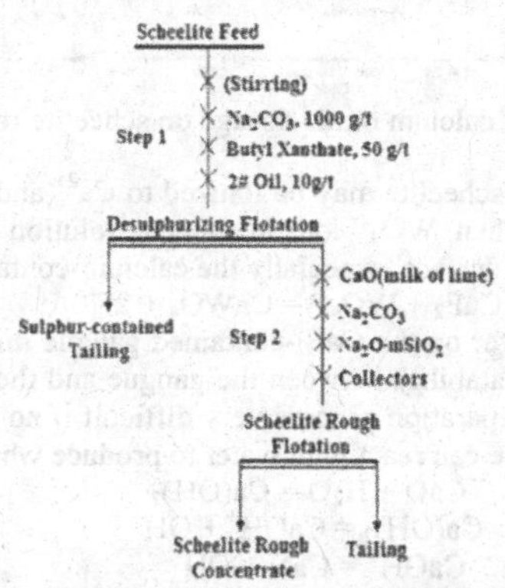

Figure 2. Scheelite rough floatation flow sheet.

Scheelite feed used for the floatation tests was produced from the ground raw ores (KNM-I high speed grinding mill machine). The particle size was controlled to 80% passing 0.074 mm to get the optimum floatation results [18, 19]. During the floatation process, all the reagents were added to the pulp as an aqueous solution for improvement the selectivity of the floatation process. After floatation tests, WO_3 contents of the scheelite rough concentrates were analyzed by ICP-AES.

Results and Discussion

Effects of Lime on Scheelite Rough Floatation

The floatation behavior of scheelite as a function of CaO dosage is shown in Figure 3. And the other set conditions of floatation were sodium carbonate 2000 g/t, sodium silicate 4000 g/t (modules 2.8), collectors 900 g/t. It was found that when no lime added, the WO_3 content in scheelite rough concentrate was minimum (1.54 *wt.*%) with least tungsten recovery (35.67%). When lime was added to 500 g/t, there was 7.77 *wt.*% WO_3 content with maximum tungsten recovery (89.43%). It means that lime should depress gangue more efficiently than sodium carbonate and strongly improve the floatability of scheelite separating from the other minerals. And with lime dosage increasing from 500 g/t to 3000 g/t, WO_3 content was upgraded to 17.39 *wt.*% at 2000 g/t lime dosage and then fallen sharply, while tungsten recovery was decreased firstly from 89.43% to 66.69% at 1000 g/t lime dosage, then increased again to 80.21% at 2000 g/t lime dosage and then lessened a little. The results suggest that lime addition can have benefit on scheelite floatation, but excessive addition may inhibit the floatation process.

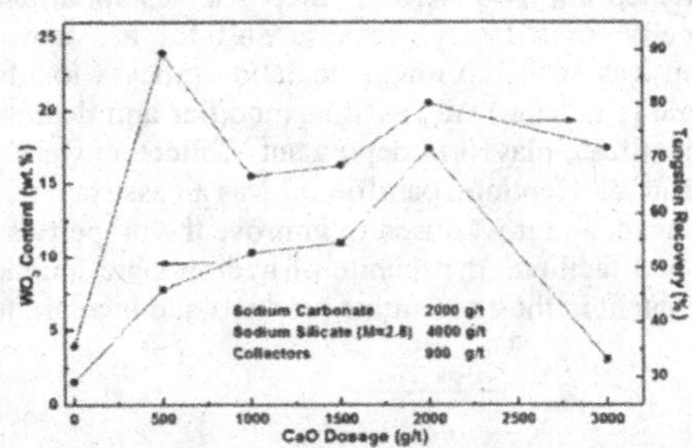

Figure 3. Effects of calcium oxide dosage on scheelite rough floatation.

As a sparingly soluble mineral, scheelite may be ionized to Ca^{2+} and WO_4^{2-} at its surface due to its ionic crystal structure [20]. Then, WO_4^{2-} ions in aqueous solution may probably react with the insoluble species on the mineral surface, especially the calcium-contained gangues. For example:

$$CaF_2 + WO_4^{2-} = CaWO_4 + 2F^- \quad (1)$$

$CaWO_4$ will precipitate and cover on the CaF_2-contained gangue mineral surface, resulting in a similar surface property and floatability between the gangue and the scheelite during floatation. This may make the floatation separation process very difficult if no depressants are added. As a pH modifier and depressant, lime can react with water to produce white lime and then ionize:

$$CaO + H_2O = Ca(OH)_2 \quad (2)$$

$$Ca(OH)_2 = CaOH^+ + OH^- \quad (3)$$

$$CaOH^+ = Ca^{2+} + OH^- \quad (4)$$

When lime is added into the pulp, the F^- on the gangue mineral surface generated by reaction (1) should absorb cations (such as $CaOH^+$ and Ca^{2+}), and then the absorbed cations will react with the added soda (Na_2CO_3) to produce $CaCO_3$ precipitation. It covers on the the CaF_2-contained gangue surface replaced $CaWO_4$ and changes the surface properties, and makes the gangue absorb sodium silicate more strongly to be more depressed [21, 22]. But the similar surface adsorption for scheelite may not happened because there is no surface anion like F^- on scheelite surface, and scheelite can scarcely be depressed by sodium silicate Therefore, the lime addition can improve the separability between scheelite and calcium-contained gangues, and the floatation separation can process more efficiently with lime addition (seen the curve of WO_3 content *vs.* CaO dosage in Figure 3). Furthermore, when CaO dosage exceeds 2000 g/t, one of the reasons that scheelite floatation was inhibited is excessive lime heightening the pulp pH above 12. In this pH condition, there is higher ionic strength between Ca^{2+} on scheelite surface, which formed by released WO_4^{2-}, and OH^- ions which will compete with the functional groups of collectors and thus decrease the separability and floatability of scheelite [2].

On the other hand, the influence of lime addition on the floatability of scheelite is indistinct (see the curve of WO_3 recovery *vs.* CaO dosage in Figure 3). According to Vazquez [14], the brittleness of scheelite makes it pulverized seriously during grinding, and absorbed more by slime during pulp conditioning. This should make a slime coating around the scheelite particle [15] so that decrease the scheelite floatability. Also the lime addition can reduce adverse effect of slime because lime may act as flocculant to coagulate slime. It was found a suitable lime dosage, such as 500-2000 g/t, can improve the floatability of scheelite, while an excessive lime dosage, such as beyond 2000 g/t, could make the slime contained scheelite particles coagulated mostly in froth and decrease the scheelite floatability (correlated to tungsten recovery) and separability as well (correlated to WO_3 content). In our floatation test, it was observed that a "froth overflow" phenomenon generated by slime coagulated in froth that became viscid and inflated [15].

Effect of Sodium Carbonate on Scheelite Rough Floatation

Floatation of scheelite as a function of sodium carbonate dosage is shown in Figure 4. The other conditions were CaO 500 g/t, sodium silicate 4000 g/t (modules 2.8), collectors 900 g/t. It is seen that WO_3 content in scheelite rough concentrate increased with Na_2CO_3 dosage increasing from 1000 g/t to 2000 g/t, and then decreased with increase in Na_2CO_3 dosage. The trend of Tungsten recovery is similar to WO_3 content, and both show a maximum at Na_2CO_3 dosage 2000 g/t with the pulp pH 10-11. The results means soda dosage significantly affects the separability and floatability of scheelite, and the best effect can appear with pH 10-11.

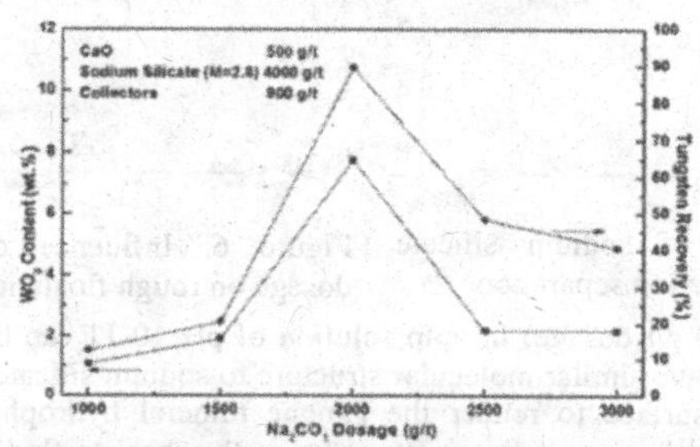

Figure 4. Influence of soda dosage on scheelite rough floatation.

The surface properties of most oxide minerals may be affected by hydrogen or hydroxyl ions, and the floatability of these oxide minerals can be governed by pH value of pulp [16]. And some modifier could adjust pH value in pulp. For soda additive, some ionization reactions may exist:

$$Na_2CO_3 = 2Na^+ + CO_3^{2-} \quad (5)$$

$$CO_3^{2-} + H_2O = HCO_3^- + OH^- \quad (6)$$

HCO_3^- is in the ascendant with pH 6-10 (soda dosage no more than 2000 g/t), while CO_3^{2-} is regnant at pH > 12 (soda dosage exceeding 2000 g/t). Both HCO_3^- or CO_3^{2-} could be absorbed on Ca^{2+}-containing surface, which formed in the first lime addition process, instead of OH^- and then react with sodium silicate to form $CaCO_3$ covering the mineral particles. If this happened on the gangue surface, the depress activity could increase, while if did on the scheelite surface, that should depress the scheelite and inhibit the absorbed collectors to worsen the scheelite separability and floatability. Only can the right soda dosage supply a suitable pH 10-11, and OH^- ions should be preponderantly absorbed on the scheelite surface for better floatation separation.

Effect of Sodium Silicate Modules and Dosage on on Scheelite Rough Floatation

The modulus (M) of sodium silicate (silica-to-soda ratio) often varied from 1.5 to 3.0 in floatation process [23]. The influence of the modules was tested using lime 1000g/t, soda 2000 g/t, sodium silicate 4000 g/t and collectors 900g/t, as shown in Figure 5. Both WO_3 content and tungsten recovery increase with higher sodium silicate modulus, showing an improved separability and floatability of scheelite. When the modulus is more than 3.0, sodium silicate can only be dissolved in hot water, which could make trouble in the floatation operation.

Figure 6 is the result of sodium silicate (M=2.8) dosage affecting on the rough floatation process with lime 500 g/t, soda 2000 g/t, and collectors 900g/t. In the test range, both scheelite separability and floatability were improved with an increase in sodium silicate dosage, maximized at 4000 g/t dosage with WO_3 content 7.77 *wt.*% and tungsten recovery 89.43%, and then worsen. It reveals that a suitable sodium silicate dosage can improve separation efficiency.

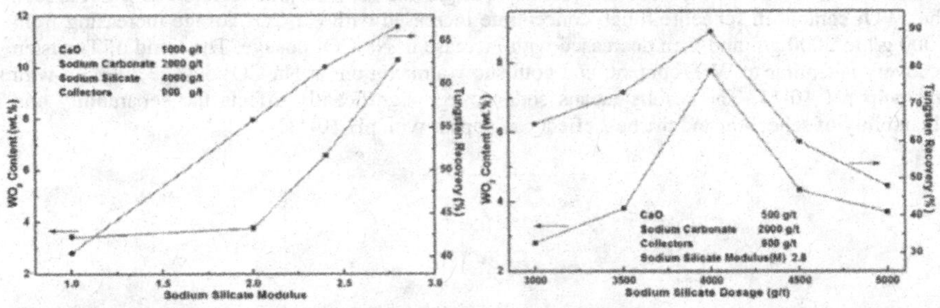

Figure 5. Influence of sodium silicate modules on rough floatation separation.

Figure 6. Influence of sodium silicate dosage on rough floatation separation.

Sodium silicate (to 4000 g/t dosage) in pulp solution of pH 10-11 can hydrolyze to $SiO(OH)_3^-$ and $SiO_2(OH)_2^{2-}$. Both have similar molecular structure to sodium silicate, and can easily adhere to the silicate gangue surface to render the gangue mineral hydrophilic, thus reducing the possibility of the unwanted minerals floating together with scheelite. Sodium silicate is also used as a dispersant, which can reduce the deleterious effects of slimes as did the lime.

The effect of sodium silicate depressing carbonate should be due to the process like following:

$$CaCO_3 + Na_2O \cdot mSiO_2 \rightarrow CaO \cdot mSiO_2 + Na_2CO_3 \quad (7)$$

This reaction could occur on carbonate gangue surface with $CaO \cdot mSiO_2$ covering the gangue surface to form a gel-like matter by the silica gel in sodium silicate, thus depressing the carbonate. When pH is above 12 (sodium silicate dosage more than 4000 g/t), a large quantity of silica gel may exist on the scheelite particles to inhibit the collectors action, which could worsens the separation efficiency.

The Final Refining Floatation Separation

A rough concentrate of WO3 7.77 *wt.*% and tungsten recovery 89.43% was obtained after the scheelite rough floatation. The cleaning floatation process was operated twice at ambient temperature (one was exactly the same as the rough process, and the other was only depressant used) and then was followed by a heating cleaning process (Petrov process). The resulting final scheelite product was examined using XRD (Rigaku D/MAX-RB12KW) quantitative analysis (K value method) to determine the content of $CaWO_4$ and WO_3. Figure 7 is a comparison of present work to a sample with 98 *wt.*% $CaWO_4$ (CP grade, Alfa Aesar (Tianjin) Chemicals Company) by XRD, and the analysis data is shown in Table II. Obviously, the WO_3 grade in final scheelite concentrate was more than 65 wt.%. This grade product can meet the requirement to produce tungsten metal and tungsten based chemicals.

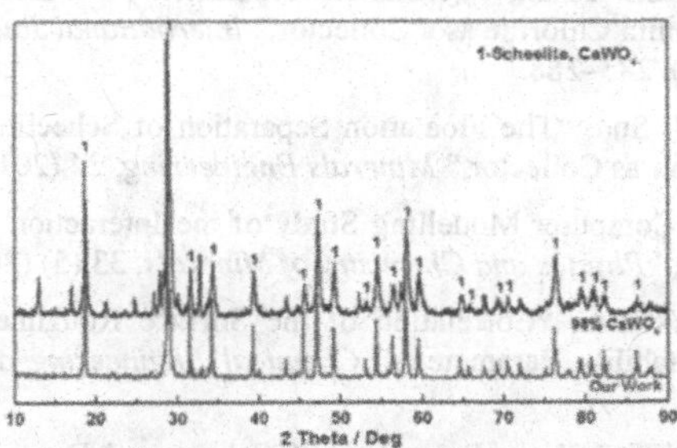

Figure 7. XRD spectra of samples in present work compared with the other reference sample containing 98 *wt*.% $CaWO_4$.

Table II. Scheelite Grade of Concentrates by XRD Quantitative Analysis

Experiment No.	$CaWO_4$ / %	WO_3 / %
1	86.57	69.71
2	84.23	67.82

Conclusions

1. A floatation separation process has been established through suitable reagents combination of depressant, modifier, and collector to upgrade the scheelite ores containing 0.50-0.77 *wt*.% WO_3 to 65% WO_3 or above to meet standard for metallurgical application.
2. Ca and Na containing additives can be used to improve selective separation during floatation process, which are effective in producing WO_3 concentrate from low sheelite value ores.

References

1. E. Lassner and W-D. Schubert, *Tungsten: Properties, Chemistry, Technology of the Element, Alloys, and Chemical Compounds* (New York, NY: Kluwer Academic/Plenum Publishers, 1999), 179-184.

2. T. G. Cooper and N. H. de Leeuw, "A Computer Modeling Study of the Competitive Adsorption of Water and Organic Surfactants at Surfaces of the Mineral Scheelite," *Langmuir*, 20 (10) (2004), 3984-3994.

3. L. M. Cabezon, M. Caballero and J. A. Perez-Bustamante, "Cofloatation Separation for the Determination of Heavy Metals in Water Using Colloidal Gas Aphrons Systems," *Separation Science and Technology*, 29 (11) (1994), 1491-1500.

4. N. K. Lazaridis et al., "Dissolved-air Floatation of Metal Ions," *Separation Science and Technology*, 27 (13) (1992), 1743-1758.

5. K. R. Kazmi et al., "Floatation Studies on Scheelite Concentrate of Chitral, NWFP, Pakistan," *Bangladesh Journal of Scientific and Industrial Research*, 46 (1) (2011), 123-126.

6. X. Tian et al., "Room Temperature Cleaner Floatation Technique for Scheelite Rough Concentrate," *Transactions of Nonferrous Metals Society of China*, 7 (2) (1997), 20-24.

7. Y. Zhang, Y. Wang and S. Li, "Floatation Separation of Calcareous Minerals Using Didodecyldimethylammonium Chloride as a Collector," *International Journal of Mining Science and Technology*, 22 (2012), 285–288.

8. Y. Hu, F. Yang and W. Sun, "The Floatation Separation of Scheelite from Calcite Using a Quaternary Ammonium Salt as Collector," *Minerals Engineering*, 24 (2011), 82-84.

9. D. Mkhonto et al., "A Computer Modelling Study of the Interaction of Organic Adsorbates with Fluorapatite Surfaces," *Physics and Chemistry of Minerals*, 33 (5) (2006), 314-331.

10. U. Ulusoy and M. Yekeler, "Correlation of the Surface Roughness of Some Industrial Minerals with Their Wettability Parameters," *Chemical Engineering and Processing*, 44 (5) (2005), 555-563.

11. U. Ulusoy et al, "Role of Shape Properties of Calcite and Barite Particles on Apparent Hydrophobicity," *Chemical Engineering and Processing*, 43 (8) (2004), 1047-1053.

12. K. I. Marinakis and G.H. Kelsall, "The Surface Chemical Properties of Scheelite ($CaWO_4$) II. Collector Adsorption and Recovery of Fine Scheelite Particles at the Iso-Octane/Water Interface," *Colloids and Surfaces*, 26 (1987), 243-255.

13. X. Tian et al., "Activation and Depression of Calcite in Calcium Minerals Floatation," *Transactions of Nonferrous Metals Society of China*, 9 (2) (1999), 374-377.

14. L. A. Vazquez et al, "Selective Floatation of Scheelite," *Floatation - A. M. Gaudin Memorial Floatation Symposium, vol. 1*, ed. M. C. Fuerstenau (New York, NY: American Institute of Mining, Metallurgical, and Petroleum Engineers, 1976), 580-596.

15. D. W. Fuerstenau, A. M. Gaudin and H. L. Miaw, "Iron Oxide Slime Coatings in Floatation," *Tran. Am. Inst. Min. Engrs.*, 211 (1958), 792-795.

16. B. J. Shean and J. J. Cilliers, "A Review of Froth Floatation Control," *International Journal of Mineral Processing*, 100 (2011), 57-71.

17. E. A. Gordon, "Scheelite Floatation Process," U.S. Patent, US4488959, 1984.9.18.

18. R. M. Rahman, S. Ata and G. J. Jameson, "The Effect of Flotation Variables on the Recovery of Different Particle Size Fractions in the Froth and the Pulp," *International Journal of Mineral Processing*, 106-109 (2012), 70–77.

19. T. V. Subrahmanyam and K. S. Eric Forssberg, "Fine Particles Processing: Shear-Flocculation and Carrier Floatation - a Review," *International Journal of Mineral Processing*, 30 (1990), 265-286.

20. M. R. Atademir, J. A. Kitchener and H. L. Shergold, "The Surface Chemistry and Floatation of Scheelite. I. Solubility and Surface Characteristics of Precipitated Calcium Tungstate," *Journal of Colloid and Interface Science*, 71 (3) (1979), 466-476.

21. W. Yin et al., "Application of NSO_H in Fluorite Ore Flotation and Its Depressing Mechanism," *Journal of Northeastern University (Natural Science)*, 30 (2) (2009), 287-290.

22. M. C. Fuerstenon and B. R. Palmer, "Anionic Flotation of Oxides and Silicates," *Flotation*, 1 (1976), 148-196.

23. C. L. Sollenberger, and R. B. Greenwalt, "Relative Effectiveness of Sodium Silicates of Different Silica-Soda Ratios as Gangue Depressants in Nonmetallic Floatation," *Mining Engineering*, 10 (1958), 691-693.

Materials Processing Fundamentals
Edited by: Lifeng Zhang, Antoine Allanore, Cong Wang, James A. Yurko, and Justin Crapps
TMS (The Minerals, Metals & Materials Society), 2013

COPPER REMOVAL FROM MOLYBDENITE BY SULFIDATION-LEACHING PROCESS

Rafael Padilla[1], Hugo Letelier[1], María C. Ruiz[1]

[1]Department of Metallurgical Engineering, University of Concepcion,
Edmundo Larenas No.285, Concepcion, 4070371, Chile

Keywords: Molybdenite purification, Sulfidation-leaching, Chalcopyrite

Abstract

Molybdenite (MoS_2) concentrates are produced by differential flotation from bulk copper–molybdenum concentrates. However, complete separation of molybdenum by flotation is very difficult, thus, molybdenum concentrates produced by this route requires a further step of chemical purification in order to reduce the copper content to a suitable level for marketing. Refractory chalcopyrite ($CuFeS_2$) is the major copper mineral impurity in these concentrates; thus the elimination of copper by leaching requires highly aggressive solutions. This paper is concerned with a novel process of chemical purification of molybdenite concentrates containing chalcopyrite as the main copper impurity. The process involves a sulfidation of the molybdenum concentrate with $S_2(g)$ at 380 °C followed by a leaching stage. H_2SO_4-NaCl-O_2 leaching of a sulfidized concentrate containing 3.4% Cu for 90 min at 100°C produced a molybdenite concentrate with less than 0.2% Cu, which is appropriate for marketing. Molybdenite dissolution was negligible in these conditions.

Introduction

Molybdenum is usually present in copper sulfide ores in very low quantities in the form of molybdenite (MoS_2). Thus in the concentration of these sulfides a bulk copper-molybdenum concentrate is produced. Differential flotation is the preferred method to separate the molybdenum from the bulk copper-molybdenum concentrates. However, the separation by this method is not highly efficient, and thus, molybdenite concentrates produced by this route requires further processing of chemical purification in order to reduce the copper content to a level suitable for marketing. Chalcopyrite is usually the major copper impurity in the molybdenite concentrates, which is a hard to leach mineral in acidic solutions. Thus, the elimination of copper from molybdenite by leaching, when the copper is in the form of chalcopyrite, requires highly aggressive solutions. Common purification methods practiced by the industry include ferric chloride leaching [1], cyanide leaching and hydrochloric acid leaching processes [2]. These processes have worked well in the particular cases for which they were developed; however, in practice they have shown some operational problems related to severe corrosion of equipment materials and/or environmental pollution.

The dissolution of the secondary copper sulfides such as covellite (CuS) and chalcocite (Cu_2S) in leaching processes is faster than the dissolution of chalcopyrite. This means that any process that converts the chalcopyrite into simple sulfides such as CuS and/or Cu_2S can be used advantageously for efficient copper removal from molybdenite concentrates. Therefore, in this paper a novel chemical purification method of molybdenite concentrate is discussed. The method involves a sulfidation stage of the molybdenite concentrates at relatively low temperature followed by a leaching stage of the sulfidized molybdenite concentrate.

The first step, the sulfidation process, has been studied for the treatment of chalcopyrite by Padilla et al. [3,4], who have shown that when chalcopyrite is heated in the presence of sulfur, a rapid transformation of its mineralogy occurs. They found that if heating is carried out in the range of 300 to 400 °C in gaseous sulfur atmosphere, chalcopyrite transforms into separate phases of covellite and pyrite, according to:

$$CuFeS_2 + 0.5S_2(g) = CuS + FeS_2 \quad (1)$$

Concerning the leaching of sulfidized material, these investigators studied the dissolution of copper in the H_2SO_4-NaCl-O_2 media, and they claimed that the covellite formed by sulfidation dissolved selectively from the pyrite according to:

$$CuS + 1/2O_2 + 2H^+ + Cl^- \rightarrow Cu^{2+} + S^o + H_2O + Cl^- \quad (2)$$

They concluded that the presence of chloride ions in the solution accelerated significantly the rate of reaction in comparison to the pure sulfuric acid media which was attributed to the fact that in the presence of chloride ions the elemental sulfur formed on the surface of the reacting particles does not form a coherent layer and, as a consequence, does not retard the rate of reaction.

Therefore, the main objective of this research was to determine the technical feasibility of removing copper from molybdenite concentrates by sulfidation and leaching to produce molybdenite concentrates with suitable levels of copper content for marketing. Thus, the leaching of the sulfidized molybdenite concentrate is crucial in this process, since the copper must be extracted rapidly without dissolving the molybdenite and producing a leaching residue of molybdenite nearly free of copper. Therefore, in this paper we present the results of the leaching of sulfidized molybdenite concentrate in H_2SO_4-NaCl-O_2 system at atmospheric pressure.

Experimental Work

The experimental work of this study was carried out using a molybdenite concentrate obtained from El Teniente mine (CODELCO-Chile). Since this material was produced by flotation unit operations, it was heat treated in nitrogen atmosphere to remove the organic reagents used in the concentration step. Subsequently, the concentrate was washed in dilute sulfuric acid solution to clean from oxides. Afterwards, this concentrate was analyzed for Mo, Cu and Fe. The results indicated that the concentrate's composition was 51.5% Mo, 3.4% Cu and 3.0% Fe. The morphology and the presence of copper as chalcopyrite in this concentrate was determined by X-ray diffraction spectroscopy (XRD) and complemented by microscopy. Figure 1 shows a micrograph of the molybdenite concentrate, where chalcopyrite particles of different size were identified and labeled. The XRD analysis also indicated that the major copper component mineral was indeed chalcopyrite since other copper minerals were not detected by this method.

Figure 1. Micrograph of the molybdenite concentrate showing several chalcopyrite particles of different sizes. The elongated particles (light to dark gray color) are molybdenite.

Sulfidation of the Molybdenite concentrate

The molybdenite concentrate was sulfidized at 380 °C for 60 min using elemental sulfur according to the methodology described by Padilla et al.[3]. Under these conditions the sulfidation conversion of the chalcopyrite in the concentrate was calculated by weight gain as 96%. Subsequently, this material was analyzed by XRD and microscopy to verify the conversion of the chalcopyrite in the molybdenite concentrate. The X-ray diffraction pattern showed strong diffraction peaks for molybdenite and peaks for covellite, and pyrite. Figure 2 shows a micrograph of a sulfidized concentrate, where we can observe clearly that the chalcopyrite present in the concentrate was transformed into covellite and pyrite according to reaction (1). The MoS_2 particles in the concentrate remained unchanged during the sulfidation. Therefore, this sulfidized material was used for the copper elimination from the concentrate by leaching.

Figure 2. Micrograph of a sulfidized molybdenite concentrate at 380 °C for 60 min showing transformed chalcopyrite particles.

Results

In the leaching of sulfidized molybdenite concentrate in sulfuric acid - sodium chloride - oxygen media at ambient pressure, the temperature, the concentration of sulfuric acid, and the concentration of sodium chloride are the main variables that could affect the dissolution of copper, and the dissolution of the molybdenum from the sulfidized material. Thus these variables were studied to determine the extent of dissolution of copper, iron, and molybdenum from the sulfidized concentrate.

Effect of the Temperature on the Copper Dissolution

Figure 3 shows the dissolution of copper as a function of time for the temperature range 70 °C to 100 °C. Other experimental conditions were 0.6 M H_2SO_4, 0.6 M NaCl, 1L/min O_2 and an agitation of 600 rpm. It can be observed in this figure that temperature has an important effect on the copper dissolution rate. Since the molybdenite concentrate must have copper content of less than 0.5% to be marketable, the dissolution of copper should be over 85%. Therefore, as can be seen in Figure 3, the desired level of copper dissolution can be achieved in 60 min of leaching at 100 °C. These results clearly indicate that the elimination of copper from the molybdenite concentrate by this route is technically feasible.

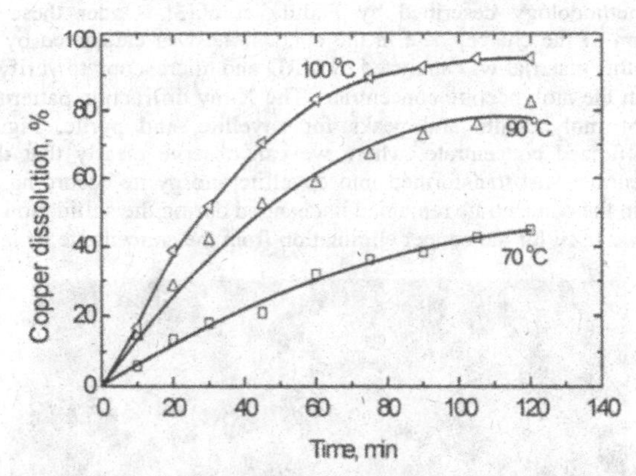

Figure 3. Dissolution of copper from the sulfidized chalcopyrite. Leaching conditions: 0.6 M H_2SO_4, 1M NaCl, 1 L/min O_2, 620 rpm.

Effect of the Concentration of Sulfuric Acid

To study the effect of the concentration of sulfuric acid on the copper dissolution, several experiments were carried out using leaching solutions containing H_2SO_4 in the molar range of 0.2 to 0.6. The results are shown in Figure 4, where one can see that the copper dissolution rate is nearly independent on the sulfuric acid concentration in the range studied. To assure independence on this variable, most of the subsequent leaching experiments were conducted using molar concentration of sulfuric acid equal to 0.6.

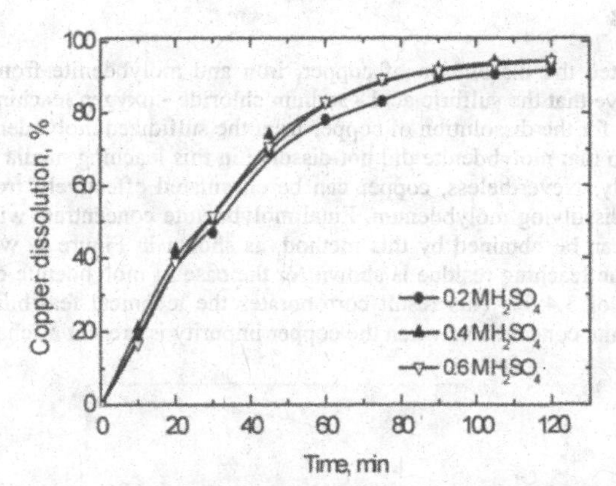

Figure 4. Effect of sulfuric acid on the copper dissolution from sulfidized molybdenite concentrates. 0.6 M NaCl, 100 °C, and 600 rpm, 1 L/min O_2

Effect of the Concentration of Sodium Chloride

Figure 5 shows the effect of sodium chloride concentration on the copper dissolution rate. It can be observed in this figure that in leaching the sulfidized concentrate without sodium chloride in the leaching solution, the dissolution of copper is low even at 2 hours of leaching time. However, the addition of sodium chloride in the leaching media drastically improves the dissolution rate of copper. It is also interesting to note that an increment in the concentration of sodium chloride over 0.6 M NaCl is not necessary since the copper dissolution rate does not improve any longer.

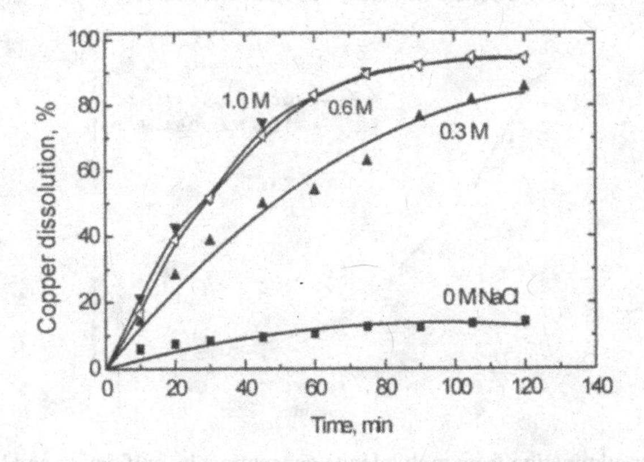

Figure 5. Effect of the concentration of NaCl on the dissolution of copper from the sulfidized molybdenite concentrate. 100 °C, 0.6 M H_2SO_4, 620 rpm, 1 L/min.

Selectivity in Leaching

In Figure 6 are depicted the dissolution of copper, iron and molybdenite from the sulfidized concentrate. We observe that the sulfuric acid - sodium chloride - oxygen leaching media is very effective and selective for the dissolution of copper from the sulfidized molybdenite concentrate. It can be observed also that molybdenite did not dissolve in this leaching media and iron can be dissolved only partially. Nevertheless, copper can be eliminated effectively from molybdenite concentrates without dissolving molybdenum. Final molybdenite concentrate with low levels of copper (< 0.2 %Cu) can be obtained by this method, as shown in Figure 7, where the copper content of copper in the leaching residue is shown for the case of molybdenite concentrate with initial copper content of 3.4 %. This result corroborates the technical feasibility of removing copper from molybdenite concentrates when the copper impurity is present as chalcopyrite.

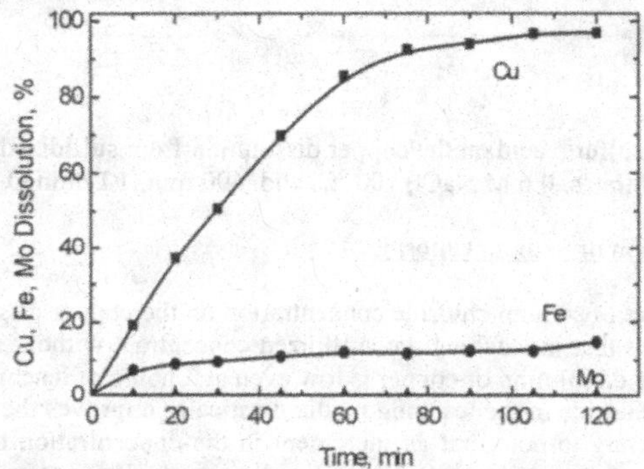

Figure 6. Copper and iron dissolution from sulfidized molybdenite concentrate, 100 °C, 0.6 M H_2SO_4, 0.6 M NaCl, 620 rpm, 1 L/min O_2.

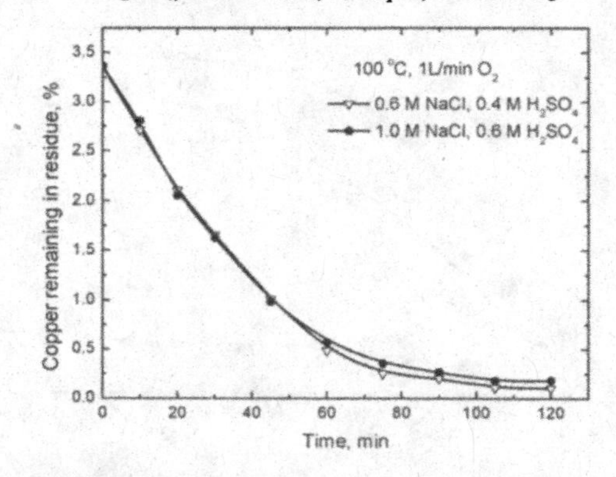

Figure 7. Copper elimination from molybdenite concentrate by sulfidation and leaching of the sulfidized concentrate in H_2SO_4-NaCl-O_2 media

Conclusions

The proposed method of sulfidation and leaching can be used effectively for copper removal from molybdenite concentrates when the copper impurity is in the form of chalcopyrite mineral, which is the general case of the Chilean molybdenite concentrates. The leaching of the sulfidized molybdenite concentrate in H_2SO_4-$NaCl$-O_2 media selectively dissolves the copper with negligible molybdenum dissolution. The copper in the concentrate can be lowered by sulfidation and leaching to less than 0.2%, which is lower than the 0.5% required for usual marketing of the molybdenite concentrates. The temperature and the concentration of sodium chloride are the two variables that influence most the copper dissolution rate in this leaching system. Finally, from the view point of the operating conditions, the sulfidation and leaching method for molybdenite concentrate purification would alleviate some of the difficulties, concerning pollution and corrosion, encountered in the traditional processes used for copper elimination from molybdenite concentrate.

Acknowledgements

The authors would like to thank the National Fund for Scientific and Technological Development (FONDECYT) of Chile for the financial support of this study through Project # 1110590.

References

1. P. H. Hennings, R. W. Stanley, and H. L. Ames, "Development of a process for purifying molybdenite concentrates," *Proceedings of Second International Symposium on Hydrometallurgy*, ed. D.J.I. Evans, AIME, New York, USA,(1973), 868.
2. R. R. Dorfler and J. M. Laferty, "Review of molybdenum recovery processes," *Journal of Metals*, 5 (1981), 48-54.
3. R. Padilla, M. Rodriguez,and M. C. Ruiz, "Sulfidation of chalcopyrite with elemental sulfur," *Metallurgical and Materials Transactions B*, Vol. 34B (2003), 15-23.
4. R. Padilla, P. Zambrano, and M. C. Ruiz, "Leaching of sulfidized chalcopyrite with H_2SO_4-$NaCl$-O_2," *Metallurgical and Materials Transactions B*, vol. 34B (2003), 153-159.

Materials Processing Fundamentals
Edited by: Lifeng Zhang, Antoine Allanore, Cong Wang, James A. Yurko, and Justin Crapps
TMS (The Minerals, Metals & Materials Society), 2013

Physical Chemistry of Roasting and Leaching Reactions for Chromium Chemical Manufacturing and Its Impact on the Environment – A Review

Sergio Sanchez-Segado, Animesh Jha

The Institute for Materials Research, Engineering Building, University of Leeds, Leeds LS2 9JT.

Keywords: Chromium, Alkali Roasting, Leaching, Kinetics, Thermodynamics, Chromite

Abstract

Chromium is mainly used for the production of stainless steel, in electroplating, as pigments and in leather and refractory industries. The natural resource of chromium is the chromite ore, which is a spinel $(Fe^{2+}, Mg)[Cr,Al,Fe^{3+}]_2O_4$, and is often present with gangue minerals such silicates. Traditionally chromium chemical extraction has been carried out with alkali and CaO roasting, which led to major environmental issues due to $CaCrO_4/Na_2CrO_4$ release.
In this review we examine the physical chemistry of both the low- and high-temperature roasting processes, which oxidizes the Cr^{3+} to Cr^{6+} state. Especially, the influence of silicates and liquid phase on oxygen transport is reviewed for chromite minerals of different origins. The physical chemistry of leaching of water soluble chromate is examined for determining the barrier to the extraction of Cr^{6+} ions. In this respect the role of spinel is examined. Finally, the precipitation of Cr^{3+} in oxide/hydroxide forms is also reviewed.

Introduction

Chromium is a transition metal with a range of oxidation states which vary from -2 to 6, however, only chromium (III) and chromium (VI) are the two most stable forms in nature. In metallic form, it has high hardness and excellent resistance to corrosion, which is why it is used as one of the main constituents of various types of stainless steels and high-temperature superalloys. Acid solutions of chromium (VI) anions are used in electroplating industry for surface coating. Chromium salts have applications in, catalyst manufacture, pigment production and in leather and refractrory industries.
Even though several chromium minerals have been identified as crocoites ($PbCrO_4$), vauquelinite ($Pb_2Cu[CrO_4][PO_4]$), uvarovite ($Ca_3Cr_2[SiO_4]_3$), merumite ($4(Cr,Al)_2O_3\ 3H_2O$), etc.[1], only chromite $(Fe^{2+}, Mg)[Cr,Al,Fe^{3+}]_2O_4$ is used as a main source for its production. Chromite minerals, belong to the spinel family of minerals which occur in ultramafic and mafic rocks. Chromite minerals and chromium are considered strategic materials, due to its relative abundance in the world geography. About 95% of chromium resources are located in South Africa and Kazakhastan [2]. Moreover, there are not many materials which can replace chromium in many of its applications [3]. The main producers of chromite ore are listed in table 1[2].

Table 1.World production of marketable chromite ore in 2009.

Country	Production (metric tons x 10^3)
India	3,760
Kazakhastan	3,330
South Africa	6,870
Other countries	5,340

Other sources of chromium might be from industrial processing wastes, which poses a serious environmental threat in the form problem due to chromium could be oxidated to chromium (VI), which is several times more toxic than chromium (III). It is known the carcinogenic effects of chromium (VI) on human health; Kristen et al. [7], have remarked some structural genetic lesions caused by the intracellular reduction of chromium (VI) such as DNA-protein crosslinking and DNA-strand breaks among others. The presence of chromium (VI) in the environment and its mobility in soils has been objective of numerous studies [4-8]. Under oxidizing conditions in soil, in the presence of Mn/Fe-ions the Cr^{6+} ions are present in the most stable forms. Becquer et al. 2003 [9] found that iron oxides, present in the soil, were able to retain chromium (VI) anions.

Crystal Structure Of Chromite Minerals

In the crystal structure chromite, shown in figure 1, there are 96 interstices between O^{2-} cubic sub-lattice where only 24 of them are occupied by cations in the following way: 8 of 64 tetrahedral interstices are occupied by divalent cations and 16 of 32 octahedral interstices are occupied by trivalent cations [10].

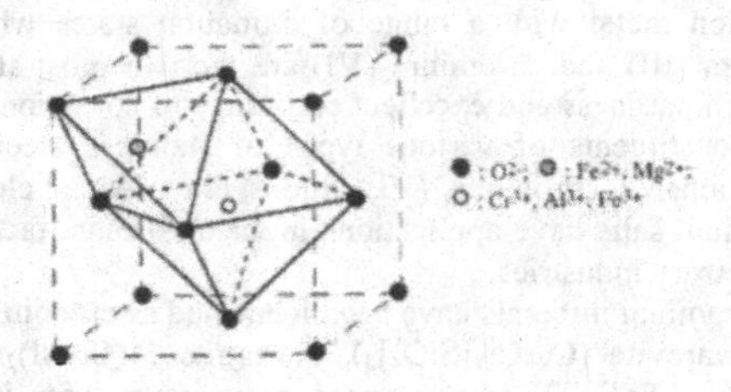

Figure 1.Chromite structure. Extracted from [1]

However, during serpentination processes can occur that divalent cations are in octahedral sites with ½ of the trivalent cations, the other half of which occupy the tetrahedral sites, rendering the structure to be an inverse spinel. Burkhard [11] reported the dependence of the degree of alteration with the Fe^{2+} and SiO_2 content in chromites. As a consequence of this alteration process, chromites are a complex solid solution of pure spinel end members ($FeCr_2O_4$, Fe_3O_4, $FeAl_2O_4$, $MgCr_2O_4$, $MgAl_2O_4$ and $MgFe_2O_4$), joint with orthopyroxene, olivine, plagioclase, chlorite, serpentine and talc as gange materials [11, 12].

Thermal Decomposition Of Chromites

The thermal decomposition of chromite ore is a complex process in which several phase changes and cations reorganization took place, moreover the gange materials associated or packed in the ore could influence these process in a significant extension.
Tiang-gui et al. [13], study the thermodynamics of the thermal decomposition of chromite ore. The free energy values (ΔG) for the thermal decomposition of the chromite in reductive and neutral atmospheres were positive which means the strength of the mineral to be decomposed in such conditions. Tathavadkar et al. [14] investigated decomposition of the chromite solid-solutions under oxidizing, inert and partially reducing 95% Argon -5% hydrogen atmospheres.
Under oxidative atmosphere, the chromite mineral undergo several changes, below 600°C the chromite starts its decomposition in two different spinels, magnesium chromite and magnesium aluminum iron oxide, as a consequence of the Fe(II) cations diffusion and oxidation to the solid-gas interface forming the rich iron phase near the surface. At temperatures higher than 600°C, these authors reported the decomposition of the magnesium aluminum iron oxide into a more stable hematite phase. Sánchez-Ramos et al [15] who work with a chromite in which chlorite was present, observed the incorporation of Fe (II) from the chlorite phase to the silicate phase until temperatures close to 800°C. At 1000°C the segregation of chromium (III) oxide were observed by these authors, this fact have also been pointed out by Parmelee and Ally [12] due to the replacement of Cr_2O_3 by Al_2O_3 or Fe_2O_3.
Other interesting observations carried out by Sánchez-Ramos et al. [15] were the decomposition of silicates in SiO_2 and the thermal reduction of iron (III) into Fe_3O_4 at 1000-1200°C.
The thermodynamic stability of solid solutions of spinels impose the upper limit for the pure phase segregation and as a result it has not been possible to produce very pure forms of chromium oxide as a result of thermal decomposition.

Alkali Roasting Of Chromites

Due to the thermodynamic limiting conditions for solid solution phase stabilities, discussed by Tathavadkar et al [14], the roasting of chromite ore with alkaline salts with or without lime were developed by le Chatelier more than 130 years ago, in which the minerals are roasted above 1100°C in air by forming water soluble sodium chromate. The presence of lime was especially adopted for reducing the alkali consumption by neutralizing the silica as silicates. However, the lime-based process yield volatile $CaCrO_4$ which is extremely water soluble and helps in spreading the pollution of Cr^{6+}-ions via air on to land and in water. Majority of developed and emerging economies have stopped the use of exogenous lime in the process for curtailing the emission of $CaCrO_4$. .
Tiang-gui et al and Tathavadkar and co-workers [13, 14] reported the thermodynamics and the physical chemistry of the lime-free alkali roasting with Na_2CO_3 of chromite ores, respectively. From the equilibrium analysis of elevated temperature reactions, the formation of $Na_2O\cdot Fe_2O_3$, $Na_2O\cdot Al_2O_3$ and $Na_2O\cdot SiO_2$ were predicted and verified experimentally. Under inert and partially reducing conditions, $Na_2Cr_2O_4$ was reported [16], which is unstable in excess oxygen. Especially important was the transmigration of cations (Mg^{2+} and Fe^{2+}) leading to formation of stable $MgO\cdot Fe_2O_3$ and $MgO\cdot SiO_2$.
Several studies, [16-19], have researched in the lime-free alkali roasting with Na_2CO_3 of chromite ores from different geographical locations with and without chromite ore processing residue (COPR). In the investigations [16-19] under oxidizing conditions, the authors claim a change in the kinetic control from low temperatue (chemical control with apparent activation energies about 200 $kJ\cdot mol^{-1}$) to high temperature (diffusion control with apparent activation

energies of 50 kJ·mol^{-1}). This change is due to the formation a low temperature eutectic liquid between Na_2CO_3 and Na_2CrO_4 at 655°C. On the first stages among 700 and 900°C, Cr^{3+} cations diffuse out of the spinel lattice and react with the alkali salt for temperatures higher than 900°C. The volume of the eutectic liquid increases, thereby posing a barrier for the oxygen diffusion to the reaction interface. The oxygen diffusion was found to be worse for high silica content chromites (e.g. China, Indonesia, Philippines) because of the high viscosity of the interfacial liquid. The presence of this liquid phase is the reason why yields higher than from 85% cannot be achieved in alkali-roasting process. Other alkali salts have also been investigated to produce soluble chromium compounds. Sun et al. [20], carried out a comparative thermodynamic study between the most common alkali salts used in chromium production, with the help of HSC 5.1 software the free energy of the following reactions for comparison were computed as a function of temperature, as shown in figure 2.

$$4/7\ Cr_2FeO_4 + 16/7\ KOH + O_2\ (g) \rightarrow 2/7\ Fe_2O_3 + 8/7\ H_2O\ (g) + 8/7\ K_2CrO_4 \quad (1)$$

$$4/7\ Cr_2FeO_4 + 16/7\ NaOH + O_2\ (g) \rightarrow 2/7\ Fe_2O_3 + 8/7\ H_2O\ (g) + 8/7\ Na_2CrO_4 \quad (2)$$

$$4/7\ Cr_2FeO_4 + 8/7\ Na_2CO_3 + O_2\ (g) \rightarrow 2/7\ Fe_2O_3 + 8/7\ CO_2\ (g) + 8/7\ Na_2CrO_4 \quad (3)$$

$$4/7\ Cr_2FeO_4 + 8/7\ K_2CO_3 + O_2\ (g) \rightarrow 2/7\ Fe_2O_3 + 8/7\ CO_2\ (g) + 8/7\ K_2CrO_4 \quad (4)$$

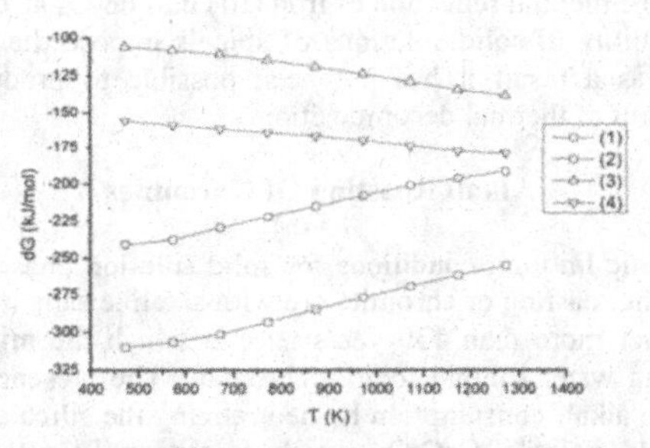

Figure 2. A comparison of the standard Gibbs energy (ΔG, kJ mol^{-1}) for different alkali salts durign chromite decomposition.

However it can be seen from figure 2, hydroxides have the most negative values of Gibbs energy which means that they are more reactive than the carbonates. Also the hydroxides reactions are exothermic and as a consequence at higher temperatures, the reactions become less feasible than that with the carbonates. From figure 2, it can be concluded that the Na_2CO_3 is the less effective alkali than for example, K_2CO_3. However, the limitations due to formation of eutectic mixtures have not been reported for the other alkali agents, leaving important gaps in their reaction mechanisms.

Arslan and Orhan [21], obtained more than 89% of chromium recovery using four times the stoichometric amount of NaOH at 650°C in 5 hours, with an air flowrate of 135 l/h and stirring speed of 210 min^{-1}. The apparent activation energy for this process was reported as 43.5 kJ·mol^{-1}. This product was treated with water at 70°C, purified by bubbling CO_2 and precipitated as CrO_3 using three times the stoichometric amount of H_2SO_4 at 180°C with low stirring speed. Under these conditions the efficiency of precipitation was 85%. Yildiz and Sengil [22], reported a chromium recovery yield of 96.2% using a weight ratio NaOH/chromite of 6/1 at 650°C for 60 min, stirring speed of 80 min^{-1} and an air flow rate of 1.65 l/h. The activation energy calculated was 94.6 $kJ \cdot mol^{-1}$. The studies carried out in references [21] and [22] were quite similar, but important differences in size and chemical composition of the ore are found. Firstly, the origin of the ores were from different locations in Turkey, the particle size of the ore was smaller in [22] than in [21] and finally the content of SiO_2 in the ore used by Arslan and Orhan [21] was higher than the one used by Yildiz and Sengil [22], 4.3 and 1.0%, respectively.

$NaNO_3$ has also been use for the roasting of rich chromium materials. Dettmer et al. [23] used this salt for processing the ash produced in the thermal treatment of leather wastes, they concluded that the use of $NaNO_3$ allow a decrease in the roasting temperature up to 750 °C with chromium recoveries of 95.50% without air injection to the system. Zhang et al.[24], studied the decomposition of chromite ore using a molten mixture of $NaOH$-$NaNO_3$, the best results achieved by these authors were a 99% of chromium extraction at 370°C, liquid phase/ore ratio of 4/1, 250 min of residence time, 60 l/h oxygen flow rate, particle size less than 58 μm and an activation energy of 106 $kJ \cdot mol^{-1}$. These authors claim the formation of a $NaOH$-$NaNO_3$ eutectic mixture below 300 °C when more than 10% w/w $NaNO_3$ is present, in which viscosity is lower than the pure molten NaOH. As a consequence the oxygen diffusion is enhance in two ways: i) best physical characteristics of the reaction media due to less melt viscosity and ii) the catalytic action of $NaNO_3$ as oxygen carrier at low temperature according to the following reaction [25]:

$$4\ NO_3^- \rightarrow 4\ NO_2 + 2O^{2-} \tag{5}$$

<u>Reactions Mechanisms Involve In Chromite Alkali Fusion</u>

Only in the works of Tathavadkar et al.[16], Tathavadkar et al. [18] and Antony et al.[19], carried out in-depth analysis about the reaction mechanism of the alkali roasting of chromites with sodium carbonate. The presence of voluminous cations like Fe^{2+} and Fe^{3+} increase the bonds length, also increase the diameter of the octahedral voids. As a consequence the energy needed for the jumps of Cr^{3+} ions to the nearest octahedral sites decrease enhancing its diffusion out of the spinel lattice. The Na^+cations diffuse inside the spinel lattice via the liquid phase formed at 650°C and react with Cr^{3+}cations to form Na_2CrO_4.

As measure as the oxidative roasting reaction takes place, the Fe^{2+} is oxidized to Fe^{3+} and the exsolution of γ-Fe_2O_3 reactive phase is observed [18]. The further reaction of this new phase with Mg^{2+} promotes the formation of $MgFe_2O_4$ which surrounds a rich chromium core, how it can be observed in figure 3.

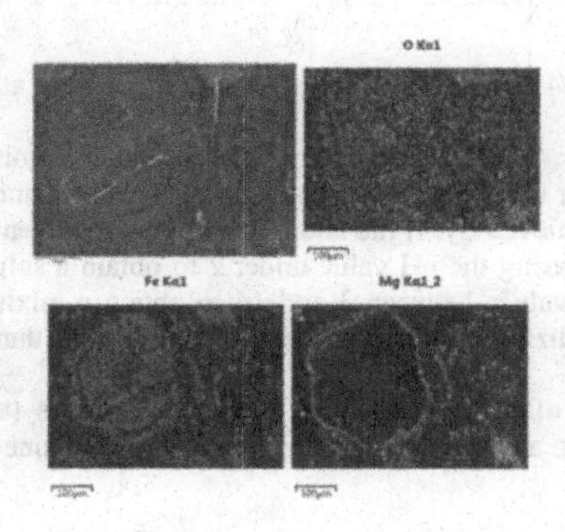

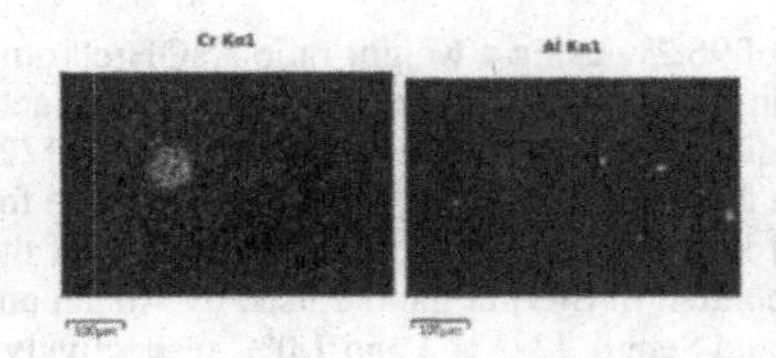

Figure 3. Alkali roasted chromite microstructure.

Similar behavior was reported by Zhang et al. [24]. They found that SiO_2 and Al_2O_3 were dissolved in the liquid phase in high proportion, a decrease in silicon concentration in the liquid was observed due to its precipitation as aluminum silicate compounds, leaving a solid residue formed mainly by $MgFe_2O_4$.
To the best of our knowledge potassium salts have not been employed in alkali fusion, however, the application of potassium hydroxide in hydrothermal treatments have been reported by several authors and will be discussed in the following section.

Leaching Of Chromite Minerals

Thermodynamic Considerations

A useful tool to design leaching, oxidation-reduction and precipitation processes is the analysis of pH-potential diagrams. Using the FACT-Sage software, the E_H (hydrogen potential,V)- pH diagram for the system Cr-Fe-H_2O was computed for 25°C and is shown in figure 4.

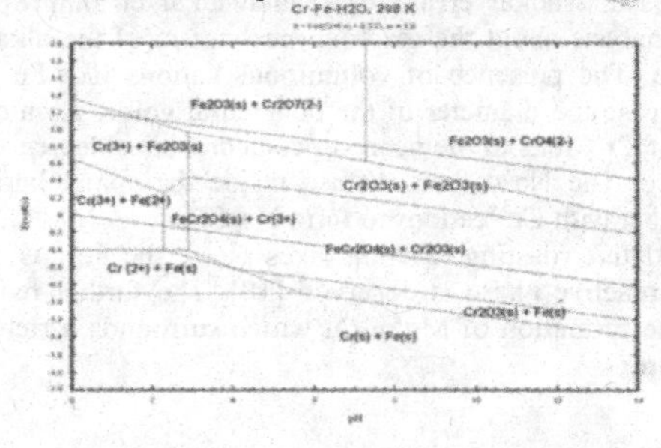

Figure 4. E_H-pH diagram of Cr-Fe-H_2O system at 25°C

It can be seen from figure 4, the specie $FeCr_2O_4$ has a large region of phase stability in the predominance area of water (dashed lines) for a wide range of pH and E_H values. The chromite decomposition could be achieved by: i) the addition of oxidizing agents at pH below 7 to obtain dichromate anion, ii) decreasing the pH value under 2 to obtain a solution of Fe^{2+} and Cr^{3+}, iii) moderate oxidation at pH values between 3 and 14 to obtain a mixture of chromium and iron oxides and iv) strong oxidizing conditions at pH values higher than 7 to obtain a chromate solution.
Leaching of chromites is often carried out at high temperatures (about 150°C) which were investigated by, Hai-xia et al [26] and the results can be explained by studying the E_H-pH

diagram, shown above. According to their analysis, they reported a decrease in the Cr^{2+}/Cr^{3+} and an increase in CrO_2^-, CrO_4^{2-} stability regions.

Acid Leaching Of Chromite Ores

Amer [27] studied the sulphuric leaching of an Egyptian chromite in a ball mill autoclave and found that working at 250°C for 30 min around 90% of chromium was leached as chromium sulphate in a 50% w/w H_2SO_4 solution for a grain size lower than 64 μm. To minimize iron dissolution MnO_2 was added to precipitate it as $Fe_2(SO_4)_3$. This author observed a decrease of aluminum leaching for temperatures higher than 180°C, probably due to the sulphate hydrolysis according to the following reaction:

$$Al_2(SO_4)_3 + 2H_2O \rightarrow 2Al(OH)SO_4 + H_2SO_4 \quad (6)$$

Vardar et al. [28], also studied the direct leaching of chromite in sulphuric media. These authors achieved chromium extraction yields around 70% for a H_2SO_4 concentration of 77% w/w, 6 hours of leaching and 210°C, when perchloric acid was added as oxidizing agent in a ratio acid/ore of ½ for 6 hours at 210°C in 82% w/w sulphuric media almost 100% of chromium was extracted as sulphate. Similar results to Vardar et al. [28] were achieved by Geveci et al. [29] using a high SiO_2 and MgO Turkish chromite concentrate. Sharma [30] employed diluted sulphuric acid solutions to increase the ratio Cr/Fe of Indian chromites, almost 70% of the iron content of the chromite was leached out at 90°C in 15 hours.
Microscopic analysis of the residues carried out by Vardar et al. [28] do not show the formation of an outer layer around the particles. The presence of perchloric acid enhances the decomposition of chromite due to the chromium oxidation to a more mobile Cr^{6+} form. The Cr^{6+} as chromic acid CrO_3 is further reduced in the presence of sulphuric acid according to the following reaction:

$$2CrO_3 + 3H_2SO_4 \rightarrow Cr_2(SO_4)_3 + 3H_2O + 3/2O_2 \quad (7)$$

Alkali Leaching Of Chromite Ores

Amer and Ibrahim [31] studied the leaching of low grades Egyptian chromites in 200 g/l NaOH solutions at 1200 rpm, 10 bar oxygen partial pressure,240°C, 1200 rpm stirring speed and 25 min of grinding in a stirred ball mill with a vertical attritor during 2 hours. Under such conditions the authors reported a chromium extraction higher than 90%. Yang et al.[32], founded that the mechanical activation of chromites by size reduction enhances greatly the chromium extraction. They reported an increase of 64% in chromium extraction yield, when the ore was ground for 10 min and then leached in >50% NaOH solutions at 400°C in the presence of bubbling oxygen for 3.5 hours.
Amer and Ibrahim [31] reported an activation energy of 6 $kJ \cdot mol^{-1}$, indicating that the process is controlled by the diffusion of oxygen and chromium ions to the boundary layer.
Xu et al. [33], carried out the leaching of Vietnamese chromites using KOH, yields higher than 90% in chromium recovery were achieved working under atmospheric oxygen pressure at 300°C, and 800 min^{-1} stirring speed. They also observed the co-dissolution of aluminum and silica at the first stage, which partly precipitation due to the reaction with KOH to form potassium aluminosilicates. Sun et al. [34], employed a new $KOH\text{-}KNO_3\text{-}H_2O$ reaction system to study the oxidative decomposition of Vietnamesse chromites ore using 50% oxygen partial pressure. The reaction media was prepared using a KOH/chromite ratio of 2/1 and KNO_3/chromite ratio of

0.8/1 in distilled water at 350°C. The stirring speed was maintained at 700 min^{-1} for 5 hours yielding a 98% of extracted chromium.
The reaction mechanism of the decomposition of chromites in KOH aqueous media follow the unreacted shrinking core model, similar to the alkali fusion at high temperature process, with activation energies in the range of 50-56 $kJ \cdot mol^{-1}$. Sun et al. [34] identified a chemical control during the first stages of the reaction and a diffusive control during the last stages due to limitations of KOH diffusion across the product layer. During the reaction oxygen dissolves in the melt to form the chromium oxidizing anion O_2^{2-}. Potassium nitrate plays the role of inorganic catalyst (similarly to alkali fusion at high temperature) for chromite oxidation according with the reactions (8) and (9) [34].

$$FeCr_2O_4 + 4KOH + 7/2KNO_3 \rightarrow 7/2KNO_2 + 2K_2CrO_4 + 1/2Fe_2O_3 + 2H_2O \qquad (8)$$

$$KNO_2 + 1/2O_2 \rightarrow KNO_3 \qquad (9)$$

Chromium Recovery

Traditionally the sodium chromate solution obtained after the water leaching of the alkali fused melt, is acidified with sulphuric acid to produce sodium dichromate, then chromium oxide (Cr_2O_3) is obtained by reaction (10) or by the path indicated by reactions (11) and (12) [35].

$$Na_2Cr_2O_7 + 2(NH_4)_2SO_4 \rightarrow Cr_2O_3 + Na_2SO_4 + N_2 + 4H_2O \qquad (10)$$

$$Na_2Cr_2O_7 + 2H_2SO_4 \rightarrow 2CrO_3 + 2NaHSO_4 + H_2O \qquad (11)$$

$$2CrO_3 \rightarrow Cr_2O_3 + 3/2O_2 \qquad (12)$$

Xu et al. [34], precipitated K_2CrO_4 due to its low solubility in concentrated KOH solutions at temperatures (120-90°C). The chromate crystals are easy of separate due to high density, allowing them to settle at the bottom. Further treatment with CO_2 and reduction with coal regenerate the alkali and permit the production of chromium (III) oxide according to reactions (13) and (14).

$$2K_2CrO_4 + H_2O + 2CO_2 \rightarrow K_2Cr_2O_7 + 2KHCO_3 \qquad (13)$$

$$K_2Cr_2O_7 + C \rightarrow Cr_2O_3 + K_2CO_3 + CO \qquad (14)$$

The reduction of Cr^{6+} by organic acids, sugars and alcohols is well known [36, 37], however the formation of Cr^{3+} organo-complexes make difficult the subsequent recovery due to the high water solubility of these complexes in a wide range of pH.
Bhattacharyya et al. [38], developed a three stages process for reducing Cr^{6+} with $NaHSO_3$ followed by the precipitation of Cr^{3+} with the help of sodium laurylsulfate. It was recorded that 97% of chromium was removed at pH values between 7.0 and 8.8. The reduction of chromium with copper smelter slag was studied by Kiyak et al. [39] who reported 100% of reduction with residence times lower than 5 min., the reduction capacity of the slag was set at 76.5 mg of Cr^{6+} per gram of slag in acid media. The chromium and other metal values were precipitated with NaOH at pH 9.
Iron compounds have been widely used to reduce Cr^{6+} to Cr^{3+}, Buerge and Hug [40], studied the reduction of Cr^{6+} at pH values between 2 and 7.2, they founded a low reaction rate at pH values

around 4, being higher for more acid and basic pH values. A maximum reduction yield of 85% was reported for pH values close to neutrality. At low pH the reaction is described by equation (15) and for pH values higher than 4 they observed the formation of hydrocomplexes of chromium and iron (equation (16)), besides the formation of a reddish-brown fine precipitate (equation (17)).

$$3Fe^{2+} + HCrO_4^- + 7H^+ \rightarrow 3Fe^{3+} + Cr^{3+} + 4H_2O \quad (15)$$

$$3Fe^{2+} + HCrO_4^- + 3H_2O \rightarrow 3Fe(OH)_2^+ + CrOH^{2+} \quad (16)$$

$$3Fe^{2+} + HCrO_4^- + 8H_2O \rightarrow Fe_3Cr(OH)_{12}(s) + 5H^+ \quad (17)$$

The rate law for pH 4.4-7.2 is a second order kinetics with the following general expression:

$$-d[Cr(VI)]/dt = k_{obs}(pH)[Cr(VI)][Fe(II)] \quad (18)$$

Where k_{obs} is a second order polynomial function of $[OH^-]$.
Pettine et al.[41] reported similar observations and realize that the presence of PO_4^{3-} anions have a negative influence in the rate of reaction due to its reaction with iron (II) and $HCrO_4^-$ to form non reactive complexes. Buerge and Hug [42], also carried out experiments to study the influence of organic ligands in the reduction of Cr^{6+} by Fe^{2+}, they established that bi and multidentate carboxylates and phenolates enhances the reaction rate meanwhile ligands such as phenantroline stop the reduction process. The same observation was pointed out by Zhou et al. [43] when they carried out the reduction of Cr^{6+} by zero-valent iron, due to the reduction mechanism involve the dissolution of Fe^0 to Fe^{2+} which reduce chromium. The presence of chelating agents is needed to avoid the precipitation of $Cr(OH)_3$, $Fe(OH)_3$ and $Cr_xFe_{1-x}(OH)_3$ which form a passivation film on the surface of Fe^0 [43,44].

Conclusions

From the literature reviewed it can be concluded that the decomposition of chromites is mainly affected by the oxygen partial pressure and the limitations of each system to its diffusion to the reaction zone. The roasting process can be carried out at high temperature with moderate alkali salt dosage or at low temperatures with high alkali dosage with high energy consumptions or costly recovery process with high dangerous waste production.
Sulphuric leaching of chromites seems to be an effective method due to the low cost of reagents used. However, the yields achieved by this process do not remove the environmental problems associated to the wastes disposal.
The reduction and precipitation of high purity $Cr(OH)_3$ from aqueous solutions is a difficult tasks due to the formation of organometallic complex when organic agents are used or chromium losses due to the formation of polymetallic precipitates as $Cr_xFe_{1-x}(OH)_3$.
At the moment the most promising solution for the treatment of chromites is the hydrothermal KOH leaching proposed by Xu et al. [33, 35], as this method employed more benign reagents, and allows the recovery of other marketable subproducts and reduces the total and hexavalent chromium content of the residue in 80 and 78%, respectively compared with the alkali fusion with a mixture of sodium carbonate and lime/dolomite traditional process.

References

[1]. N. Kanari, I. Gaballah and E. Allain, "A study of chromite carbochlorination kinetics", *Metall Mater Trans B,* 30 (B) (1999), 577-587.

[2]. K. Salazar and M.K. McNutt. *"Minerals Yearbook"*, U.S. Geological Survey (2011), 1-198.

[3]. R.H. Nafziger et al., "A review of the deposits and beneficiation of lower grade chromite", *Journal of the South African institute of mining and metallurgy* (1982), 205-226.

[4]. A.B. Mukherjee."Chromium in the environment of Findland", *The science of the total environment* 217 (1998), 9-19.

[5]. C.H. Weng et al., "Chromium leaching behavior in soil derived from the chromite ore processing waste", *The science of the total environment* 154 (1994), 71-86.

[6]. R.Mattuck and N.P. Nikolaidis."Chromium mobility in freshwater wetlands", *Journal of contaminat hydrology* 23 (1996), 213-232.

[7]. G.R.C. Cooper."Oxidation and toxicity of chromium in ultramafic soils of Zimbabwe", *Applied Geochemistry* 17 (2002), 981-986.

[8]. M.K. Banks, A.P. Schwab and C. Henderson."Leaching and reduction of chromium in soil as affected by soil organic content and plants", *Chemosfere* 62 (2006), 255-264.

[9]. T. Becquer et al.,"Chromiumaviability in ultramafic soils of New Caledonia", *The science of the total environment* 301 (2003), 251-261.

[10]. K.E. Sickafus and J.M. Wills, "Structure of Spinel", *Journal of the American Ceramic Society* 82 (12) (1999), 3279-3292.

[11]. D. J.M. Buckhard, "Accessory chromium spinels: Their coexistence and alteration in serpentinites", *GeochimicaetCosmochimicaActa*57 (1993), 1297-1306.

[12]. C.W. Parmelee and A. Ally. "Some properties of chrome spinel",(Paper presented at the Refractories Division of the American Ceramic Society Annual Meeting, Cleveland, Ohio 2 January 1932).

[13]. Q. Tian-gui et al., "Thermodynamics of chromite ore oxidative roasting process",*Journal of Central South University of Technology* 18 (2011), 83-88.

[14]. D. Vilas Tathavadkar, M.P. Antony and A. Jha, "The physical chemistry of thermal decomposition of South African chromite minerals", *Metallurgical and Materials Transactions B* 36 (B) (2005), 75-84.

[15]. S. Sánchez-Ramos et al. "Thermal decomposition of chromite spinel with chlorite admixture", *ThermochimicaActa* 476 (2008), 11-19.

[16]. V.D. Tathavadkar, M.P. Antony and A.Jha. "The soda-ash roasting of chromite minerals: Kinetics considerations", *Metallurgical and Materials transactions B* 32 (B) (2001), 593-602.

[17].M.P. Antony et al.,"The soda-ash roasting of chromite ore processing residue for the reclamation of chromium", *Metallurgical and Materials transactions B* 32 (B) (2001), 987-995.

[18]. V.D. Tathavadkar, M.P. Antony and A.Jha. "The effect of salt phase composition on the rate of soda-ash roasting of chromite ores", *Metallurgical and Materials transactions B* 34 (B) (2003), 555-563.

[19].M.P. Antony, A.Jha and V.D. Tathavadkar. "Alkali roasting of Indian chromite ores: Thermodynamic and kinetic considerations", *Mineral processing and extractive Metallurgy* 115 (2) (2006), 71-79.

[20]. Z. Sun, S.L. Zheng and Y. Zhang. "Thermodynamics study on the decomposition of chromite with KOH", *ActaMetallurgyca* 20 (3) (2007), 187-192.

[21]. C. Arslan and G. Orhan. "Investigation of chrome (VI) oxide production from chromite concentrate by alkali fusion", *International Journal of Mineral Processing* 50 (1997), 87-96.

[22]. K. Yildiz and I.A, Sengil, "Investigationof efficient conditions for chromate production from chromite concentrate by alkali fusion", *Scandinavian Journal of Metallurgy* 33 (2004), 251-256.

[23]. A. Dettmer et al.,"Obtaining sodium chromate from ash produced by thermal treatment of leather wastes", *Chemical Engineering Journal* 160 (2010), 8-12.

[24]. Y. Zhang et al., "Decomposition of chromite ore by oxygen in molten NaOH-$NaNO_3$", *International Journal of Mineral Processing* 95 (2010), 10-17.

[25]. A.K. Tripathy, H.S. Ray and P.K. Pattnayak, "Kinetics of roasting of chromium oxide with sodium nitrate flux", *Metallurgical and Materials Transactions B* 26 (B) (1995), 449-454.

[26].Y. Hai-xia, X. Hong-bin, Z. Yi, Z. Sheng-li, G. Yi-ying, "Potential-pH diagrams of Cr-H_2O system at elevated temperatures", *Transactions of Nonferrous Metals Society of China* 20 (2010), s26-s31.

[27]. A.M. Amer, "Processing of Ras-Shait chromite deposits", *Hydrometallurgy* 28 (1992), 29-43.

[28]. E. Vardar, R.H. Eric and F.K. Letowski, "Acid leaching of chromite", *Minerals Engineering* 7 (5/6) (1994), 605-617.

[29]. A. Geveci, Y. Topkaya and E. Ayhan, "Sulfuric leaching of Turkish chromite concentrate", *Minerals Engineering* 15 (2002), 885-888.

[30]. T. Sharma, "The kinetics of iron dissolution from chromite concentrate", *Minerals Engineering* 3 (6) (1990), 599-605.

[31]. A.M. Amer, I.A. Ibrahim, "Leaching of a low grade Egyptian chromite ore", *Hydrometallurgy* 43 (1996), 307-316.

[32] Z. Yang et al., "Effect of mechanical activation on alkali leaching of chromite ore", *Transactions of Nonferrous Metals Society of China* 20 (2010), 888-891.

[33]. H.B. Xu et al., "Oxidative leaching of a Vietnamesse chromite ore in highly concentrated potassium hydroxide aqueous solution at 300ºC and atmospheric pressure", *Minerals Engineering* 18 (2005), 527-535.

[34]. Z. Sun et al., "A new method of potassium chromate production from chromite and KOH-KNO_3-H_2O binary submolten salt system", *American Institute of Chemical Engineers Journal* 55 (10) (2009), 2646-2656.

[35]. H.B. Xu et al., "Development of a new cleaner production process for producing chromic oxide from chromite ore", *Journal of Cleaner Production* 14 (2006), 211-219.

[36]. E. Remoudaki et al., "The role of metal-organic complexes in the treatment of chromium containing effluents in biological reactors", (Paper presented at the 15 th International Biohydrometallurgy Symposium, Hellas, Athens, 14-19 September 2003).

[37]. T. Banda, S. Koihe and T. Hara, "Organic acid chromium (III) salt aqueous solution and process of producing the same" (United States patent publication US2009/0194001 A1, 6 August 2009).

[38]. D. Bhattacharyya, J.A. Carlton and R.B. Grieves, "Precipitate flotation of chromium", *American Institute of Chemical Engineers Journal* 17 (2) (1971), 419-424.

[39]. B. Kiyak et al., "Cr(VI) reduction in aqueous solutions by using copper smelter slag", *Waste Management* 19 (1999), 333-338.

[40]. I.J. Buergue and S.J. Hug, "Kinetics and pH dependence of chromium (VI) reduction by iron (II)", *Environmental science & technology* 5 (31) (1997), 1426-1432.

[41]. M. Pettine et al. "The reduction of chromium (VI) by iron (II) in aqueous solutions", *Geochimica et Cosmochimica Acta* 9 (62) (1998), 1509-1519.

[42]. I.J. Buergue and S.J. Hug, "Influence of organic ligands on chromium (VI) reduction by iron (II)", *Environmental science & technology* 32 (1998), 2092-2099.

[43]. H. Zhou et al. "Influence of complex reagents on removal of chromium (VI) by zero-valent iron", *Chemosphere* 72 (2008), 870-874.

[44]. M. Rivero-Huguet and W.D. Marshall, "Influence of various organic molecules on the reduction of hexavalent chromium mediated by zero-valent iron", *Chemosphere* 76 (2009), 1240-1248.

Materials Processing Fundamentals
Edited by: Lifeng Zhang, Antoine Allanore, Cong Wang, James A. Yurko, and Justin Crapps
TMS (The Minerals, Metals & Materials Society), 2013

ELECTROPHORETIC CLASSIFICATION OF ULTRAFINE SILICA PARTICLES IN DILUTE AQUEOUS SUSPENSION

Ryan D. Corpuz[1,2] and Lyn Marie Z. De Juan[2]

[1]Mindanao State University-Iligan Institute of Technology
[2]University of the Philippines-Diliman

Keywords: EPD, stainless steel, silica, size classification, electrophoresis, electrophoretic mobility

Abstract

This study aims to classify ultrafine silica particles in a dilute aqueous suspension using their electrokinetic properties. Initially, the electrophoretic mobility of three (3) silica particle with mean diameters 2.39 µm, 3.74 µm and 5.31 µm are determined from pH 3 to 10 using two (2) electrolyte concentrations (0.01M and 0.001 M KNO_3). To evaluate classification, electrophoretic deposition experiments are performed using a modified electrophoretic/ deposition (EPD) cell. The experiments are conducted at pH 10 with an electrolyte concentration of 0.001M KNO_3 in four (4) time interval levels (15, 20, 25 and 30 min).

Results show that the electrophoretic mobility (µm.cm/V.s) of ultrafine silica particles in aqueous suspension decreased with increasing particle size. Electrophoretic mobilities also increased with increasing pH, and decreased with a corresponding increase in electrolyte concentration. Furthermore, classification experiments through electrophoretic deposition (EPD) show a gradient in both mean particle diameter and particle size distribution with increasing deposition time. The result are as follows: a) At 15 minutes, the mean particle diameter is 2.24 b) at 20 minutes, the mean particle diameter is 3.01 µm c) at 25 minutes, the mean particle diameter is 3.9 µm for EPD and 4.06 µm for electrophoresis; and, d) at 30 minutes, the mean particle diameter is 4.88 µm.

Based on the observed gradient in particle size and particle size distribution, it shows that the presence of smaller particles in the early stage of migration is probably due to their higher electrophoretic mobility as compared to the larger particles, thus resulting to their classification.

Introduction

Silica is the most abundant mineral on the earth's crust. It has a lot of uses especially in ceramic industries where it is considered as a primary component for the manufacture of traditional ceramic articles such as table wares, pot, vases, bathroom fixtures, roofing tiles, floor tiles, walls tiles, terracotta products, paving bricks and etc. Aside from its significant impact in traditional ceramic industries, silica also gains importance in the production of advanced materials used in cosmetics, electronics, biomedical, photonics, and semiconductor industries. Even today, in the promising field of nanoscience and nanotechnology, the potential application of silica is considerable. However, along with the sophisticated advancement introduced by nanoscience and nanotechnology is its difficulty in producing miniature materials from big entities. Truth be told, that silica is not an exception of this challenge.

Classification methods that will address the issue of producing micro and nano particles with uniform sizes and narrow size distribution beyond the range of ordinary sieves therefore is necessary for effective and efficient harnessing of the most economical and abundant mineral on earth for its utilization in the production of advanced micro and nanomaterials.

There is a lot of classification methods used for collecting fine and coarse particles. These methods include the use of hydrocyclone [1], elutriators [2], fluidized beds [3], centrifuge [4] and microsieves [5]. Commonly used force for these type of classification includes, hydrodynamic, aerodynamic, magnetic, centrifugal, and the force due to gravity in exploiting properties of materials like mass, density, shape, size, and magnetic properties. However these existing techniques could only classify material within the size range 5 to 1000 µm.

This study however demonstrates the relationship of electrophoretic mobility to particle size classification. The primary objective is to classify ultrafine silica particles in the size range 1-10µm which is an important size range for superhydrophobicity of silica particle in a metal substrate and where ordinary sieves and hydrodynamic classification methods are not capable of.

Experimental

Electrophoretic Mobility Measurement

The particle size and particle size distribution of silica powder collected from a 400 mesh sieve was analyzed first using coulter counter. Out of the collected powder, 5 grams of which was poured in a 300 ml distilled water and then thoroughly mixed for 10 minutes, ultrasonicated for 30 minutes and then poured to the cylindical sedimenting vessel with a capacity of 4 L. The particles were then allowed to settle for 2.5 hours. After which, the suspension were then collected and partitioned into three levels (top, middle and bottom) each having a 1 L capacity. Out of the collected suspensions, 250 ml from each group was allotted for particle size and particle size distribution analysis using Malvern zetasizer nano particle size analyzer. Addition of Potassium Nitrate (KNO3) and distilled water was then done to achieve a concentration of 0.01M and 0.001 M. The suspensions were then set aside overnight before the adjustment of pH using Nitric Acid and Potassium Hydroxide. Before measurement of electrophoretic mobility, the suspension was ultrasonicated for 6 minutes and stirred for 2 minutes after each pH adjustment. Measurements of electrophoretic mobility were then done using Malvern zetasizernano ZS90.

Electrophoretic Classification Experiment

Electrophoretic classification of micron sized silica particles was done in four time intervals (15, 20, 25, 30) using an electrophoretic deposition cell by applying a constant voltage of 30 V between two 316 stainless steel parallel plates 40 mm apart, with 10mm x 40 mm dimensions. The working electrode however is covered with a non conducting adhesion tape to expose a 10mm x 10mm surface area only. The sample was then image using Scanning Electron Microscopy to determine the particle size and particle size distributions.

Results and Discussion

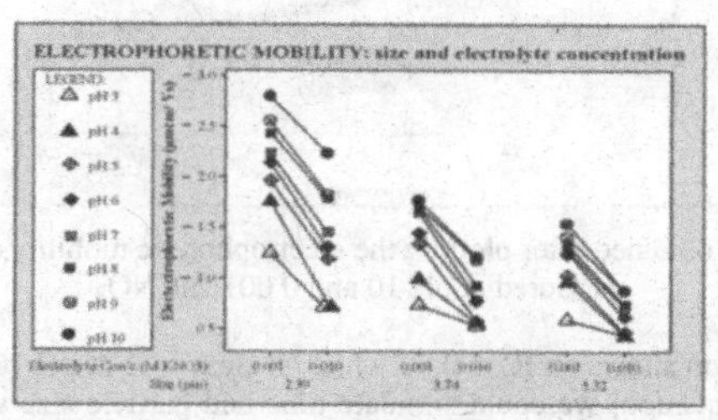

Figure 1. electrophoretic mobility of silica particles with respect to particle size, electrolyte concentration and pH

Figure 1 shows the electrophoretic mobility of silica particles with respect to particle size, pH and electrolyte concentrations. In this graph, it is very evident that silica particles with small particle radius have higher electrophoretic mobility compared to the larger ones. This behavior is due to the fact that electrophoretic mobility is inversely proportional to the radius and directly proportional to the charge of the particle. The ratio of these two parameters normally determines the electrophoretic mobility of the particle. However, since a large particle, is very much influenced by the force due to gravity compared to the small one, the competing forces such as electric field, gravitational force and viscous force that may act on the particle will determine the overall effect or net electrophoretic mobility of this particle as it traverse from one point to another rather than the electric field alone.

Moreover, it is also noticeable that electrophoretic mobility increases with increasing pH. This behavior is attributed to the increasing surface charge of silica particles at higher pH. Every time the pH is increased more and more OH- ions are introduced into the system (in cases wherein KOH is used for pH adjustment) thus making the silica particle more negative. Aside from the OH-ions, more and more K+ ions are also introduced to support the double layer thus making it more stable and more mobile since this will lead to an increase in repulsive interaction among particles that are present in the system due to the larger electrical double layer that each of these silica particles are carrying.

Furthermore, it is also observed that electrophoretic mobility decrease with respect to the addition of electrolyte, this behavior is due to the shielding effect that these ions introduced that may hinder the long range effect of the surface charge. Additions of excess ions will compressed the double layer and will also increase the conductivity of the system. When this happens, the electrophoretic mobility will decrease since other than the particle that is supposedly the main carrier of charge, the excess ions present will also carry charges that will hinder the movement of the particles.

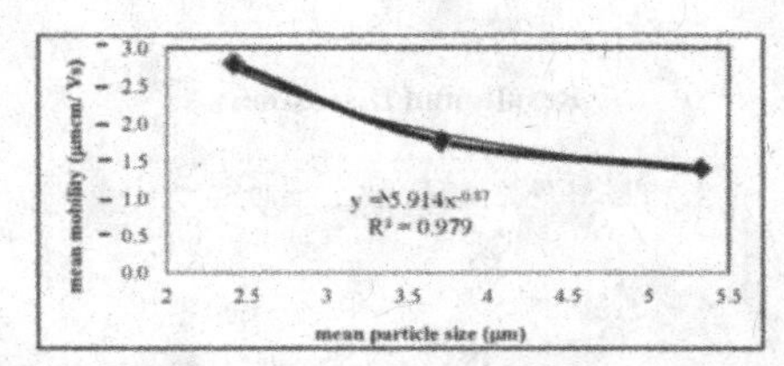

Figure 2. Relationship obtained after plotting the electrophoretic mobility and mean particle size measured at pH 10 and 0.001M KNO_3

The obtained relation shows an $R^2 = 0.98$ which is close to 1 at pH 10 and 0.001M KNO_3. This indicates that, using the equation, we could simulate time and particle size values to generate a graph that will precisely describe time-mean particle size relationship for a given electrophoretic mobility.

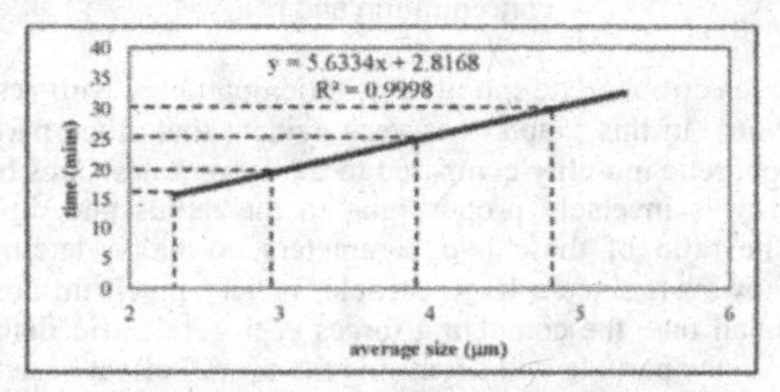

Figure 3. Relationship between time and average particle size deposited.

Figure 3 shows the linear dependence of the mean average particle size with respect to time with $R^2 = 1$. Using this relationship, the mean particle size deposited and deposition time could now be correlated.

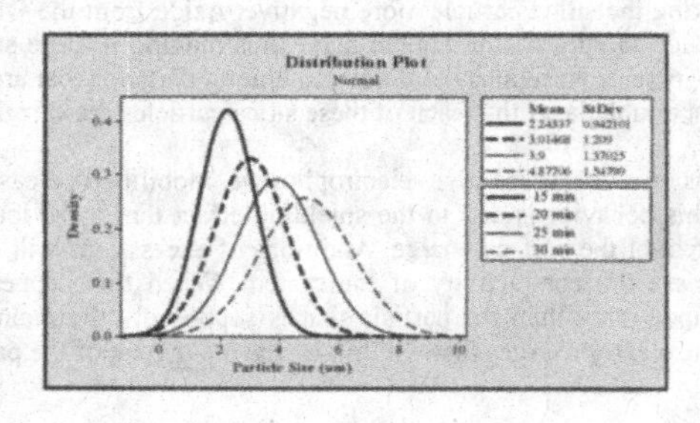

Figure 4. Particle size and particle size distribution after Electrophoretic deposition

Figure 4 shows the particle size diameter and particle size distribution with respect to deposition time considered. At 15 minutes, the mean particle diameter is around 2.24 microns, at 20 minutes, the mean particle diameter is around 3.01 microns, and 25 minutes, the mean particle diameter is around 3.9 and at 30 minutes, the mean particle diameter is around 4.88 microns. This result implies that the mean particle size diameter increases with respect to deposition time and the particle size distribution also become wider with time.

Table 1. Comparison between theoretical and EPD

	average particle size (μm)		
time (mins)	Calculated	EPD	%difference
15	2.16	2.24	3.7314
20	3.05	3.01	1.1657
25	3.94	3.90	0.9599
30	4.82	4.88	1.0714

Table 1 is the comparison of values of mean particle size obtained using the relationship in figure 3 and the mean particle sizes measured after subjecting the samples for image analysis. It is obvious from this table that the values that are experimentally determined are close to the values calculated. This implies that there is a strong correlation between the radius of electrophoretically classified silica particles and its electrophoretic mobility. This is due to the fact that the deposit yield is directly proportional to electrophoretic mobility with respect to time in a dilute solid concentration and low electric field direct current electrophoretic deposition (DC-EPD).

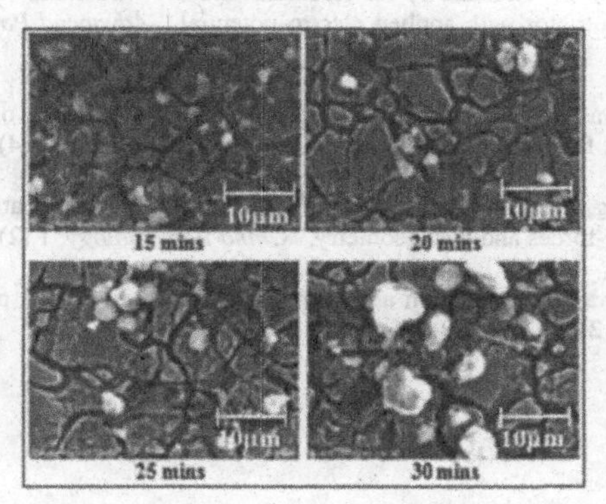

Figure 5. Scanning electron micrographs take at x3500 magnitude for the four time intervals considered.

Figure 5 shows the scanning electron micrographs of the deposited silica particles after deposition. It is evident in this figure that the particle sizes deposited increases with increasing deposition time.

Conclusion

Based on electrophoretic mobility measurement, it shows that the electrophoretic mobility of silica particles is very much influenced by pH, electrolyte concentration and its size. It was observed, that electrophoretic mobility is higher at basic pH, higher in a more dilute concentration and higher in small particles.

Furthermore, it was also observed that small silica particles are deposited first in early times of deposition and only at later time did the large particle become prominent. In addition, there is a linear dependence between the mean diameters of these deposited silica particles and deposition time as verified by electrophoretic deposition experiment and Scanning Electron Micrographs gathered.

It is therefore concluded, that the presence of small silica particles in the early times of deposition is probably due to its higher electrophoretic mobility compared to the larger particles. Thus, the difference in electrophoretic mobility of these silica particles leads to its classification.

References

[1] L. Zhao, M. Jiang, and Y.Wang, "Experimental study of a hydrocyclone under cyclic flow conditions for fine particle separation," *Separation and Purification Technology,* 59 (2) (2008), 183-189.

[2] H.Yoshida, K. Fukui, T. Yamamoto, A. Hashida, and N. Michitani, "Continuous fine particle classification by water elutriator with applied electro-potential," *Advanced Powder Technology*, 20 (4) (2009), 398-405.

[3] T. Nakazato, J. Kawashima, T. Masagaki, and K. Kato, "Penetration of fine cohesive powders through a powder-particle fluidized bed," *Advanced Powder Technology*, 17 (4) (2006), 433-451.

[4] Y. Wang, E. Forssberg, J. Li, and Z. Pan, "Continuous ultra-fine classification in a disc-stack nozzle centrifuge — effects of G-forces and disc geometry," *China Particuology*, 1 (2) (2003), 70-75.

[5] J. Hidaka and S. Miwa, "Fractionation and particle size analysis of fine powders by micro sieve," *Powder Technology,* 24, (2) (1979), 159-166.

Materials Processing Fundamentals
Edited by: Lifeng Zhang, Antoine Allanore, Cong Wang, James A. Yurko, and Justin Crapps
TMS (The Minerals, Metals & Materials Society), 2013

QEMSCAN ANALYSIS OF WADI-SHATTI IRON ORE (LIBYA)

Ali M. Tajouri[1], Patrick R.Taylor[2], Corby Anderson[3] ,Moftah M.Abozbeda[4]

1 Ali M. Tajouri, Materials &Metallurgical Eng., Faculty of engineering, U T, Libya
2 Patrick R.Taylor, Metallurgy and Materials Eng., CSM, Golden, CO 80401, USA
3 Corby Anderson, Metallurgy and Materials Eng., CSM, Golden, CO 80401, USA
4 Moftah M. Abozbeda, Mining Eng., U T, Libya

Keywords: Iron ore analysis, QEMSCAN analysis, Mineral Characterization

Abstract

Libyan Iron ores reserve estimated to be > 5.0 billion tons, with 48-55% Fe& 1.0% P& Libyan Previous detailed investigation by FSG& IRC was not sufficient& did not adopt any technology to remove P), (Si) & alkaline. Due to price increasing of iron ores in global markets, gave potentials to treat high (P) iron ores in economical ways. Recent global treatment of high (p) iron ores results may be adopted to treat the Libyan ores & the predication may be fruitful .A Preliminary QEMSCAN analysis of Libyan iron ores samples was conducted at(advanced mineralogy research center- CSM), to determine, phosphorous phases, locking and liberation characteristics of the phosphorous phases and mineral associations. Determination shows that 61 % Fe as Oxide/Hydroxide, 31 % quartz, 5 % apatite, and other traces. Fe occurs in fraction -300/+150 μm. Highest concentrations of apatite occur in the coarsest fraction assay shows 1.0 % P and 39 % Fe & most of Fe found in coarsest size fraction. Size fractions of P < 0.5 mass %& elemental deportment of (P) is present in apatite, in fraction -75 μm &0.3 mass % P occur in monazite.68 % of apatite is liberated; the majority of liberated apatite 62 % can be observed in the -75 μm fraction. The fruitful findings may give an excellent encouragement to perform preliminary floatation& leaching experiments as long as the phosphorous existed as apatite phase& it can be separated by many different promising techniques as well.

Introduction

The rise of iron ores prices in international market, made the global steel manufacturing companies to search for high (P) cheap ores, the phosphorous is one of the harmful element to ferrous metallurgy. LISCO (Libya) importing the pre-fluxed iron ore pellets from Sweden& Brazil, As reported by LISCO, that the price increase of imported pellets is 300% in years of 2002-2007 &the price increase of transportation in such years is 1000%, such price increase issue made LISCO in hard position to compete with the international steel production companies. Due to price increase issue LISCO reports encouraging the decision makers in ministry of industry to invest the local ores. Research in USA ,Australia, Canada, Sweden & China concerning treatment of high(p) iron ores which is in close fit to the Libyan iron ore; may open the gate of investment. The Iron ore of wadi shatti (Libya) has been known long time ago& the investigation done by French study group (FSG) & IRC in 1971-1976, estimated the ore to be >5.0 billion tons with average grade of 48-55% Fe, 1.0% P, 4%Si &4%Al2O3 [1,2].Based on FSG report Both laboratory & pilot plant scale tests were not capable to remove (P)&reduce the other impurities to the level of steel making standers. A Canadian work was performed back to 1974, on Snake River iron ore deposit; Successful beneficiation was achieved, where the concentrate was 69% Fe, <0.03%P& they recommend the same treatment may be successful to be applied to similar iron ores [3].The LKAB concentrator in Kiruna, Sweden treated high (P) magnetite ore, influence &operational variables on kinetics of apatite floatation was performed. The authors made the statistical tests on (P) recovery; the maximum recovery of (P) was attained, with minor losses of magnetite [4].Recently acid leaching considered as a possible alternative to floatation techniques, the acid consumption can be reduced by selecting the proper scheme. Swedish group performed leaching technique on high Swedish phosphorous iron ore with Nitric acid, they listed the advantages, disadvantages of other leaching solutions, the leaching process were capable to reduce (P) content from 0.2% to 0.05%&the leached ore was used for production of pellets [5,6].In another work they treated Swedish ore ,where the apatite content was 1%, the (P) removal extent to 95%, iron loss was<0.05%, while the alkali reduced to 60%.China research group made comprehensive& comparative work, where alkali/acid leaching were performed for the dephosphorization of China high (P) iron ore, which contains 1.12 %(P). Sodium hydroxide, sulpheric acid, hydrochloric acid & nitric acid were used, they showed that (P) in apatite phase could be removed by alkali leach, but (P) occurring in iron phase could not [7].

Sulpheric acid was the most effective three kinds of acid, 92% of (P) was removed with 1% of sulpheric acid, Fe loss <0.25%.STU &mining companies in Sweden summarized their work in several possible ways to produce high quality steel from high (P) apatite iron ores. Australian researchers conducted an experimental work on an Australian 0.15% (P) iron ore, an effective dephosphorization was performed. They chosen sulpheric acid to be the leachant on basis of availability& low cost, leached (P) reduced to 0.04%, with minimum acid cost& minimum iron loss [8, 9]. This review of different countries experience that developed for, high (P) iron ores treatment may play potential & gives encouragement to re- investigate Libyan high (P) iron ores, mean while the economics of treatment is very promising as well.

Samples for QEMSCAN analysis

Libyan Fe ore sample was submitted for QEMSCAN analysis. The aim was to determine the sample's mineralogy, phosphorous-bearing phases, locking& liberation characteristics of the phosphorous-bearing phases and mineral associations. The sample was separated into three size fractions: -75 μm, -150/+75 μm and-300/+150 μm, each fraction was split into eight representative sub-splits. One split per sample was mixed with graphite of the same size range, mounted into epoxy resin. The cured sample blocks were ground and polished using water-based lubricants and suspensions following standard procedures. QEMSCAN analyses were conducted in Particle Mineral Analysis mode at15 μm, 10μm and 5 μm resolutions depending on fraction size.

Results

Mineralogy

The Fe ore sample contains 61 mass % Fe Oxide/Hydroxide, 31 mass % quartz, 5 mass % apatite,1 mass % ilmenite, 1 mass % clay minerals and other traces (Table 1).By mass, the majority of Fe Oxide/Hydroxide can be found in fraction 300/+150 μm (37 mass % of total sample weight),the least amount is present in the -75 μm fraction (7 mass %)(Fig. 1; Table 2). The highest concentrations of apatite occur in the coarsest fraction and the least amount can be observed in the medium and fine size fractions. In contrast, the highest quartz concentrations occur in the medium size fraction150/+75 μm)

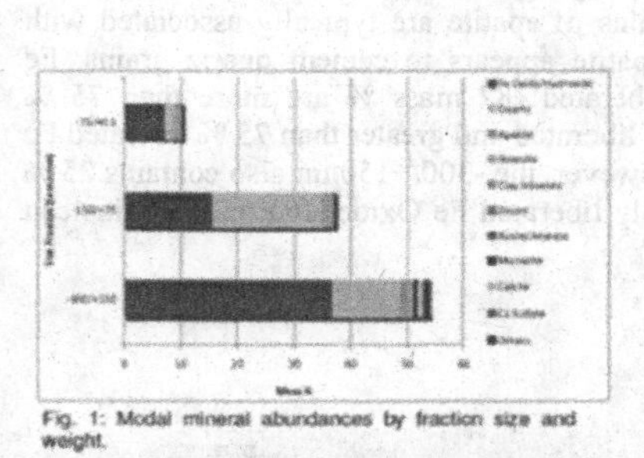

Fig. 1: Modal mineral abundances by fraction size and weight.

Table 1: Modal mineral abundances (Mass %)

Minerals	Wadi-Shatti
Fe Oxide/Hydroxide	61
Quartz	31
Apatite	5
Ilmenite	1
Clay minerals	1
Zircon	tr
Rutile/Anatase	tr
Monazite	tr
Calcite	tr
Ca Sulfate	tr
Others	1

*tr = <0.5 mass %

Elemental Assay and Elemental Deportment of Phosphorous

QEMSCAN shows 1 mass % P and 39 mass % Fe in the sample (Table 3). As is reflected in the modal mineral abundances, most of the Fe can be found in the coarsest size fraction (23 mass %; Table 4), and the least amount of Fe occurs in the smallest size fraction. In all size fractions P makes up less than 0.5 mass %.Elemental deportment of P by mass % illustrates that the vast majority of P is present in apatite (99.7 to 100 mass %); in fraction -75 μm 0.3 mass % P occur in monazite (Table 5).

Table 3: Elemental Assay

Elements (Mass %)	Wadi-Shatti
Al	tr
Ca	2
Fe	39
K	tr
Mg	tr
Na	tr
O	41
P	1
S	tr
Si	15

*tr = <0.5 mass %

Table 4: Elemental Assay by Fraction Size and Weight

Elements (Mass %)	-300/+150	-150/+75	-75/+0.1
Al	tr	tr	tr
Ca	1	1	tr
Fe	23	10	5
K	tr	tr	tr
Mg	tr	tr	tr
Na	tr	tr	tr
O	21	17	4
P	tr	tr	tr
S	tr	tr	tr
Si	6	9	1

*tr = <0.5 vol. %

Table 5: Elemental Deportment of P by Mass %

Minerals	-300/+150	-150/+75	-75/+0.1
Apatite	100.0	100.0	99.7
Monazite	0.0	0.0	0.3

Mineral Associations and Locking/Liberation Characteristics

Analysis of the locking and liberation characteristics of apatite shows that 42 mass % of Apatite are 100 % liberated, 26 mass % are more than 75 % liberated and the remaining Apatite grains are more than 25 % liberated Fig. (2). Taking the fraction sizes and Weights into consideration, the majority of apatite (62 mass %) can be observed in fraction-75 µm where it occurs as liberated or greater than 75 % liberated particles. The majority of less than 75 % liberated apatite occurs in the coarsest fraction. Partially locked grains of apatite are typically associated with quartz and Fe Oxide/Hydroxide; in some cases apatite appears to cement quartz grains. Fe Oxide/Hydroxide grains are also predominantly liberated (87 mass % are more than 75 % liberated; Fig. (3).The majority (52 mass %) of fully liberated and greater than 75 % liberated Fe Oxide/Hydroxide grains occur infraction -75 µm. However, the -300/+150µm also contains 25 % mostly liberated Fe Oxide/Hydroxide grains. Partially liberated Fe Oxide/Hydroxide grains can be found forming cements around quartz grains.

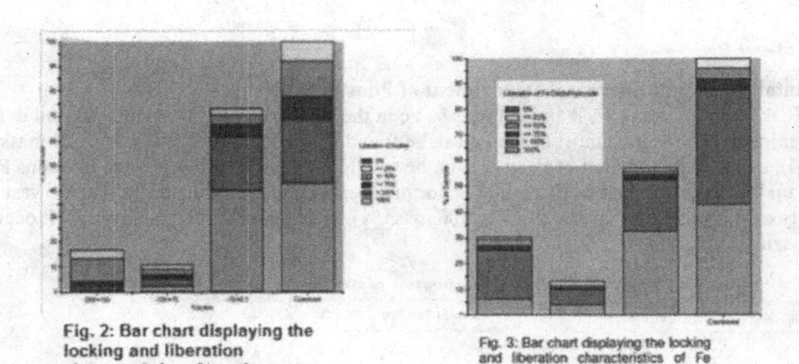

Fig. 2: Bar chart displaying the locking and liberation characteristics of apatite

Fig. 3: Bar chart displaying the locking and liberation characteristics of Fe oxide/Hydroxide.

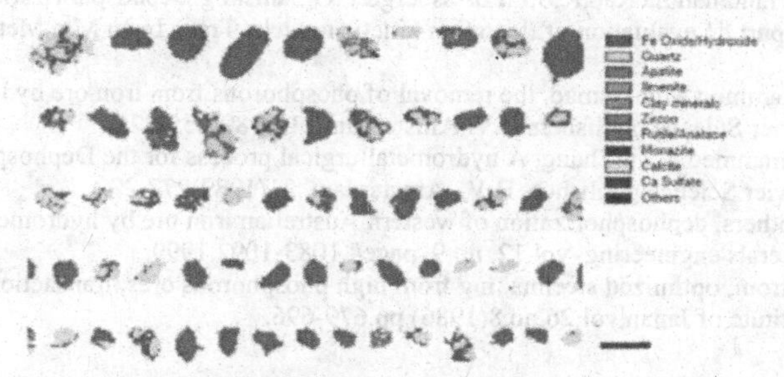

Fig. 4: QEMSCAN images of representative particles showing liberated Fe Oxide/Hydroxide and apatite grains as well as both minerals cementing quartz. Scale bar = 300 μm.

Conclusions

The Wadi-Shatti Fe ore sample contains 61 mass % Fe Oxide/Hydroxide, 31 mass % Quartz, 5 mass % apatite, 1 mass % ilmenite, 1 mass % clay minerals and trace amounts of Zircon, rutile/anatase, monazite, calcite and Ca Sulfate. By mass, the majority of Fe Oxide/Hydroxide occurs in the coarsest fraction and the least amount in the finest fraction. The same trend can be observed for apatite. QEMSCAN elemental assay shows 1 mass % P and 39 mass % Fe in the sample. In all size fractions P makes up less than 0.5 mass %.

Elemental deportment of P by mass % illustrates that the vast majority of P is present in Apatite (99.7 to 100 mass %); in fraction -75 μm 0.3 mass % P occur in monazite. Both Fe Oxide/Hydroxide and apatite are predominantly greater than 75 % liberated. Partially Liberated Fe Oxide/Hydroxide and apatite commonly cement quartz grains. Today's technology is capable o treat high (P) iron ores that occurs as apatite phases, their development may play key role of encouragement to invest Libyan high (P) iron ores, mean while the economics of treatment is very promising as well. The most promising economic parameters that may open the door of investment in Libya are; importance of international location of Libya, Low labor cost, cheap natural gas& energy, moderate climate& environment, international increase of iron ore cost, international shortage of natural high iron grad, un-used reserve of Waddi-shatti ore.

References

1. French study group (FSG) report, joint contract with Libyan industrial research (IRC) Contract#44045/05800, 1977.

2. Private technical report, IRC-Libya, 2002

3. B.D.Sparks&A.F.Sirianni, Beneficiation of a phosphorous iron ore by Agglomeration methods, journal of mineral processing, 1(1974), Page /231-241

4. Fenwei Su,K.Hanumantha Rao,K.S.E.Forsssberg&P.O.Samskog,Dephosphorization of magnetite fines, part #2,influence of chemical variables on floatation kinetics,Trans.Instn Min.Metall,sec C:,Pag#107,1998.

5. Fenwei Su,K.Hanumantha Rao,K.S.E.Forsssberg&P.O.Samskog,Dephosphorization of magnetite fines, part #1,evaluation of floatation kinetic models ,Trans.Instn Min.Metall,sec C:, 1998.
6.Yu Zhang& Mammon Muhammad, the removal of phosphorous from iron ore by leaching with nitric acid, Elsevier Science publishers B.V.,Amsterdam,21(1989)255-275.
7. Mammon Mohammed &Yu Zhang, A hydrometallurgical process for the Dephosphorization of iron ore, Elsevier Science publishers B.V., Amsterdam, 21(1989)277-292.
8. C.Y.Cheng&others, dephosphorization of western Australian iron ore by hydrometallurgical process, j.of minerals engineering, vol.12, no.9, page# 1083-1092,1999.
9.John Olaf Edstrom, optimized steelmaking from high phosphorous ores, transactions of the iron & steel ,institute of Japan,vol.26,no.8(1986) pp.679-696.

Materials Processing Fundamentals
Edited by: Lifeng Zhang, Antoine Allanore, Cong Wang, James A. Yurko, and Justin Crapps
TMS (The Minerals, Metals & Materials Society), 2013

REMOVAL OF ARSENIC FROM ENARGITE RICH COPPER CONCENTRATES

Maria C. Ruiz[1], Ricardo Bello[2] and Rafael Padilla[1]

[1]Department of Metallurgical Engineering, University of Concepción,
[2]Department of Chemical Engineering, University of Concepcion
Edmundo Larenas 285, Concepcion, 4070371, Chile

Keywords: Arsenic removal, Enargite, Alkaline baking

Abstract

Arsenic is a troublesome impurity in copper concentrates where it is usually present as enargite (Cu_3AsS_4). When the enargite content is high in the concentrates, they cannot be treated by direct smelting processes because of the high risk of arsenic emissions to the atmosphere. In this work, some experimental results are presented on a novel alternative process to remove selectively the arsenic from an enargite rich copper concentrate. The process includes an alkaline (Na_2S-NaOH) baking step followed by leaching with water. The experimental results showed that temperature and Na_2S concentration are important variables in the baking step for efficient arsenic removal. In experiments using a chalcopyritic copper concentrate containing 2.2% arsenic as enargite, 94% of the arsenic was removed when the concentrate was baked at 85 °C for 10 min by using 2.5 times the stoichiometric Na_2S requirement for arsenic dissolution. Copper dissolution was negligible under these conditions.

Introduction

Arsenic, usually present as enargite is a toxic impurity in Chilean copper concentrates. When the concentrates contain high levels of arsenic (>0.5% As) they cannot be treated by conventional smelting processes, because of the high risk of arsenic emissions to the atmosphere [1].

There are two main routes to lower the arsenic content of a "dirty" copper concentrate to levels appropriate for the concentrate to be safely treated by a conventional smelter:

(1) Roasting to volatilize the arsenic from the concentrates
(2) Selective dissolution of the arsenic from the enargite by a hydrometallurgical treatment.

Because of the high partial pressure of most arsenic sulfide and oxide compounds, the roasting process is an effective method for the arsenic elimination from copper concentrates [2]. However, this alternative possesses a major risk of atmospheric contamination.

On the other hand, hydrometallurgical processes are potentially less risky to the environment than roasting. Thus alkaline leaching using sodium hypochlorite (NaClO) or sodium sulfide (Na_2S) can be used advantageously to dissolve the arsenic from copper concentrates. The leaching using NaClO-NaOH would dissolve the arsenic from the enargite to form arsenate ions (AsO_4^{3-}), while the copper from the enargite is converted into solid CuO [3-5]. Although, this process has a rapid kinetics at ambient pressure, it is not very selective since other copper sulfides also react to a certain extent in this media, particularly covellite (CuS) whose reactivity is similar to enargite. Another drawback of sodium hypochlorite leaching is that the antimony minerals, which are usually present in copper concentrates, do not react appreciably [6].

In the case of sodium sulfide leaching (Na_2S-NaOH), the dissolution of enargite can be represented by the simplified reaction (1), where arsenic from enargite is dissolved as thioarsenate ions (AsS_4^{3-}) and the copper remains in the solid residues as sulfide [7-9].

$$2Cu_3AsS_4(s) + 3S^{2-} = 3Cu_2S(s) + 2AsS_4^{3-} \quad (1)$$

In this leaching system, the main role of the NaOH apparently is to prevent the hydrolysis of the S^{2-} to HS^- [7]. High dissolution of arsenic can be obtained at temperatures near the boiling point of the solution or at higher temperatures with large excesses of Na_2S [7, 9].

Since in alkaline sodium sulfide media the copper from the enargite remains in the residues as a solid copper sulfide, the solid residues are appropriate for the treatment in the conventional pyrometallurgical technology of smelting/converting. Other common copper sulfide minerals usually present in copper concentrates, such as chalcopyrite, bornite, covellite, chalcocite, etc., should not dissolve appreciably in this leaching media. However, most of the other minor element sulfides such as tetrahedrite, stibnite and bismuthinite, will be dissolved in this system [10, 11].

Thus, the sodium sulfide leaching has several advantages over hypochlorite leaching for the purification of contaminated copper concentrates with arsenic, namely:

i) A higher selectivity for the dissolution of enargite over other copper minerals.
ii) The solid leaching residues are suitable for direct treatment by smelting.
iii) Other noxious impurities which are usually present in the copper concentrates, such as antimony and bismuth minerals will be dissolved as well.

The main disadvantage of the alkaline sodium sulfide leaching which has limited its application for the treatment of dirty copper concentrates is the slow kinetics of the dissolution reaction (1). Thus, for an efficient arsenic removal from the concentrates, extended leaching (up to 8 hours) may be required.

In this work, we discuss a novel hydrometallurgical alternative for rapid removal of the arsenic from enargite containing copper concentrates which consists of baking the arsenic containing concentrate with a concentrated Na_2S and NaOH solution followed by a leaching stage with water.

Experimental Work

The primary materials used in the experimental work were a chalcopyrite rich copper concentrate and an enargite concentrate. Both materials were classified by sieving using a USA sieve series. Narrow size fractions of the chalcopyrite concentrate were mixed with size fractions of the enargite concentrate to obtain copper-arsenic concentrate samples to be used in the experiments. The experimental results shown here were obtained with two of these mixtures: Concentrate C1 was prepared mixing 80% of a chalcopyrite concentrate with size fraction 270/400 mesh and 20% enargite concentrate with size fraction 140/200 mesh, while concentrate C2 was prepared using the size fraction 140/200 mesh of the chalcopyrite concentrate and the size fraction

100/140 mesh of the enargite concentrate mixed in the same proportions as C1. The arsenic content of C1 and C2 were 2.23% and 2.21%, respectively.

The experiments were carried out with 20 g of the chalcopyrite-enargite concentrate samples (C1 or C2) prepared as described. The solid was charged into a beaker and heated to the desired temperature in a water bath. The alkaline baking was then started by wetting the solid sample with a small amount (5 or 4 ml) of a preheated solution containing the desired amounts of Na_2S and NaOH. During baking, the wet solid was constantly mixed using a rod for the predetermined time. Once the baking period ended, the solids were leached with water to dissolve the sodium thioarsenate produced. It should be pointed out that the experiments presented here were focused in the baking step; therefore, the leaching was performed under standard conditions for a period of 60 min in order to assure that all the soluble arsenic from the baked concentrate went into the solution. Thus, the entire leaching tests were carried out in a 2-liter glass reactor with a mechanical agitation system using 500 ml of water at 80 °C and 370 rpm. Afterwards, the solids were filtered, washed and dried at 80 °C for chemical analysis. In some experiments, the filtrate was also analyzed for copper and arsenic.

Results

In the experiments, the amount of Na_2S used in the baking was expressed as a multiple of the stoichiometric amount necessary to react with all the enargite present in the concentrate, according to reaction (1), i.e. 1.5 moles of Na_2S per mole of enargite in the concentrate sample. The water leaching experiments of both baked concentrates (C1 and C2) were always carried out in the standard conditions already described. The NaOH concentration used as a standard value in most baking tests was 2 molar. The experimental results obtained with both concentrate samples were very similar. The effect of the main baking variables is described in the following.

Effect of Baking Temperature and Time

To study the effect of baking temperature and time on the arsenic removal (dissolution), experiments were carried out with concentrate C1, at temperatures in the range of 70 to 95 °C for baking times of 5 and 10 min. The results are summarized in Figure 1, where we can see that at the highest baking temperature of 95 °C, the solubilizing of arsenic reaction is fast since 95.7% of the arsenic was removed after 5 min of baking by the standard water leaching, producing a leaching residue with 0.1 % As. On the other hand, when the baking was done at 70 °C, for 10 min, the arsenic removal was only 86.1% and the arsenic content of the residues was 0.32%. It should be pointed out; that considering that the maximum arsenic content of copper concentrates to be treated in a smelter is about 0.5% [1], the baking at 70°C would still produce a saleable copper concentrate.

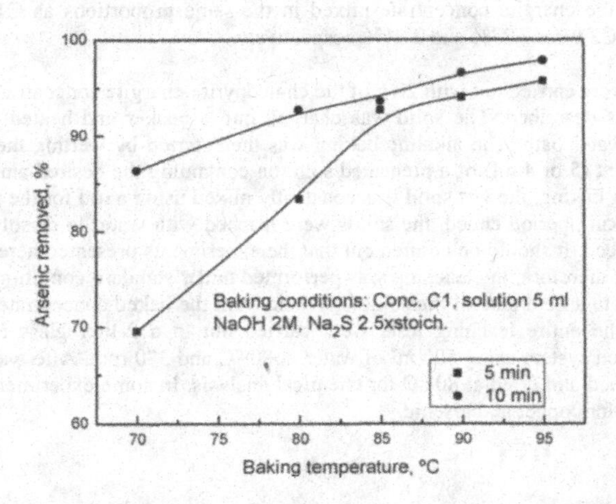

Figure 1. Removal of arsenic from concentrate C1 after 5 and 10 min of baking.

Effect of the Amount of Baking Solution

Experiments were performed with concentrate C1 where the volume of the solution added to the solids during baking was decreased from 5 to 4ml. In these experiments the amount of NaOH was also decreased proportionally, in order to maintain a NaOH concentration of 2M. However, the amount of Na_2S added was kept constant at 2.5 times the stoichiometric requirement. The results are presented in Figure 2.

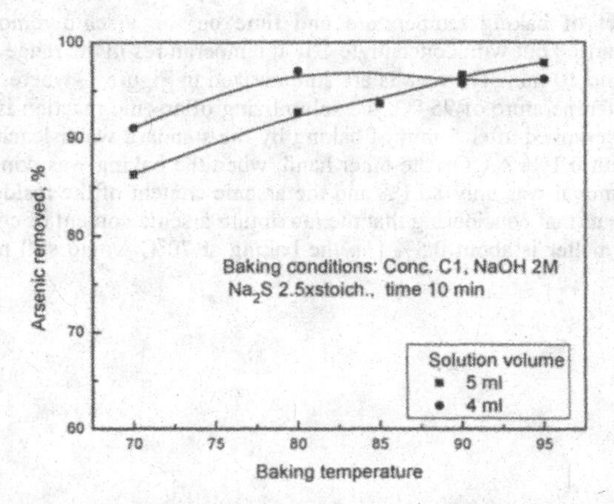

Figure 2. Effect of the volume of baking solution on arsenic removal from copper-arsenic concentrate C1.

As seen in this figure, at the lower temperatures of 70 and 80 °C, a decrease in the volume of the baking solution added to the concentrate produced a substantial increase in the fraction of arsenic removed from the concentrate C1. This is probably due to the increase in Na_2S concentration in the solution, since as indicated earlier the dosage of reagent added was 2.5 times the stoichiometric amount for both solution volumes. At the highest temperatures (90 and 95°C) the tendency of arsenic removal appears to be the opposite, although the difference for 4 and 5 ml of solution is small. This result could be attributed to excessive loss of moisture of the solids because of faster water evaporation at the higher temperatures.

Effect of an Increase in the Amount of Na_2S

Additional experiments were carried out at various temperatures using a constant volume of baking solution of 5 ml and increasing the amount of Na_2S to 3.5 times the stoichiometric amount. The results are presented in Figure 3 and compared with the experiments when the addition of Na_2S was 2.5 times the stoichiometric amount. In these conditions, the experiments with higher amounts of Na_2S gave better arsenic removal in the whole range of temperatures tested, although the difference was less significant at high temperatures.

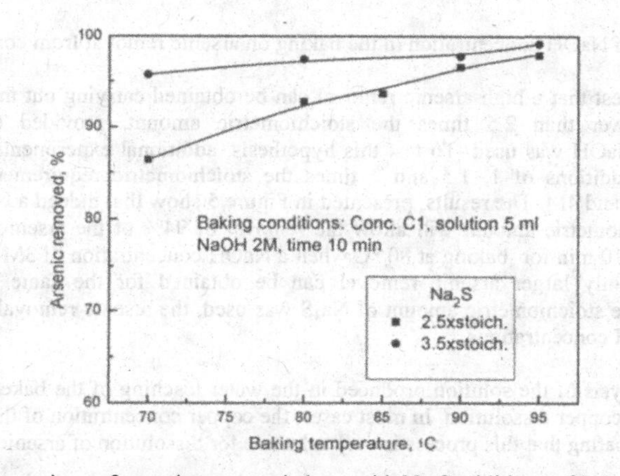

Figure 3. Comparison of experiments carried out with Na_2S additions of 2.5 and 3.5 times the stoichiometric amount using concentrate C1.

Effect of the NaOH Concentration in the Baking Solution.

The effect of the NaOH concentration in 4 ml of baking solution on the arsenic removal was investigated in the range of 0.5 to 4 M. The experiments were carried out using the concentrate sample C2, baked at 80°C for 5 min with Na_2S equal to 2.5 times the stoichiometric requirement. The results presented in Figure 4 show that a higher NaOH concentration in the baking improves the extent of arsenic removal from the concentrate. For NaOH equal to 4M 97.8% of the arsenic was removed in the standard water leaching, while for NaOH of 0.5M the arsenic removal decrease to 83.2%.

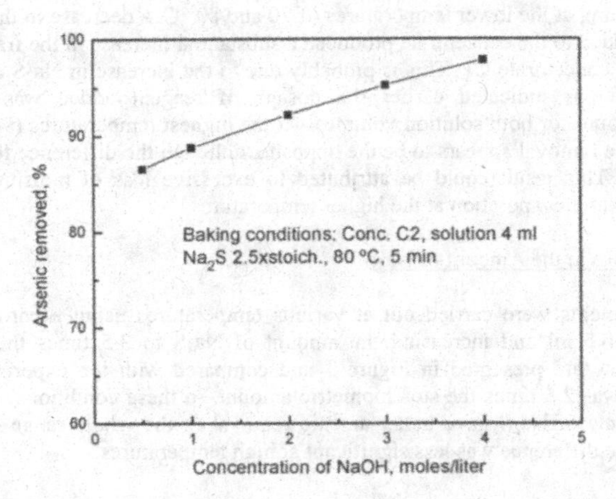

Figure 4. Effect of NaOH concentration in the baking on arsenic removal from concentrate C2.

These results suggest that a high arsenic removal can be obtained carrying out the baking with Na_2S dosages lower than 2.5 times the stoichiometric amount, provided that a higher concentration of NaOH was used. To test this hypothesis, additional experiments were carried out using Na_2S additions of 1, 1.5 and 2 times the stoichiometric requirement and NaOH concentration of 3 and 4M. The results, presented in Figure 5 show that indeed a Na_2S dosage of 2 times the stoichiometric amount will allow the removal of 94% of the arsenic content from concentrate C2 in 10 min for baking at 80 °C when a NaOH concentration of 3M was used. For NaOH 4M a slightly larger arsenic removal can be obtained for the same Na_2S dosage. However, when the stoichiometric amount of Na_2S was used, the arsenic removal obtained was low for both NaOH concentrations.

The chemical analysis of the solution produced in the water leaching of the baked concentrates showed very little copper dissolution. In most cases, the copper concentration of the solution was below 2 ppm, indicating that this process is very selective for dissolution of arsenic over copper.

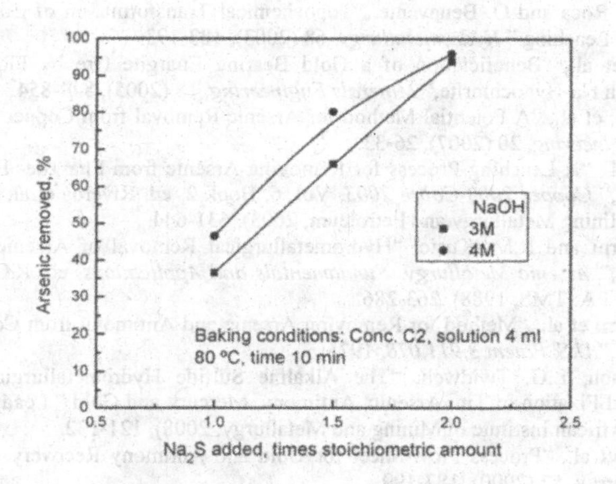

Figure 5. Arsenic removal as a function of the Na_2S dosage used in baking for NaOH concentration of 3 and 4 M.

Conclusions

From the experimental results on the alkaline baking step, it can be concluded that the baking-leaching process is very effective method for the arsenic removal from an enargite-rich chalcopyritic copper concentrate.

The reactions that take place in the baking of the concentrate are fast and a baking time of 10 min is sufficient to solubilize most of the arsenic from the enargite.

The main variables that affect the arsenic dissolution from the concentrate are the baking temperature and the dosage of Na_2S.

High arsenic removal can be obtained keeping to a minimum the volume of the baking solution. Increasing the NaOH concentration in the baking solution improves the arsenic removal, too.

Acknowledgments

This work was supported by the National Fund for Scientific and Technological Development (FONDECYT) of Chile under project N° 1120585.

References

1. K. Baxter, H. Scriba and I. Vega, "Treatment of High-Arsenic Copper-Gold Concentrates-An Options Review," *Proceedings of Copper 2010*, Hamburg, Germany,vol. 5, 1783-1802.
2. E.H. Smith, "Metallurgy and Mineral Processing Plant at St, Joe's El Indio Mine in Chile," *Mining Engineering*, October 1986, 971-979.

3. J. Viñals, A. Roca and O. Benavente, "Topochemical Transformation of Copper Oxide by Hypochlorite Leaching," *Hydrometallurgy*, 68 (2003), 183-193.
4. L. Currelli et al., "Beneficiation of a Gold Bearing Enargite Ore by Flotation and As Leaching with Na-Hypochlorite," *Minerals Engineering*, 18 (2005), 849-854.
5. I. Mihajlovic, et al., "A Potential Method for Arsenic Removal from Copper Concentrates," *Minerals Engineering*, 20 (2007), 26-33.
6. A. Roca et al., "A Leaching Process for Removing Arsenic from Enargite- Bearing Copper Concentrates," *Copper 2003-Cobre 2003*, Vol. 6, Book 2, ed. Riveros et al. (The Canadian Institute of Mining Metallurgy and Petroleum, 2003), 631-644.
7. R.M. Nadkarni and R.M. Kusic, "Hydrometallurgical Removal of Arsenic from Copper Concentrates," *Arsenic Metallurgy, Fundamentals and Applications*, ed. R.G. Reddy et al. (Warrendale, PA: TMS, 1988), 263-286.
8. R.M. Nadkarni et al., "Method for Removing Arsenic and Antimony from Copper Ores and Concentrates," *U.S. Patent 3,911,078*, 1975.
9. C.G. Anderson, L.G. Twidwell, "The Alkaline Sulfide Hydrometallurgical Separation, Recovery and Fixation of Tin, Arsenic, Antimony, Mercury and Gold," Lead and Zinc 2008 (The South African Institute of Mining and Metallurgy, 2008), 121-132.
10. S. Ubaldini et al., "Process Flow-Sheet for Gold and Antimony Recovery from Stibnite," Hydrometallurgy, 57 (2000), 187-199.
11. S.A Awe, A. Sandström, "Selective Leaching of Arsenic and Antimony from a Tetrahedrite Rich Complex Sulphide Concentrate using Alkaline Sulphide Solution", Minerals Engineering 23 (2010), 1227-1236.

Materials Processing Fundamentals
Edited by: Lifeng Zhang, Antoine Allanore, Cong Wang, James A. Yurko, and Justin Crapps
TMS (The Minerals, Metals & Materials Society), 2013

EFFECT OF CALCIUM ON THE SOLUBILITY OF ZINC OXIDE IN THE SODIUM HYDROXIDE SOLUTION

Chen Ai-liang, Xu Dong, Chen Xing-yu, Liu Xu-heng, Zhu Wei-xiong

(School of Metallurgical Science and Engineering, Central South University, Changsha 410083, China)

Abstract: Effect of calcium on the solubility of zinc oxide in the sodium hydroxide solution was studied in this paper. Results showed that ZnO equilibrium concentration was different at different temperatures when CaO or $Ca(OH)_2$ was added to the saturated sodium zincate solution of Na_2O-ZnO-H_2O system. ZnO equilibrium concentrations had little change with $Ca(OH)_2$ added at 100°C. But the concentrations became complicated at the temperature of 25°C, 50°C and 75°C, which decreased at $Na_2O \leq 20\%$ and then had no change at $Na_2O > 20\%$ with $Ca(OH)_2$ added at 25°C, 50°C and 75°C. The curves of the decreased equilibrium concentration formed arc shape when CaO or $Ca(OH)_2$ added. Moreover, the curves of the concentration added CaO more crooked than that added $Ca(OH)_2$. The dissociated Ca from $Ca(OH)_2$ bonded to oxygen of ZnO_2^{2-} and produced calcium zincate precipitation($Ca(Zn(OH)_3)_2 \cdot 2H_2O$), which leads to zinc loss in the solution and decreases the balance concentration of zinc oxide. Therefore, in order to avoid the loss of zinc when CaO or $Ca(OH)_2$ added to sodium zincate solution, Na_2O concentration would be controlled above 20% or temperature be about100°C.

Key Words: Calcium hydroxide, Solubility, Zinc oxide, Sodium zincate, Calcium zincate

1. Introduction

With the increase of China's demand of zinc source and the increasingly shortage of zinc sulfide, zinc oxide ore resources have gradually developed and utilized [1,2]. But hemimorphite zinc oxide ore is a kind of refractory minerals which is difficult to be treated by mineral processing and smelting [3-5]. Especially, the silica would be inevitably formed in acid aqueous process[2,6]. Based on these reasons, a new process that zinc oxide ore is treated by alkali leaching has been invented in our precious research. This method mainly contains the two processes of alkaline leaching zinc oxide ore and desilication using CaO or $Ca(OH)_2$ like the metallurgy of aluminum oxide [7-9]. During the process, there are a small amount of zinc (about 4% Zn) in the desiliconization residues in the leaching solution[10]. Moreover, alkaline leaching method was also widely used in other zinc oxide ore containing carbonate and silicates, such as smithsonite ($ZnCO_3$), hydrozincite ($2ZnCO_3 \cdot 3Zn(OH)_2$), zincite (ZnO) and willemite ($ZnSiO_4$)[2,5]. These treating process contain the technology of desilication using CaO or $Ca(OH)_2$ [10]. During the desilication process, more or less of zinc would lose. The reason of zinc loss was that $ZnFe_2O_4$ was produced by iron ball from the XRD pattern of the leach residue during the process of mechanical activation. But we found zinc also lose without iron ball by adding CaO or $Ca(OH)_2$. So, we should make it clear that how calcium affects the solubility of zinc oxide in the sodium hydroxide solution and how calcium reacts with zinc oxide or sodium zincate.

There was less reports on it except a few relevant references as follows. Johnson C. A. and Kersten M.[11]investigated the binding of Zn (II) to the cement mineral calcium silicate hydrate in a well-defined laboratory system. They found calcium silicate hydrate (Ca:Si 1:1) was synthesized by coprecipitation with varying contents of Zn(II). Dissolved Zn (II) concentrations were below the solubility of ZnO. They postulated that the mechanism is Zn (II) bounded in the calcium silicate hydrate matrix. The solubility of calcium zinc samples and ZnO were compared and the result showed that the solubility of calcium zincate is smaller than ZnO[12-14]. The solubility product, $\log K_{SP}$, of calcium zincate ($CaZn_2(OH)_6 \cdot 2H_2O$) was 43.9 at 25°C under the condition of an ionic strength of 0.1 M in a pH range between 11.4 and 12.7[11,15]. From the above, calcium zincate ($CaZn_2(OH)_6 \cdot 2H_2O$) has certain solubility but people did not interpret the effect of calcium on the stability of ZnO in the sodium hydroxide solution. Thus, in this paper, a series of experiments were used to clarify it as follows.

2. Experimental Procedure

2.1. Materials

All chemicals were AR (analytical reagent) -grade. Solutions were generally prepared using ultrapure water. Before use, the ultrapure water was boiled and then cooled to 20°C under argon. In order to prevent CO_2 contaminate the samples during preparation, all preparation procedures involving alkaline solutions and solids were performed in a glovebox.

2.2. Equipment And Procedure/ Analysis

HDPE flasks were used for the experiments. Firstly, sodium zincate solution was synthesized. A superfluous amount of ZnO was added into 500 ml of NaOH solution. Stirring was stopped till the dissolved ZnO concentration had no change for 24h, whose solid phase contain white powder of ZnO. The clear liquid, sodium zincate solution, was taken out for analysis. Successively, a superfluous amount of $Ca(OH)_2$ or CaO was added to the liquid-solid mixed system of Na_2O-ZnO-H_2O under continuous stirring. The bottle was sealed tightly and shaken for the next 48 h (the dissolved ZnO concentration had no change within 24h) on a rotary shaker at 200 rpm. After that, the clear liquid was decanted for analysis. The solid phase was washed with water on a vacuum filtration unit until the solution attained a pH value of about 7. The solid was freeze-dried and then examined by XRD. The concentration of ZnO and Na_2O in the clear liquid was respectively analyzed by complexometric titration and neutralization titration.

3. Results And Discussion

3.1 Effect Of Different Source Of Calcium On ZnO Equilibrium Concentration

Saturated sodium zincate solution was obtained from the clear liquid of the balance system of Na_2O-ZnO-H_2O at 25°C. And CaO or $Ca(OH)_2$ was added to the system of Na_2O-ZnO-H_2O and its dosage was 1.5 mole times as theoretical value. Fig.1 showed that the equilibrium concentration of zinc oxide changed when CaO or $Ca(OH)_2$ was added to the saturated sodium zincate solution of Na_2O-ZnO-H_2O system. The equilibrium concentration of zinc oxide decreased when Na_2O concentration was below 20% and the curve formed arc shape. In the Na_2O-ZnO-H_2O system, ZnO reacted with NaOH and produced dissolvable Na_2ZnO_2 as chemical reaction equations (1):

$$ZnO+2NaOH = Na_2ZnO_2 + H_2O \tag{1}$$

CaO reacted with H_2O and produced $Ca(OH)_2$ as equations (2):

$$CaO+H_2O = Ca(OH)_2 \tag{2}$$

The solubility product of $Ca(OH)_2$ was 5.5×10^{-6} and a small amount of $Ca(OH)_2$ can be dissolved to water.

Ca^{2+} ions can be ionized out from the dissolvable $Ca(OH)_2$ as equations (3):

$$Ca(OH)_{2(aq)} = Ca^{2+} + OH^- \quad (3)$$

The ionized Ca^{2+} ions reacted with ZnO_2^{2-} and produced calcium zincate ($Ca(Zn(OH)_3)_2{\bullet}2H_2O$) as equations (4):

$$2ZnO_2^{2-} + Ca^{2+} + H_2O = Ca(Zn(OH)_3)_2{\bullet}2H_2O + 4OH^- \quad (4)$$

Although calcium zincate has its own solubility in aqueous solution[9,11], its solubility (lgK_{sp}= 43.9, 25°C) is so lower in the alkaline solution that calcium zincate easily precipitate and zinc stays in the solid, which decreases the balance concentration of zinc oxide in the solution and leads to zinc loss during the process of zinc extraction. But the equilibrium of chemical reaction equations (3) moved to the left if OH^- increases. Then the concentration of Ca^{2+} ions decreased and the concentration of ZnO_2^{2-} that can be combined with Ca^{2+} ions accordingly decreased. So the equilibrium concentration of zinc oxide increased again with CaO or $Ca(OH)_2$ added. When Na_2O concentration was above 20%, the equilibrium concentration of zinc oxide had little change with the addition of CaO or $Ca(OH)_2$. This is because OH^- increases and the equilibrium of chemical reaction equations (3) moved to the left. Then the concentration of free Ca^{2+} ions decreased so much that there was too little Ca^{2+} in the solution. Little Ca^{2+} can reaction with ZnO_2^{2-} and ZnO concentration had no change with the addition of CaO or $Ca(OH)_2$, which leads to the three curves overlapped.

Moreover, Fig.1 also showed that the curve with the addition of CaO more crooked than that of $Ca(OH)_2$ when Na_2O concentration was less than 20%. This is because CaO reaction with H_2O and water consumed, which is equivalent to increase alkali concentration. That is to say, the concentration of OH^- increased and the equilibrium of chemical reaction equations (3) move to the left. The concentration of free Ca^{2+} ions that combined with ZnO_2^{2-} decreased. Nevertheless, $Ca(OH)_2$ didn't consume water, which don't affect the OH^- concentration by itself. Thus, the equilibrium concentration of zinc oxide added with $Ca(OH)_2$ was higher than that added with CaO.

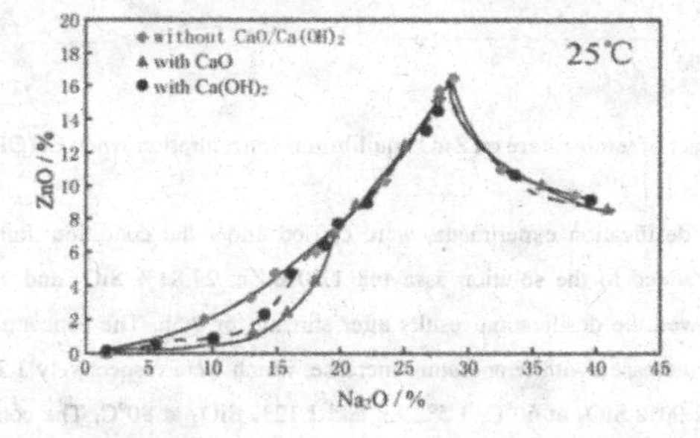

Fig.1. Effect of different calcium material on the composition of sodium zincate solution

3.2 Effect Of Temperature On ZnO Equilibrium Concentration

Xu, et al[16] reported the ZnO equilibrium concentration in the balance system of Na_2O-ZnO-H_2O at different temperature. Fig. 2 showed the change of ZnO equilibrium concentration at different temperature when $Ca(OH)_2$ was added to the balance system of Na_2O-ZnO-H_2O. It was shown that ZnO equilibrium concentration decreased at $Na_2O \leq 20\%$ with the addition of $Ca(OH)_2$ at the temperature of 25°C, 50°C and 75°C. The higher the temperature was, the less the concentration of ZnO decreased. The curve of decreased ZnO concentration at higher temperature crooked less than that at lower temperature. In other words, the reduced scope of ZnO equilibrium concentration decreased with temperature increase. At 100°C, ZnO equilibrium concentration had no change with $Ca(OH)_2$ added. This is because that calcium hydroxide has certain solubility in the water liquor. However, the solubility decreases with temperature increase. In reference [13] the value is 0.16g at 20°C, 0.15g at 30°C, 0.13g at 50°C, 0.105g at 70°C, 0.095g at 80°C and 0.07g at 100°C, respectively. Similarly, $Ca(OH)_2$ solubility in the alkaline system of Na_2O-ZnO-H_2O decreases with temperature increase. Thus, there is little zinc loss when $Ca(OH)_2$ was added to sodium zincate solution at 100°C.

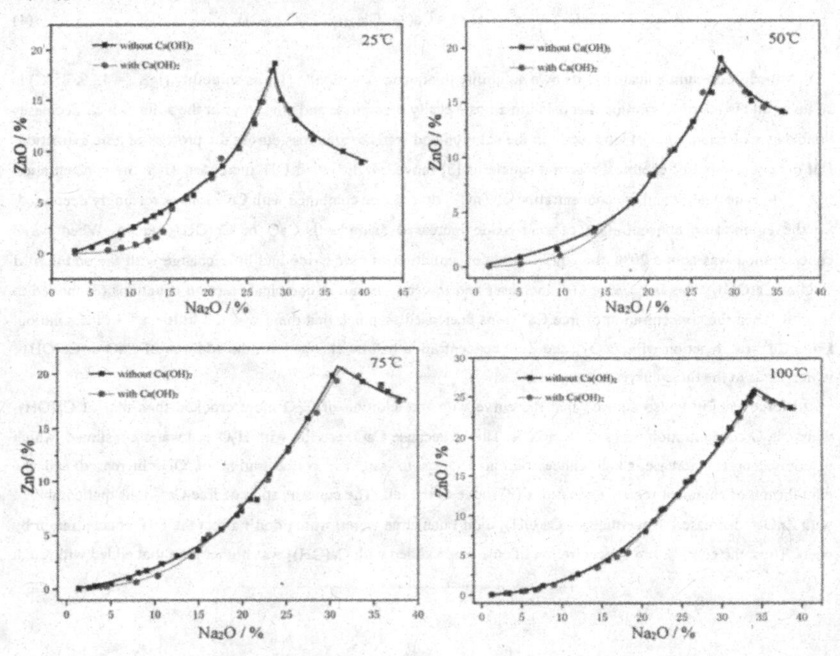

Fig. 2. Effect of temperature on ZnO equilibrium concentration when $Ca(OH)_2$ added

Additionally, a few desilication experiments were carried under the condition that $Ca(OH)_2$, 1.2 times as theoretical amount, was added to the solution assaying 1.50% Zn, 27.81% SiO_2 and Na_2O 9.81% at different temperature. Table I showed the desilication results after stirring for 0.5h. The concentration of SiO_2 obviously decreased but that of Zn increased with temperature increase, which were respectively 1.21% Zn and 4.22% SiO_2 at 25°C, 1.35% Zn and 2.90% SiO_2 at 60°C, 1.5% Zn and 1.12% SiO_2 at 80°C. The concentration of Na_2O was about 9.80 %, which had little change.

Table I Desilication results after stirring for 0.5h with the addition of $Ca(OH)_2$ (wt%)

Temperature/ °C	Zn	SiO_2	Na_2O
25	1.21	4.22	9.77
60	1.35	2.90	9.90
80	1.50	1.12	9.84

3.3 Description Of Calcium Zincate

Owing to abundant of ZnO and $Ca(OH)_2$ solid stayed in the solid phase of Na_2O-ZnO-H_2O system, calcium zincate produced are difficultly found by XRD when $Ca(OH)_2$ directly added to the equilibrium of Na_2O-ZnO-H_2O. This is because a lot of powder of ZnO and $Ca(OH)_2$ masked a small amount of calcium zincate in the XRD patterns. In order to gain more calcium zincate without other materials interference, we had to add directly $Ca(OH)_2$ to sodium zincate solution at 25°C and stirred for 48h. Fig. 3 showed the XRD patterns of the solid phase. It was seen that the solid phase was basically calcium zincate ($Ca(Zn(OH)_3)_2 \cdot 2H_2O$) and $Ca(OH)_2$. This exactly explained Ca^{2+} ions can reaction with ZnO_2^{2-} and produce calcium zincate precipitate, which further confirmed the experiment results of mechanochemically leaching refractory hemimorphite that zinc would lose when $Ca(OH)_2$ added to the leaching solution for silicon removal[10].

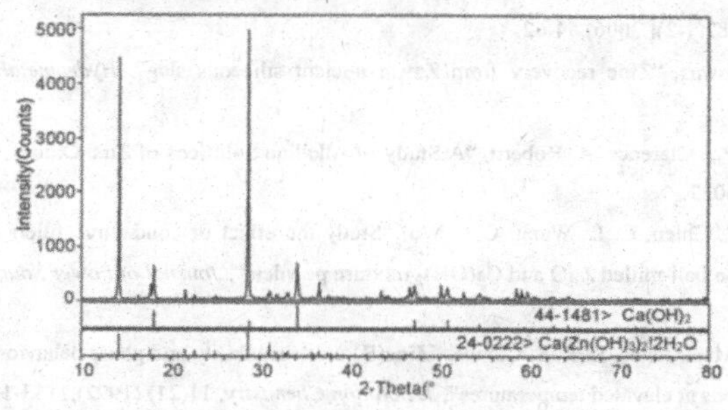

Fig. 3. XRD patterns of the solid phase of sodium zincate solution when $Ca(OH)_2$ added

Moreover, Xavier et al [12] showed the crystalline $Ca(Zn(OH)_3)_2 \cdot 2H_2O$ structure. $Ca(Zn(OH)_3)_2 \cdot 2H_2O$ is characterized by the presence of Ca atoms bonded to six oxygen, resulting in $[CaO_6]$ clusters with octahedral coordination. On the other hand, Zn atoms are coordinated to four oxygens in a tetrahedral configuration, i.e., forming $[ZnO_4]$ clusters. H_2O molecules are bonded by hydrogen bridges to the hydroxyl groups $[OH^-]$ for the charge balance of the lattice. This just explained that dissociated Ca from $Ca(OH)_2$ bonded to oxygens of ZnO_2^{2-} and produced calcium zincate, which leads zinc to lose from sodium zincate solution.

4. Conclusions

The solubility of zinc oxide decreased at $Na_2O \leq 20\%$ when CaO or $Ca(OH)_2$ was added into the saturated sodium zincate solution of Na_2O-ZnO-H_2O system and the curve formed arc shape. The higher the temperature was, the less the concentration of ZnO decreased. ZnO equilibrium concentration had no change with $Ca(OH)_2$ added at 100°C. The dissociated Ca from $Ca(OH)_2$ bonded to oxygens of ZnO_2^{2-} and produced calcium zincate precipitation($Ca(Zn(OH)_3)_2 \cdot 2H_2O$) which lead to zinc loss from sodium zincate solution and zinc oxide balance of the concentration decreased. Thus, the desilication condition would control Thus, the desilication condition would control the conditions that Na_2O concentration was above 20% or the temperature was 100°C with the addition of $Ca(OH)_2$ to sodium zincate solution.

Acknowledgements

This work was financially supported by the National Basic Research Program of China (No. 2007CB613603).

References

[1] R.X.Wang, M. T.Tang, S.H.Yang, W.H. Zhang, C.B.Tang, J. He, J.G.Yang, " Leaching kinetics of low grade zinc oxide ore in NH_3-NH_4Cl-H_2O system", *J. Cent. South Univ Techno.*, 15(2008), 679-683.

[2] A.L.Chen, Z.W.Zhao, X.J. Jia, "Alkaline leaching Zn and its concomitant metals from refractory hemimorphite zinc oxide ore", *Hydrometallurgy*, 97(3-4)(2009), 228-232.

[3] E.A.Abdel-aal, "Kinetics of sulfuric acid leaching of low-grade zinc silicate ore", *Hydrometallurgy*, 39(2) (2000),247-254.

[4] Z.L. Yin, Z.Y. Ding, H.P. Hu, K. Liu, Q.Y. Chen, "Dissolution of zinc silicate (hemimorphite) with ammonia–ammonium chloride solution", *Hydrometallurgy*, 103(1-4) (2010), 215-220.

[5] S. Espiari, F. Rashchi, S.K. Sadrnezhad, "Hydrometallurgical treatment of tailings with high zinc content", *Hydrometallurgy*, 82(1-2)(2006),54-62.

[6] R. Kumar, A.Biswas, "Zinc recovery from Zawar ancient siliceous slag", *Hydrometallurgy*, 15(3)(1986), 267-280.

[7] P. D. Thedford, P.J. Clarence, A. Robertt, "A Study of Alkaline Solutions of Zinc Oxide", *J. Am. Chem. Soc.*, 76(1954), 6022-6023.

[8] C.C. Yang, W.C. Chien, C. L. Wang, C.Y. Wu, "Study the effect of conductive fillers on a secondary Zn electrode based on ball-milled ZnO and $Ca(OH)_2$ mixture powders", *Journal of Power Sources*, 172 (1) (2007), 435-445

[9] S. E. Ziemniak, M. E. Jones, K. E. S. Combs, "Zinc(II) oxide solubility and phase behavior in aqueous sodium phosphate solutions at elevated temperatures", *J. Solution Chemistry*, 11(21) (1992),1153-1196.

[10] Z.W. Zhao, S. Long, A.L. Chen, G.S. Huo, H.G. Li, X.J. Jia, X.Y. Chen, "Mechanochemical leaching of refractory zinc silicate (hemimorphite) in alkaline solution",*Hydrometallurgy*, 99(3-4) (2009), 255-258.

[11] C.A. Johnson, M. Kersten, "Solubility of Zn(II) in Association with Calcium Silicate Hydrates in Alkaline Solutions", *Environmental Science & Technology*, 13(33) (1999), 2296-2298.

[12] C.S. Xavier, J.C. Sczancoski, L.S. Cavalcante, C.O. Paiva-Santos, J.A.Varela, E. Longo, M.S. Li, "A new processing method of $CaZn_2(OH)_6{\cdot}2H_2O$ powders: Photoluminescence and growth mechanism", *Solid State Sciences*, 11(12)(2009),2173-2179

[13] F. Zhao, "The study of the method and mechanism their formation of Calcium Zincate and zinc dendrites", Zhengzhou University, 2010.

[14] J. X. Yu, H.X. Yang, X.P. Ai, X.M. Zhu, "A study of calcium zincate as negative electrode materials for secondary batteries", *Journal of Power Sources*, 103 (2) (2001), 93-97

[15] F. Ziegler, C.A. Johnson, "The solubility of calcium zincate ($CaZn_2(OH)_6{\cdot}2H_2O$)", *Cement and Concrete Research*, 31 (9)(2001),1327-1332

[16] D. Xu, Z.W. Zhao, A.L. Chen, X.H. Liu, X.Y. Chen, "Leaching zinc oxide in alkaline solutions", *Journal of University of Science and Technology Beijing*, 33(7) (2011), 812-816.

Supplementary data

Attached Table Ⅰ Original Data of Fig.1

Effect of different calcium material on the composition of sodium zincate solution (wt/%)

Without CaO or $Ca(OH)_2$		With CaO		With $Ca(OH)_2$	
Na_2O	ZnO	Na_2O	ZnO	Na_2O	ZnO
2.98	0.25	2.98	0.04	2.93	0.22
6.13	0.83	6.02	0.77	6.12	0.77
8.76	1.86	10.67	0.85	10.81	0.98
13.91	4.00	13.25	1.55	14.32	2.35
15.31	4.70	15.86	2.87	15.93	4.69
18.14	5.70	18.09	4.88	19.04	5.82
20.48	7.74	20.55	7.39	20.55	7.89
24.07	10.27	21.73	9.28	22.13	9.38
28.67	18.61	25.95	12.78	26.39	13.20
29.47	15.01	27.59	15.77	27.61	15.07
33.57	11.11	28.25	17.85	33.53	11.00
39.71	8.86	35.53	10.23	39.75	9.56
		40.05	9.23		

Attached Table Ⅱ Part of Original Data of Fig.2

Effect of temperature on ZnO equilibrium concentration when $Ca(OH)_2$ added (wt/%)

25 °C		50 °C		75 °C		100 °C	
Na_2O	ZnO	Na_2O	ZnO	Na_2O	ZnO	Na_2O	ZnO
2.99	0.04	1.54	0.04	3.86	0.22	3.00	0.20
7.15	0.32	5.18	0.28	4.53	0.24	5.48	0.50
8.88	0.54	9.77	1.06	7.88	0.61	7.14	0.80
10.67	0.86	13.64	3.21	10.32	1.22	8.96	1.23
12.24	1.24	18.11	5.90	14.23	3.03	11.15	2.13
13.26	1.56	20.33	8.40	16.16	5.04	14.42	3.45
14.19	2.10	22.67	10.54	19.83	7.63	16.86	4.78
15.36	4.37	24.97	12.72	22.91	10.13	18.23	5.33
17.17	5.60	26.45	15.88	25.20	12.92	21.46	9.34
20.56	7.39	28.45	17.72	28.47	16.39	24.53	12.61
21.74	9.28	31.33	15.96	30.78	19.33	26.07	14.41
24.04	10.37	33.49	14.50	33.25	19.75	27.68	16.66
25.95	13.00			35.78	18.64	29.85	19.91
27.10	14.77			37.86	17.44	31.11	21.01
28.25	17.85			--	--	32.23	22.53
30.53	14.00					33.59	24.38
33.26	11.53					35.43	24.90
						36.19	24.45
						37.48	24.03

Note: Xu, et al, (2011) listed ZnO equilibrium concentration without the addition of $Ca(OH)_2$ at different temperature, which need not repeat again here.

MATERIALS PROCESSING FUNDAMENTALS

Poster Session

Materials Processing Fundamentals
Edited by: Lifeng Zhang, Antoine Allanore, Cong Wang, James A. Yurko, and Justin Crapps
TMS (The Minerals, Metals & Materials Society), 2013

LUMINESCENCE ENHANCEMENT OF SKY-BLUE ZnS:Tm PHOSPHOR BY PROMOTER DOPING

Su-Hua Yang, Yin-Hsuan Ling

Department of Electronic Engineering, National Kaohsiung University of Applied Sciences, Kaohsiung, Taiwan, R.O.C.

Keywords: Phosphor, Luminescence, Promoter, Energy transfer

Abstract

Luminescence of ZnS:Tm phosphor was enhanced by the addition of KCl promoter. High synthesis temperature and TmF_3 concentration promoted the substitution of Tm^{3+} ions for Zn^{2+} ions, leading to the growth of Tm_2S_3 phase. The ionic defects were increased with the increase of Tm^{3+} concentration, which influenced the luminescence characteristics of phosphor when the energy transferred to the defects was not effectively cascaded to the luminescence center. KCl promoter doping facilitated the growth of hexagonal phase and decreased the phase transformation temperature of ZnS. The emission pattern and peaks of the ZnS:Tm,KCl phosphor were similar to those of ZnS:Tm phosphor. This shows that the variation in the ligand field of Tm^{3+} because of KCl doping was insignificant. The luminescence center concentration, the excitation energy absorption, the effective energy transfer, and the charge balance were improved by the addition of KCl promoter. Consequently, the luminescence of phosphor was effectively enhanced.

Introduction

ZnS-based phosphors are important II-VI group luminescence materials and are suitable for applications related to electroluminescence (EL) device [1] and field emission display (FED) [2]. Usually, transition metals and rare-earth materials are doped to modify the emission color of the ZnS-based phosphor. Tm is one of the preferred choices for the luminescence centers of the blue ZnS phosphor. The Tm^{3+} ion has a $4f^{12}$ electron configuration, and its energy states can be perturbed and shifted by the variation of crystal field.

In order to improve the luminescence characteristic of phosphor, donor-acceptor pairs are usually formed [3] and a halide sensitizer is generally added [4]. According to the literatures, uniform grains were obtained by chlorine flux doping and large grains with irregular morphology were obtained when synthesized with fluoride fluxes [5]. In addition, the doped alkaline and halide ions substantially and interstitially present in the host lattice will induce a variation in the luminescence of phosphors.

The objective of this study is to prepare sky-blue ZnS:Tm phosphor by solid state reaction and doped it with a KCl promoter to enhance the luminescence of the phosphor. The crystalline and luminescence properties of the phosphors were investigated in detail.

Experimental Procedure

ZnS (99.99%) and TmF_3 (99.9%) were used as source materials to synthesize ZnS:Tm phosphor using the conventional solid-state reaction method. The TmF_3 doping concentration was varied from 1 to 7 mol%. Moreover, in order to improve the luminescence of phosphor, KCl promoter was added, and its concentration was changed in the range of 1–7 mole%. To synthesize phosphor, first, the source powders were weighed in stoichiometric ratios, mixed with deionized water, and milled for 24 h. Subsequently, the mixed solution was dried in an oven at temperature of 80°C for 8 h. The resulting dry powder was then sintered in a tube furnace at temperatures of 700–1200°C for 0.5–2.5 h in a N_2 atmosphere. The flow rate of N_2 was set at 20 sccm. Redox reactions proceeded during sintering. After synthesis, the furnace was cooled naturally to the room temperature. The synthesized phosphor was screen-printed on an indium-tin oxide (ITO)-coated glass to measure the luminescence property.

The surface morphology of the phosphor was evaluated with a Philips XL-40 FEG scanning electron microscope (SEM). The elemental compositions were determined using an energy dispersive X-ray spectroscope (EDX) attached to the SEM system. Furthermore, the crystalline phase was analyzed using a Siemens D5000 X-ray diffraction (XRD) system with Cu Kα radiation ($\lambda = 0.1541$ nm) and recorded in the range of $2\theta = 10°–70°$ with an increment of 0.1°. A Hitachi F-7000 fluorescence spectrophotometer equipped with a 150 W xenon lamp and a monochrometer was used to evaluate the photoluminescence (PL) and PL excitation spectra of the phosphor. The excitation wavelength for PL measurement was set at 325 nm, and the emission wavelength was scanned from 300 to 700 nm. A 6 W (1350 $\mu W/cm^2$) ultraviolet (UV) lamp was used to measure the luminance intensity of the phosphor, and the Commission Internationale de l'Eclairage (CIE) coordinates were measured using the Minolta chroma meter CS-100A.

Results and Discussion

The crystalline property of the phosphor prepared by a solid-state reaction is significantly influenced by the synthesis parameters, especially in the case of a phosphor that crystallizes to form different structures under different synthesis conditions (e.g., ZnS:Tm phosphor). The as-synthesized ZnS:Tm had a cubic (C) structure when it was prepared at low synthesis temperatures (< 1000°C), and it had a hexagonal (H) structure when preparing at higher temperatures, which was consistent with the phase transformation temperature of 1020°C for the ZnS-based phosphors reported in the literatures [6]. The average particle size of the ZnS:Tm phosphor measured using SEM was approximately 1–2 μm.

For the ZnS:Tm phosphor, it is observed that the growth of hexagonal phase was increased with the increase of synthesis temperature, in the meantime, the growth of the Tm_2S_3 phase was enhanced as well. This growth can be attributed to the low solid solubility of Tm^{3+} ions in the ZnS:Tm lattice, and the substitution of Tm^{3+} ions for Zn^{2+} ions and the reaction of Tm^{3+} ions with S^{2-} ions were enhanced at higher synthesis temperature. The substitution resulted in an increase in the ionic defects in the ZnS:Tm, which influenced the luminescence characteristics of the phosphor when the energy transferred to the defects was not effectively cascaded to the luminescence center [7].

In order to expedite the solid-state reaction at low synthesis temperatures, alkali halide KCl promoter was doped because it has an unique characteristic of low melting point, approximately 771°C [8]. Figures 1(a)–(b) show the crystalline analyses of the phosphors synthesized at 1000°C for 1 h; Fig. 2 exhibits the FWHM values of XRD for ZnS:Tm,KCl doped with different concentrations of KCl promoter, where the doping concentration of Tm was 3 mol%.

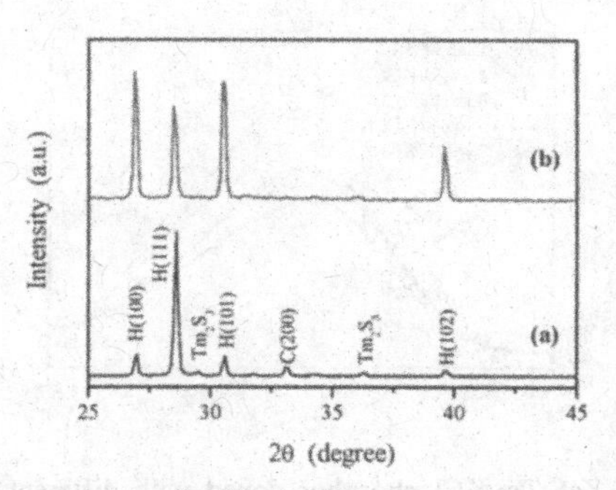

Figure 1. XRD patterns of the (a) ZnS:Tm phosphor doped with 1 mol% TmF_3 and (b) ZnS:Tm,KCl phosphor doped with 3 mol% TmF_3 and 7 mol% KCl.

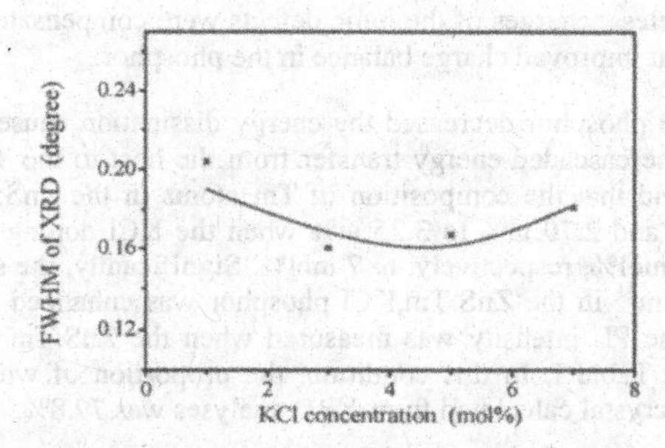

Figure 2. FWHM values of XRD for ZnS:Tm,KCl doped with different concentrations of KCl promoter.

The proportions of wurtzite to sphalerite phases, calculated from XRD analyses [9], in the ZnS:Tm,KCl crystal were 28.1%, 49.3%, 79.8%, and 90.9% for KCl doping concentrations of 1, 3, 5, and 7 mol%, respectively. Hence, the KCl promoter doping facilitated the growth of hexagonal phase. It is expected that the ZnS:Tm,KCl phosphor will be almost completely transformed to the hexagonal phase as the KCl concentration is increased to more than 7 mol%, and thus, doping with KCl could decrease the phase transformation temperature of ZnS (which is typically 1020°C). The proportion of hexagonal to cubic phases in the crystal will influence the luminance property of phosphor [6].

The emission pattern and peaks of the ZnS:Tm,KCl phosphor were similar to those of ZnS:Tm phosphor, as shown in Fig. 3. The emission peaks, located at approximately 460, 470, and 480 nm, were attributed to the $1D2 \rightarrow {}^3F_4$ and ${}^1G_4 \rightarrow {}^3H_6$ electron transitions of the Tm^{3+} ions [10]. The electrons were also transited in the defect states, leading to a wide-band emission. The emission pattern and peaks of the ZnS:Tm,KCl phosphor were similar to those of ZnS:Tm phosphor. This shows that the variation in the ligand field of Tm^{3+} in the crystal because of KCl

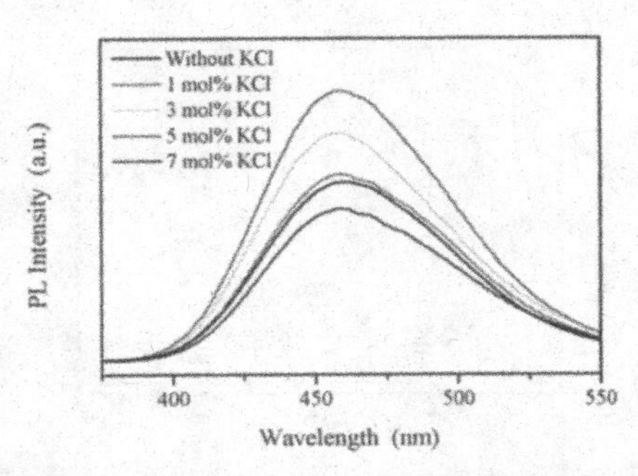

Figure 3. PL spectra of ZnS:Tm,KCl phosphor doped with different concentrations of KCl promoter.

doping was small. Nevertheless, charges of the ionic defects were compensated; thus, KCl acted as a charge compensator that improved charge balance in the phosphor.

The balanced charge in the phosphor decreased the energy dissipation caused by ionic defects and effectively enhanced the cascaded energy transfer from the host to the Tm^{3+} luminescence center. Further, it was found that the composition of Tm atoms in the ZnS:Tm,KCl phosphor increased from 1.93, 2.52, and 2.70 at% to 3.25 at% when the KCl doping concentration was increased from 1, 3, and 5 mol%, respectively, to 7 mol%. Significantly, the solid solubility and the oscillator strength of Tm^{3+} in the ZnS:Tm,KCl phosphor was enhanced by KCl doping. A 1.5-fold enhancement in the PL intensity was measured when the ZnS:Tm was doped with 5 mol% of KCl, as listed in Table I. In this condition, the proportion of wurtzite to sphalerite phases in the ZnS:Tm,KCl crystal calculated from XRD analyses was 79.8%.

Table I. Relative PL intensity of ZnS:Tm,KCl phosphor doped with different concentrations of KCl promoter

KCl concern. (mol%)	0	1	3	5	7
Relative PL intensity	1	1.05	1.28	1.51	0.86

Figure 4 shows the PL excitation (PLE) spectra, monitored at emission wavelength of 460 nm, of the ZnS:Tm phosphor with and without KCl doping. Two excitation energies at wavelengths of approximately 338 and 373 nm were absorbed, which were attributed to the intrinsic absorption of ZnS host (bandgap = 3.67 eV) [9] and the electron absorption transition in ${}^3H_6 \rightarrow {}^1D_2$ of Tm^{3+} activator, respectively. Hence, the excitation mechanism of the Tm^{3+} activator was attributed to the energy absorption by the Tm^{3+} activator along with nonradiative energy transfer from the ZnS host to the Tm^{3+} activator. The energy absorption was significantly enhanced by doping with the KCl promoter, attributing to the decrease in energy absorption by the defect states. It was also found that the host-to-activator absorption ratio was increased from 0.77 to 1.70 when the ZnS:Tm phosphor was doped with KCl. The improved host absorption increased the energy transfer to the Tm^{3+} activator.

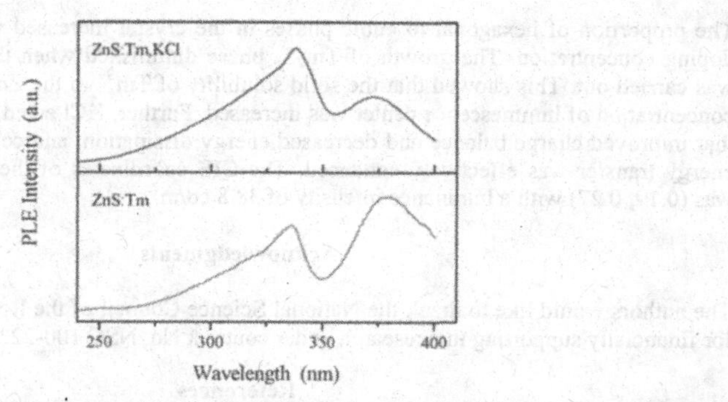

Figure 4. PLE spectra of the ZnS:Tm and ZnS:Tm,KCl phosphors, which were monitored at emission wavelength of 460 nm,.

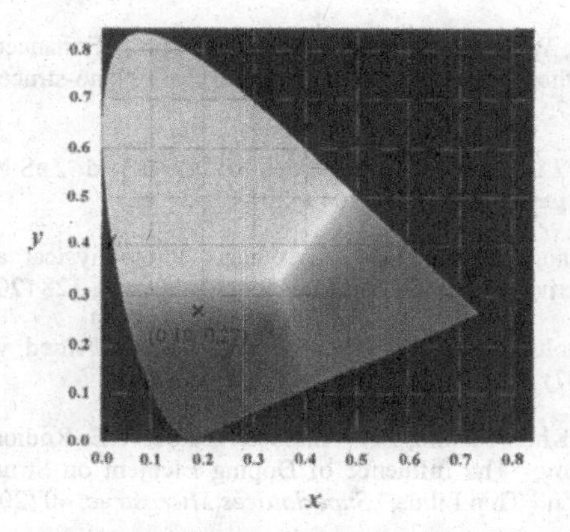

Figure 5. CIE coordinates of the ZnS:Tm,KCl phosphor.

From the investigations above, it was found that the luminescence center concentration, the excitation energy absorption, the effective energy transfer, and the charge balance in the phosphor were improved by KCl promoter doping, as a result, an increased PL intensity was measured. Figure 5 shows the CIE coordinates of the ZnS:Tm,KCl phosphor doped with 3 mol% TmF_3 and 5 mol% KCl and synthesized at 1000°C for 1 h. It was sky-blue emission; a luminance intensity of 38.8 cd/m^2 with CIE coordinates of (0.19, 0.27) was measured under the excitation of a 6 W (1350 μW/cm^2) ultraviolet lamp.

Conclusions

In this study, it is found that the ionic defects influenced the luminescence characteristics of the phosphor when the energy transferred to the defects was not effectively cascaded to the luminescence center.

The proportion of hexagonal to cubic phases in the crystal increased with the increase in KCl doping concentration. The growth of Tm_2S_3 phase diminished when the KCl promoter doping was carried out. This showed that the solid solubility of Tm^{3+} in the ZnS was improved and the concentration of luminescence center was increased. Further, KCl acted as a charge compensator that improved charge balance and decreased energy dissipation, and consequently, the cascaded energy transfer was effectively enhanced. The CIE coordinates of the ZnS:Tm,KCl, phosphor was (0.19, 0.27) with a luminance intensity of 38.8 cd/m^2.

Acknowledgments

The authors would like to thank the National Science Council of the Republic of China, Taiwan, for financially supporting this research under contract No. NSC 100-2221-E-151-031.

References

[1] D.K. Sinha and Y.N. Mohapatra, "Charge Trapping and Electroluminescence at Quantum Dots Embedded in a Polymer Matrix," *Org Electron*, 13 (2012) 1456–1462.

[2] S.I. Ahn, S.E. Lee, W.H. Le, S.H. Park, and K.C. Choi, "Enhanced Cathodoluminescence from ZnS Based Phosphors with Self-assembled ZnO Nano-structures," *Chem Phys Lett*, 493 (2010) 113–117.

[3] G. Murugadoss, "Luminescence Properties of Co-doped ZnS:Ni,Mn and ZnS:Cu,Cd Nanoparticles," *J Lumin*, 132 (2012) 2043–2048.

[4] W. Wang, F. Huang, Y. Xia, and A. Wang, "Photophysical and Photoluminescence Properties of Co-activated ZnS:Cu,Mn Phosphors," *J Lumin*, 128 (2008) 610–614.

[5] T.-P. Tang, "Photoluminescence of ZnS:Tb Phosphors Fritted with Different Fluxes," *Ceram Int*, 33 (2007) 1251–1254.

[6] T. Kryshtab, V.S. Khomchenko, J.A. Andraca-Adame, V.E. Rodionov, V.B. Khachatryan, and Y.A. Tzyrkunov, "The Influence of Doping Element on Structural and Luminescent Characteristics of ZnS Thin Films," *Superlattices Microstruc*, 40 (2006) 651–656.

[7] B. Dong, L. Cao, G. Su, and W. Liu, "Synthesis and Characterization of Mn Doped ZnS d-dots with Controllable Dual-color Emissions," *J Colloid Interface Sci*, 367 (2012) 178–182.

[8] H.-L. Li, Z.-N. Du, G.-L. Wang, and Y.-C. Zhang, "Low Temperature Molten Salt Synthesis of SrTiO3 Submicron Crystallites and Nanocrystals in the Eutectic NaCl–KCl," *Mater Lett*, 64 (2010) 431–434.

[9] S.-H. Yang, Y.-J. Lial, N.-J. Cheng, and Y.-H. Ling, "Preparation and Characteristics of Yellow ZnS:Mn,Ce Phosphor," *J Alloys Compd*, 489 (2010) 689–693.

[10] W. Qin, G. Qin, Y. Chung, Y.-I Lee, C. Kim, K. Jang, "Strong Enhancements of Infrared-to-Ultraviolet Upconversion Emissions in Yb^{3+} and Tm^{3+} Co-Doped Sub-Micron Fluoride Particles Prepared by Using Pulsed Laser Ablation," *J Korean Phys Soc*, 44 (2004) 925–929.

Materials Processing Fundamentals
Edited by: Lifeng Zhang, Antoine Allanore, Cong Wang, James A. Yurko, and Justin Crapps
TMS (The Minerals, Metals & Materials Society), 2013

Effect of Thermal History on the Hot Ductility and Fracture Mechanisms of Low Carbon Peritectic Steel

Zhihua Dong, Dengfu Chen*, Xing Zhang, Mujun Long

Laboratory of Metallurgy and Materials, College of Materials Science and Engineering, Chongqing University, Chongqing 400030, China

Keywords: Hot ductility, thermal history, fracture mechanism, continuous casting.

Abstract

Hot tensile test was employed to measure the hot ductility of peritectic steel in the paper. In order to understand the effect of thermal history on slab hot ductility, tensile samples were subjected to two different thermal cycles prior to the deformation, including the experience of cooling to testing temperatures directly (C-cycle) and reheating to testing temperatures after cooling (R-cycle). The results indicated that the R-cycle led to lower yield strength in high temperature region as well as lower tensile strength in the whole temperature region. With the thermal history from C-cycle to R-cycle, the ductility trough was deeper and occurred at higher temperature zone. The fracture surfaces and microstructures of the specimens were examined at the embrittlement zone. Results indicated that equiaxed (Fe,Mn)S particles as well as pro-eutectoid ferrite were precipitating at γ gain boundary, leading to inter-granular brittle fracture. The precipitations were enhanced by R-cycle at high temperature.

Introduction

Cracking in continuously cast steel slabs has been one of the main problems in casting for decades [1]. The majority of cracks found in slabs are known to be initiated in the mushy or embrittlement zone, both regions being characterized by low ductility. In order to manufacture excellent products, the relation of high-temperature mechanical properties examined on the isothermal tensile test and the occurrence of cracks, especially in the bending and straightening operations, has been extensively studied [2-5]. The findings have mitigated the cracking problem to some degree.

The thermal cycling to the test temperature and cooling rate have been found to largely influence the ductility of casting steels. Mintz et al. [6] found that the cyclic oscillation of temperature led to a wider and deeper ductility trough for C-Mn-Al-Nb steel, and this was worsened by increasing the amplitude of oscillation. According to the experiments of Cardoso et al. [7] on a C-Mn-Al steel, the temperature oscillation prior to isothermal tensile encouraged the precipitation of AlN, therefore, the ductility trough apparently moved to high temperature zone. Likewise, on the basis of studying on weather resistant steel, Chen et al. [8] concluded that the fasting cooling rate likely caused the ductility trough to extend to low temperature region, resulting in a wide embrittlement zone. The finds were in reasonable agreement with the results of Mintz [9]. In the regression equation of Ar3 temperature, the increase of cooling rate

*Corresponding author. Dengfu Chen is a professor of College of Materials Science and Engineering, Chongqing University of Chongqing, China. E-mail: chendfu@cqu.edu.cn.

obviously decreased the temperature of Ar3, which is fundamentally important to the ductility restoration at low temperature zone for casting steels.

However, the correlation between laboratory and industrial results is far from precise. Part of the problem may be because that many tests, in fact, do not accurately simulate the thermal history occurring in continuous casting prior to bending or straightening. The present work focuses on investigating the influence of thermal history on low carbon peritectic steel which is sensitive to cracks for coarse grains formed in mold. At embrittlement zone, the high-temperature mechanical properties were measured after directly cooling thermal cycle (C-cycle) or reheating thermal cycle (R-cycle) respectively. The fracture surface and microstructures were also examined to understand the mechanisms of cracking.

Experimental

Materials

The chemical composition of the steel studied is given in Table I. The steel was supplied as pieces sectioned from continuously cast slab. C content strongly affects the cracking susceptibility of continuously cast slabs. According to Maehara et al. [10], the maximum susceptibility of cracking was in 0.10~0.15% C region due to its coarse grains of austenite.

Table I. Chemical composition of the steel (mass %)

C	Si	Mn	P	S	Alt
0.10-0.17	0.10-0.35	0.80-1.10	≤0.030	≤0.030	≥0.008

Experimental procedures

Cylindrical tensile test specimens of 10mm in diameter and 120mm length were machined with the specimens axes parallel to the casting direction and fixed/loose side. All the hot tensile tests were performed on the Gleeble 1500-D thermo-mechanical simulator. Specimens were heated in a vacuum atmosphere of ~1.33×10^{-5}Mpa at the rate of 20℃/s. After solution at 1300℃ for 60s, two types of thermal cycles were used to test the specimens, shown in Figure 1. Tensile tests were carried out at a constant strain rate of $5\times10^{-3}s^{-1}$ until complete failure. The reduction of area, yield strength and tensile strength were taken as the ductility criteria and strength criteria respectively.

The heat treatments prior to testing (Figure 1) were as follows:

(1) Cooling thermal cycle (C-cycle): after solution treatment at 1300℃, the specimens were cooled down to the desired test temperature directly at a cooling rate similar to that of a continuously cast slab (100℃/min), and then holding for 30s prior to deformation.

(2) Reheating thermal cycle (R-cycle): the specimens were firstly cooled down to a low temperature (T_{min}) at a same cooling rate with the C-cycle, after a hold time of 30s, specimens were reheated to the test temperature (T_{test}) from T_{min} at the heating rate of 60℃/min. To simulate the oscillation of temperature occurring in continuous casting, T_{min} was taken as 100℃ lower than T_{test} at each test. A hold time of 30s before deformation was also applied to R-cycle.

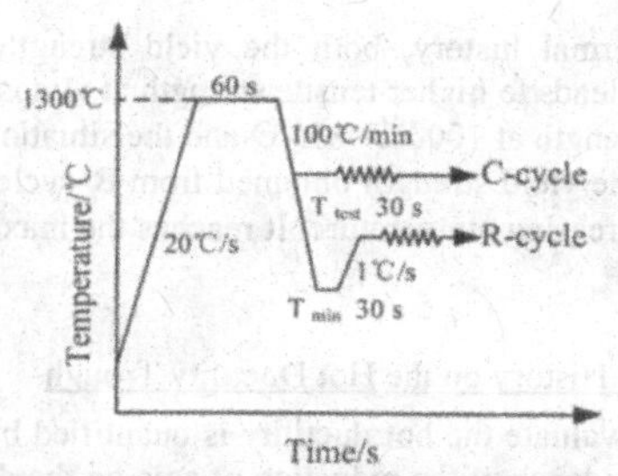

Figure 1. Temperature time schedules followed with the C-Mn peritectic steel

After fracture, fractography was performed by using a scanning electron microscope (TESCAN TS 5136XM). Cross sections parallel to the fracture surface were taken from the broken specimens and prepared for metallographic examination in the usual manner. The microstructure was revealed by a 3% nital solution.

Results and Discussions

Influence of Prior Thermal History on the Strength

The variation of strength for C-Mn peritectic steel as a function of temperature is shown in Figure 2. The results indicate that both temperature and thermal history strongly affect the strength of studied steel. With decreasing the temperature from 1000℃ to 700℃, the yield strength and tensile strength for the peritectic steel subjected to C-cycle prior to deformation gradually increase from 29Mpa and 52Mpa to 85Mpa and 131Mpa, respectively. During the growth, the rising rate of yield strength as well as tensile strength apparently varies at some typical temperatures (e.g. 800℃). This behavior is considered to associate with the austenite-ferrite transformation during cooling [11,12]. The dependence of strength obtained from R-cycle on temperature is similar to that of C-cycle. The yield strength and tensile strength rapidly increase from 27Mpa and 50Mpa to 103Mp and 138Mpa, respectively, when the temperature drops from 1000℃ to700℃.

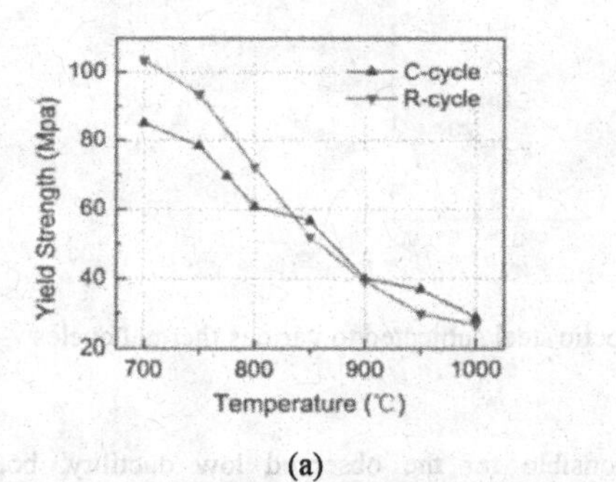

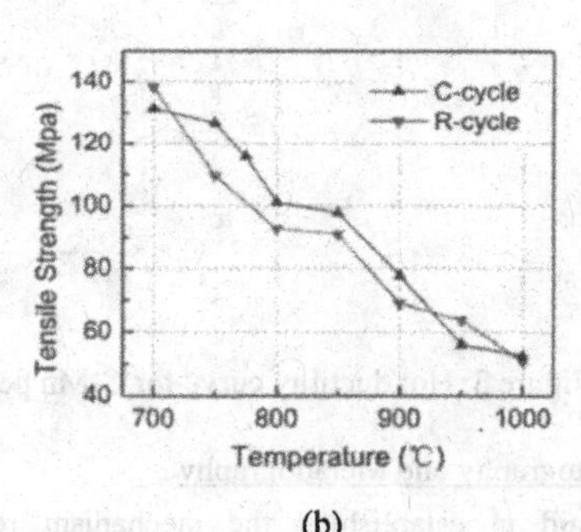

(a) (b)

Figure 2. Strength curve of C-Mn peritectic steel: (a) yield strength; (b) tensile strength

With the variation of thermal history, both the yield strength and tensile strength change apparently. While C-cycle leads to higher tensile strength at almost the whole temperature range, R-cycle has lower yield strength at 1000℃~835℃ and the situation is subverted at 835℃~700℃, where the gap between the yield strength obtained from R-cycle and that of C-cycle is getting bigger and bigger with decreasing temperature. It reaches the maximum at 700℃ with a value of 19Mpa approximately.

Influence of Prior Thermal History on the Hot Ductility Trough

A more effective way to evaluate the hot ductility is quantified by the reduction of area (%RA) of the specimen. The dependence of the reduction of area on the deformation temperature for the C-Mn peritectic steel is presented in Figure 3.

As can be seen from Figure 3, each curve illustrates a hot ductility trough, however, the depth and the temperature zone of the trough vary with thermal history obviously. In terms of the rule that the zone with the reduction of area less than 60% is named the hot embrittlement zone for continuously cat slab [13,14], the hot embrittlement zone for the peritectic steel subjected to C-cycle is 715℃~775℃ whit minimum ductility value of approximately 48% at 750℃. When the thermal history changes to R-cycle, the hot embrittlement zone apparently shifts to higher temperature region. The hot ductility trough varied to 766℃~826℃ with the minimum ductility growing up to 800℃ at a value of 34%. However, little difference between the width for the hot ductility trough obtained from C-cycle and R-cycle is detected, both of them steady at 60℃ approximately. At high temperature range from 850℃ to 1000 ℃, the reduction of area obtained after R-cycle is closed to that of C-cycle, which is more than 80%.

The findings are of importance for the design of parameters such as unbending position and temperature at the secondary cooling zone to enhance the quality of casting steels. When the temperature oscillation frequently occurs in continuous casting, the unbending temperature should be controlled at higher temperature zone.

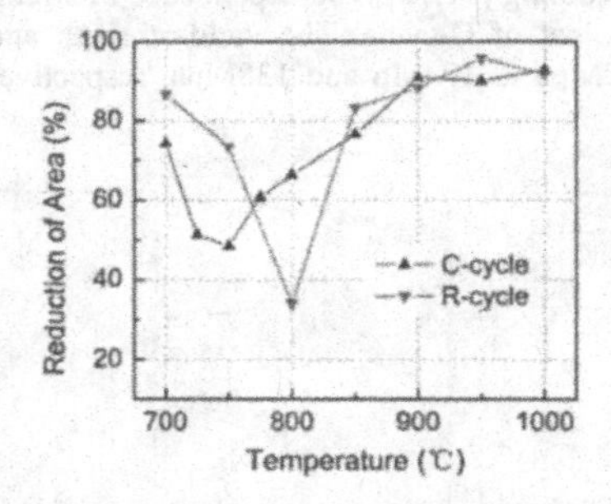

Figure 3. Hot ductility curve for C-Mn peritectic steel subjected to various thermal cycles

Fractography and Metallography

To aid in establishing the mechanism responsible for the observed low ductility, both fractographic and metallographic examination of the structures resulting from the various thermal histories were carried out by SEM and optical microscope. Scanning electron microscope fractographs of representative specimens are presented in Figure 4(a) and Figure 4(b) for correlation with the embrittlement behavior shown in Figure 3. As can be seen from Figure 4(a) and Figure 4(b), the fracture surfaces of the specimens for C-cycle and R-cycle have almost the same features, showing intergranular fracture essentially. Detailed examination shows that intergranularly fractured surfaces consist of network dimples tens of microns in diameter and that equiaxed precipitates a few microns in diameter are present in each dimple. With the thermal history varies to R-cycle, the dimples become shallower, while the equiaxed participates apparently lose in diameter but increase in its amount. EDS analysis of these individual precipitates indicated by the arrows, detects Fe-Mn-S-Al-Ti and Fe-Mn-S-O-Mg signals. Although some of the Fe signals are believed to come from the iron matrix, these precipitates are presumably compounded of (Fe, Mn)S, Al2O3 and MgO. According to the density of signals, (Fe, Mn)S should be the main component for equiaxed precipitates.

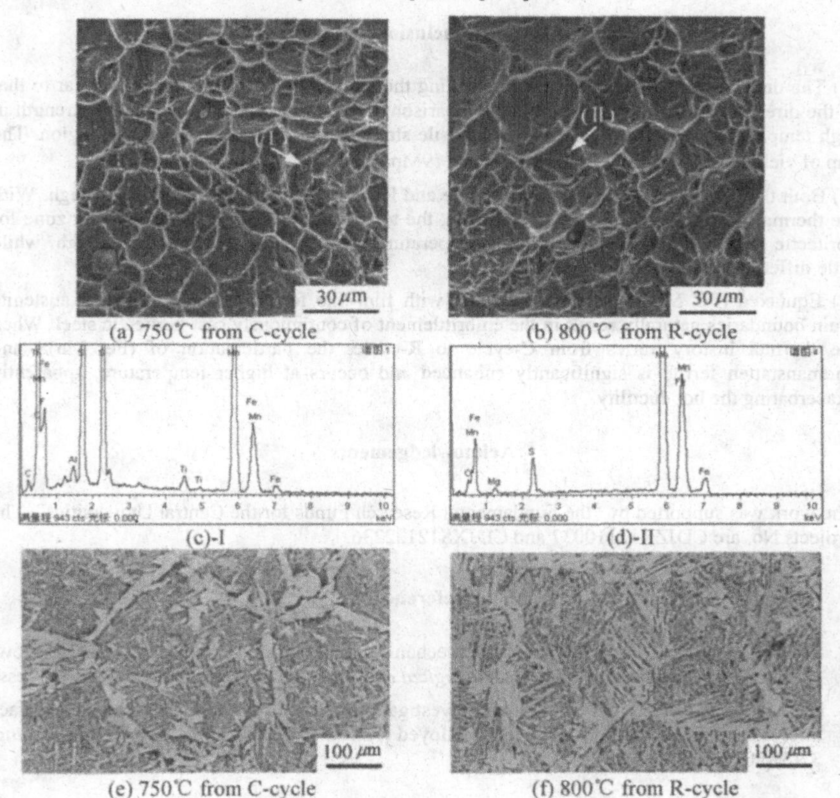

Figure 4 Fractographs and metallographs of C-Mn peritectic steel from various thermal histories

Optical microstructures of the cross section near to the fracture surface are shown in Figure 4(e) and Figure 4(f). The results indicate that the low ductility at 750℃ and 800℃ for C-cycle and R-cycle is basically related to the pro-eutectoid ferrite presented in prior austenite grain boundaries. When the thermal history varies to R-cycle, the film-like ferrite obviously loses in its thickness and a great number of widmanstatten ferrite participate in prior austenite grain boundaries, further exacerbating the hot ductility of specimens.

All mentioned above indicate that voids coalescence along grain boundary induced by (Fe,Mn)S participates and stress concentration in grain boundaries due to the film-like ferrite has occurred for C-cycle and R-cycle. It is also the nature of the embrittlement in the temperature region of 900℃ to 700℃. With the thermal history from C-cycle to R-cycle, the participation of (Fe, Mn)S and widmanstatten ferrite is significantly enhanced and occurs at higher temperature, apparently exacerbating the hot ductility for the continuously cast peritectic steel.

Conclusions

(1) The dependence of strength for the reheating thermal cycle on temperature is similar to that of the directly cooling thermal cycle. In comparison, the R-cycle leads to lower yield strength in high temperature region as well as lower tensile strength in the whole temperature region. The gap of yield strength reaches the maximum of 19Mpa at 700℃.

(2) Both the hot ductility curves from C-cycle and R-cycle illustrate a hot ductility trough. With the thermal history from C-cycle and R-cycle, the temperature region of embrittlement zone for peritectic steel apparently shifts to high temperature with a deeper hot ductility trough, while little difference of its width is detected.

(3) Equiaxed (Fe, Mn)S participates couple with film-like ferrite presented in prior austenite grain boundaries naturally result in the embrittlement of continuously cast peritectic steel. When the thermal history varies from C-cycle to R-cycle, the participation of (Fe, Mn)S and widmanstatten ferrite is significantly enhanced and occurs at higher temperature, apparently exacerbating the hot ductility.

Acknowledgements

The work was supported by “the Fundamental Research Funds for the Central Universities”. The Projects No. are CDJZR12110037 and CDJXS12132236.

Reference

1. B. Santillana, et al, “High-temperature mechanical behavior and fracture analysis of a low-carbon steel related to cracking,” *Metallurgical and Materials Transactions A*, 2012.In press.
2. F.J. Ma, et al, “In situ observation and investigation of effect of cooling rate on slab surface microstructure evolution in microalloyed steel,” *Ironmaking and Steelmaking*, 37(3)(2010),211-218.

3. M. Sawada, et al, "Effect of V, Nb and Ti addition and annealing temperature on microstructure and tensile properties of AISI301L stainless steel," *ISIJ International*, 51(6)(2011),991–998.

4. D.F. Chen, et al, "Thermal simulation on mechanical properties of steel Q345 for continuous casting slab," *Materials Science Forum*, 575-578(2008), 69-74.

5. B. Mintz, S. Yue, and J.J. Jonas, "Hot ductility of steels and its relationship to the problem of transverse cracking during continuous casting," *International Materials Reviews*, 36(5)(1991),187-220.

6. B. Mintz, J.M. Stewart and D.N. Crowther, "The influence of cyclic temperature oscillations on precipitation and hot ductility of a C-Mn-Nb-Al steel," *Transactions ISIJ*, 27(1987),959-964.

7. A.M. EL-Wazri, et al, "The Effect of thermal history on the hot ductility of microalloyed steels," *ISIJ International*, 39(3)(1999),253-262.

8. D.F. Chen, et al, "Effect of cooling rate on high temperature mechanical properties of weathering steel," *Journal of Chongqing University*, 34(11)(2011),50-55.

9. B. Mintz, J.R. Banerjee and K.M. Banks, "Regression equation for Ar_3 temperature for coarse grained as cast steels," *Ironmaking and Steelmaking*, 38(3)(2011),197-203.

10. Y. Maehara, et al, "Effect of carbon on hot ductility of as-cast low alloy steels," *Transactions ISIJ*, 25(1985),1045-1052.

11. M.J. Long, et al, "Simulation and investigation on physical of continuous casting slab AH36 at high temperature," *Materials Science Forum*, 575-578(2008),75-79.

12. H.G. Suzuki and D. Eylon, "Hot ductility of titanium alloys-a comparison with carbon steels," *ISIJ International*, 33(12)(1993),1270-1274.

13. B. Mintz, Z. Mohamed and R. Abu-shosha, "Influence of calcium on hot ductility of steels," *Materials Science and Technology*, 5(7)(1989),682-688.

14. B. Mintz and J.M. Arrowsmith, "Hot-ductility behavior of C-Mn-Nb-Al steels and its relationship to crack propagation during the straightening of continuously cast strand," *Metals technology*, 6 (1)(1979),24-32.

Materials Processing Fundamentals
Edited by: Lifeng Zhang, Antoine Allanore, Cong Wang, James A. Yurko, and Justin Crapps
TMS (The Minerals, Metals & Materials Society), 2013

INFLUENCE OF CORIOLIS FORCE ON THE FLOW FIELD OF COMBINED TOP AND BOTTOM BLOWN CONVERTER

Haiyan Tang[1,2], Tongbo Zhang[2], Jingshe Li[1,2], Yongfeng Chen[2]
[1]State Key Laboratory of Advanced Metallurgy, University of Science and Technology Beijing, No.30 Xueyuan Road, Haidian District, Beijing, 100083, P.R.China
[2]School of metallurgical and ecological engineering, University of Science and Technology Beijing, No.30 Xueyuan Road, Haidian District, Beijing, 100083, P.R.China

Keywords: Coriolis force, Converter's flow field, Water model, Bottom tuyere, Mixing time

Abstract

Based on the principle of Coriolis force and the structure of combined top and bottom blown converter, the relationship between Coriolis force and the converter was analyzed theoretically. It is concluded that Coriolis force affects converter's flow field by the action on top and bottom blown gas. It makes top gas clockwise deflection and bottom gas counterclockwise deflection, and the deflection angle of the former smaller than the latter. According to the theoretical analysis, the water model experiments were designed to primarily test and verify the influence of Coriolis force on converter's flow field. It is shown from the investigation that the counterclockwise trend of bottom gas caused by Coriolis force can be superimposed with the counterclockwise trend of flow field caused by the arrangement of bottom tuyeres, further to reduce the mixing time of bath.

Introduction

Coriolis force is generated by the earth's rotation. It reflects on making those objects which are moving horizontal on the Earth's surface deflected. An object having horizontal velocity at Northern hemisphere will deflect to the right whereas at Southern Hemisphere to the left. A case which can prove the existence of Coriolis force is the whirling effect while water running through a funnel. The hot cyclone caused by the non-uniform heat flow in the atmosphere and the erosion of right bank of the river in the Northern hemisphere can also prove the existence of Coriolis force [1].

Due to its widespread existence, its influence on construction, water conservancy, and industrial production has to be taken into consideration[2,3]. In the currently steelmaking process, the Corilolis force's effect on gas flow created by the top and bottom blown gas in the converter has to be considered. The currents generated from the top lance to decarburize and the bottom lance to stir the bath, both of the two flows have a certain horizontal velocity, which provides the initial conditions for Coriolis force to act. If the flow field of converter can be superimposed with the action of Coriolis force, the circulation of the flow field will be strengthened, and also with the mixing time reduced, productivity increased.

Many researchers had investigated the flow field of the converter using the water modeling [4-9], however to our knowledge; no study has been done on the effect of the Coriolis force on the flow field in combined top and bottom blown converter. Although Coriolis force cannot dominate the flow field in a converter, its action can't be ignored. In our previously experiments, the results disagreed with the expected without considering the Coriolis force. Many experimental phenomena can be rationally explained since the Coriolis force was applied in the system. In this paper, theoretical calculations and water modeling experiment were combined together to investigate the influence law of Coriolis force on the flow field of the converter.

Parameter Analysis of Coriolis force

Assuming that there is an object whose mass is m on the Earth's surface, its force is different at moving state and stationary state. When the object is stationary, it will subject to the gravity force F and the centrifugal force C, where $C=\omega^2 R\cos\varphi$. The C and F are decomposed into axial and tangential directions, which are expressed as C_Z、C_H和 and F_Z、F_H, respectively.

$$C_Z = m(\omega^2 R\cos\varphi)\cos\varphi \tag{1}$$

$$C_H = m(\omega^2 R\cos\varphi)\sin\varphi \tag{2}$$

Where m is the mass of the object, kg; ω= 7.292×10^{-5} rad·s^{-1} which is the angular velocity of the Earth's rotation; R=6.371×10^3 km which is Earth's average radius; φ is the latitude of the object located; φ' is the geocentric latitude. As the earth's radius is very big, φ is quite close to φ'.
C_H and F_H are in the opposite direction, equivalent value and offset each other. Fz is far more than Cz and their resultant force is gravity force. For the stationary object, its centrifugal force which caused by rotation of the Earth is negligible. While an object in motion along longitude or latitude will deflect. When the object is moving along latitude, it will produce extra centrifugal force along the earth's rotation direction. For example, if an object at Northern hemisphere is moving along latitude from west to east at speed V, at this moment the angular velocity of the moving object to the Earth is $V/(R\cos\varphi)$ and the centrifugal force acting on the object is C' which can be expressed as:

$$C' = m(\omega + \frac{V}{R\cos\varphi})^2 \cdot R\cos\varphi \tag{3}$$

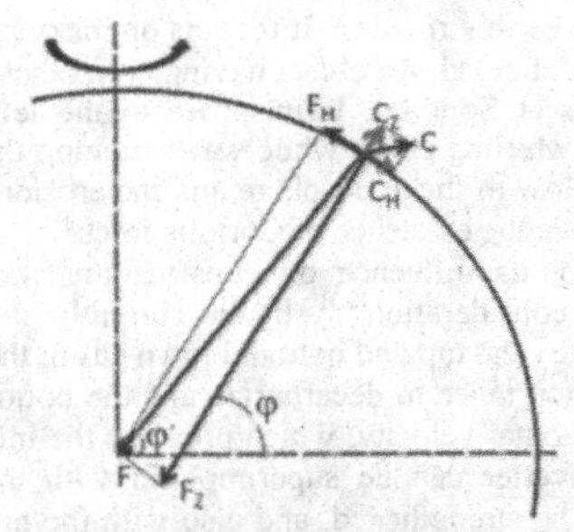

Figure 1. Force analysis of stationary object on the earth's surface

The component force at tangential direction can be expressed as:

$$\begin{aligned} C_H' &= C'\bullet\sin\varphi = m(\omega + \frac{V}{R\cos\varphi})^2 \cdot R\cos\varphi\sin\varphi \\ &= C_H + m(2\omega + \frac{V}{R\cos\varphi})\bullet V\sin\varphi \end{aligned} \tag{4}$$

Let

$$A = m(2\omega + \frac{V}{R\cos\varphi}) \bullet V\sin\varphi \quad (5)$$

The equation (6) is obtained from (4) and (5)

$$C_H^{'} = C_H + A \quad (6)$$

Because C_H and F_H are balanced each other, there is an extra force A drives the object moving to the South.

In the equation (5),

$$2\omega >> \frac{V}{R\cos\varphi}$$

So,

$$A \approx 2m\omega V\sin\varphi$$

Similarly, when an object moving from east to west, its tangential force is expressed as $C_H^{''} = C_H - A$. It shows that, in addition to gravity force, there is another force which drives the object to the North.

The above analysis shows that when an object moves along the latitude, there is a force perpendicular to its motion direction acting on it. At Northern hemisphere, direction of the force is on the right of moving velocity, while at Southern hemisphere, it is on the left. The force is called horizontal Coriolis force with its value $2m\omega V\sin\varphi$. The similar results can be obtained for the moving object along longitude [3].

So, it is deduced that any object with the horizontal velocity will be affected by Coriolis force, whose intensity is decided by the mass, velocity and position of the object.

Theoretical analysis of the effect of Coriolis force on converter flow field

Effect on the jet flow of top blowing oxygen lance

When the gas is injected from the nozzle, which has an inclined angle for multi-pore oxygen lance, the velocity out of the nozzle can be decomposed in horizontal and perpendicular directions. Thus, the jet flow will be affected by Coriolis force. **Figure 2** shows the model of lance head designed with five holes, in which four opening holes distributed around the head with an inclined angle of 13°, and one center hole without any inclined angle. Since there is no inclined angle in the center hole, only the four around holes were calculated with the effect of Coriolis force. The nozzle calculation parameters are shown in **Table1.**

Table 1 Nozzle Parameters

Parameter	Value
Distance from nozzle to liquid surface	0.2m
Nozzle flow rate	$80Nm^3 \cdot h^{-1}$
Nozzle outlet velocity	$121.43\ m \cdot s^{-1}$
Diameter of nozzle outlet	0.007m
Divergence angle of Laval nozzle	3.4°
Inclined angle of four around holes	13°

The outlet velocity of injection(V) can be decomposed into the horizontal velocity V_0 and the perpendicular velocity V_2, the distance of gas flow traveling along V direction from outlet to liquid surface can be calculated by l=0.2m /cos13° = 0.2053 m

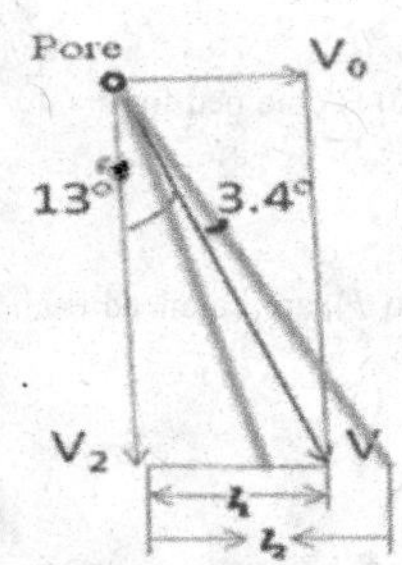

Figure 2. The physical model of lance head(a) and gas injection diagram of around hole (b)

Assuming there is no natural diffusion on the jet flow, and then it can be regarded fluctuating between -3.4° and +3.4° as in Fig.2 (b), the rate which gas reaches to the liquid surface can be obtained as $V' \approx$ 9.38 $m \cdot s^{-1}$.

Assuming the uniform attenuation gas flow speed from outlet to liquid surface, according to formulas (7) and (8) [8], the time for this process can be calculated, which is $t = 3.05 \times 10^{-3}$ s.

$$V' = V + at_0 \tag{7}$$

$$l = 0.5at_0^2 + Vt_0 \tag{8}$$

Where V' is the velocity which gas reaches the liquid surface, $m \cdot s^{-1}$; V is the gas outlet velocity, $m \cdot s^{-1}$; a is gas flow acceleration, $m \cdot s^{-2}$; t_0 is the time of gas traveling from outlet to liquid surface, s.

The horizontal velocity of gas at the nozzle outlet is calculated by $V_0 = V \times \cos 77° = 27.3$ $m \cdot s^{-1}$, and the horizontal velocity of gas traveling to the surface after time t is $V_0' = V' \times \cos 77° = 2.11$ $m \cdot s^{-1}$. The rotational angular velocity of the earth, $\omega = 7.29 \times 10^{-5}$ $rad \cdot s^{-1}$, and the local latitude, $\varphi = 40°$. According to the equation (1), the Coriolis force of gas flow at nozzle outlet and liquid surface is $A_0 = 2.55 \times 10^{-3} m$ and $A_0' = 1.98 \times 10^{-4} m$, respectively.

The path of gas flow will travel in a curve way due to the Coriolis force always perpendicular to the gas motion direction and the velocity is varying continuously.

The attenuation rate of Coriolis force can be expressed as:

$$\Delta A = \frac{A_0' - A_0}{t_0} = -0.77m$$

Then the Coriolis force at any time (t) from outlet to liquid surface is obtained by:

$$A_t = A_0 - \Delta A \cdot t$$

According to the centripetal formula: $F = m\omega^2 r = \frac{mV_t^2}{r}$, A_t provides the centripetal force. The velocity of object at time t is given by: $V_t = V + at$

The deflection angle of jet flow is integrated as below

$$\theta_1 = \int_0^{3.05\times10^{-3}} \omega' dt = \int_0^{3.05\times10^{-3}} \frac{F}{mV_t} dt = 6.02\times10^{-8}(^\circ) \quad (9)$$

Where θ_1 is the deflection angle of jet flow of top lance from outlet to liquid surface caused by Coriolis force.

Effect on gas flow of bottom lance

Figure 3 shows that there is an inclined angle of 13° when the gas exits from bottom lance of converter. The experiment showed the gas velocity would decrease to a certain value rapidly after entering into the bath, and then kept a horizontal velocity and the buoyancy velocity in bath.

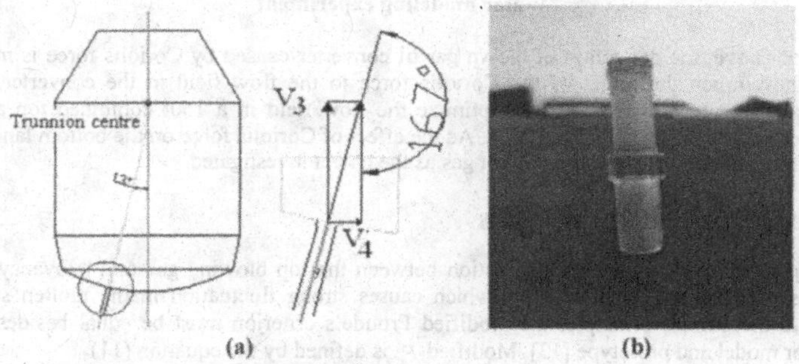

Figure 3 Illustration of bottom lance gas flow of converter (a) and the physical model of bottom lance (b)

The bubble buoyancy velocity mainly depends on the bubble diameter, temperature of molten steel [11]. Shi et al [12] had investigated the relation between the buoyancy velocity and bubble diameter showing that the acceleration of bubbles was negligible when the bubble diameter was smaller than 500μm. In this work, the maximum bubble diameter reached to 1.25mm which meant that acceleration of the bubble cannot be neglected. To simplify the model, assuming the horizontal velocity after attenuation is independent on the acceleration; bubbles will deflect in a certain angle before escape from the molten steel surface due to the effect of Coriolis force. The flotation time of bubbles was about 1s by experimental observation.

The total gas flow rate in four bottom lances is $1.8 Nm^3 \cdot h^{-1}$ and four holes in each bottom lance with 1mm of diameter, the initial velocity of gas at nozzle outlet can be calculated as : $v' = 39.8\ m \cdot s^{-1}$..

$v^{\prime}$ can be decomposed into vertical velocity V_3 and horizontal velocity V_4 ; $V_3 = v^{\prime}\cos 13°$ = 38.7 m·s^{-1}, V_4 = 8.9 m·s^{-1}. The depth of liquid is 0.288 m, then the horizontal velocity of bubble while floating to the liquid surface $V_4^{\prime}$ = 0.064 m·s^{-1}. According to the formula (1), Coriolis force can be calculated as

$$A'' = 6\times10^{-6}\times m$$

The deflection angle of bubble caused by Coriolis force from the bottom of converter rising to liquid surface is obtained by:

$$\theta_2 = \int_0^1 \omega'' dt = \int_0^1 \frac{A''}{mV_4} dt = 9.4\times10^{-5}(°) \tag{10}$$

Comparing the equations (9) and (10), the deflection angle is three orders of magnitude larger than top lance gas flow, which shows that the effect of Coriolis force on the bottom lance gas flow is much more important.
According the right hand rule of Coriolis force, the top gas flow will get clockwise deflection and the bottom gas flow counterclockwise deflection.

Water modeling experiment

As calculated above, the deflection of blown gas of converter caused by Coriolis force is minimal. To investigate how much the effect of the Coriolis force to the flow field in the converter, the water modeling experiment was performed to optimize the flow field in a 150t combined top and bottom blown converter from a steel plant in China. As the effect of Coriolis force on the bottom lance gas flow is larger than on the top gas, take the bottom gas as the object investigated.

Experimental principle of water modeling

In melting pool of a converter, the interaction between the top blowing gas and buoyancy of bottom blowing gas produces asymmetric flow, which causes strong fluctuation in the molten steel cavity. According to the similar principle, the modified Froude's criterion must be equal besides geometric similarity for model and prototype [13]. Modified $F_r^{\prime}$ is defined by the equation (11)

$$Fr' = \frac{u^2 \rho_g}{gL(\rho_l - \rho_g)} \tag{11}$$

Where u is gas flow speed, m·s^{-1}; L is feature size, m; ρ_l and ρ_g are densities of liquid and gas, respectively, kg·m^{-3}; g is gravity acceleration, m·s^{-2}.
The equation (15) is obtained according to the equations (12)~(14)

$$(F_r^{\prime})_m = (F_r^{\prime})_p \tag{12}$$

$$\frac{u_m}{u_p} = (\frac{L_m}{L_p})^{\frac{1}{2}} \times (\frac{\rho_{gp}}{\rho_{gm}})^{\frac{1}{2}} \times (\frac{\rho_{lm} - \rho_{gm}}{\rho_{lp} - \rho_{gp}})^{\frac{1}{2}} \tag{13}$$

$$Q = u \times \frac{\pi}{4} L^2 \times n \times 3600 \tag{14}$$

$$\frac{Q_m}{Q_P} = (\frac{L_m}{L_p})^{\frac{5}{2}}(\frac{\rho_{gp}}{\rho_{gm}})^{\frac{1}{2}}(\frac{\rho_{lm} - \rho_{gm}}{\rho_{lp} - \rho_{gp}})^{\frac{1}{2}} \quad (15)$$

Where script m and p denote model and prototype, respectively; n is the number of nozzle; Q is gas flow rate, $Nm^3 \cdot h^{-1}$. The converter in water model was made by plexiglass at the similarity ratio 1:6(model to prototype). Lava nozzle lance was made by copper with the inclination of 13° as shown in Fig.2(a). Nitrogen was used to simulate bottom gas and air to simulate top gas in the experiment. More details are given by **table 2.**

Table 2 Main parameters of prototype and model of the converter

Parameters	Steel case	Water Modeling case
Diameter of bath/m	5.4438	0.9073
Height of bath/m	7.790	1.2983
Height of liquid surface/m	1.728	0.288
Gas density/$kg \cdot m^{-3}$	1.43(top)/1.23(bottom)	1.29
Liquid density/ ($kg \cdot m^{-3}$)	7200	1000
Gas flow rate of top lance/$Nm^3 \cdot h^{-1}$	17000	76.6
Distance of top lance from outlet to liquid surface/m	1.680	0.280
Diameter of top lance outlet/m	0.0423	0.00705
Mach number of top lance	1.98	1.98
Gas flow rate of bottom lance/ ($m^3 \cdot h^{-1}$)	230~500	1.0~2.2

Experimental method and apparatus

The mixing in a converter under different gas blown conditions was studied by conductivity method. In the course of experiment, two electrodes were fixed at the bottom of the pool at different positions and saturated KCl solution was added into the bath as tracer. The mixing time τ was obtained by detecting the change of conductivity after addition of tracer. Take the average of three times of all the experiments as the measured value for the mixing time.The experimental apparatus is shown in **figure 4**.

To optimize the flow pattern and decrease the mixing time, four different cases of bottom lance position were implemented in the system, where the bottom tuyeres were set at the position of 0.4R and 0.6R (R is the radius of the converter bottom) with different combinations as in **figure 5**. Counterclockwise flowing trend of liquid occurred in the converter in case a, whereas, clockwise flow in case b which is opposite to the effect of Coriolis force on bottom gas. Based on case a and case b, the number of bottom lances were increased in case c and case d. The flow in case c made a counter-clockwise pattern, however there was no chang in flow pattern in case d.

The layout of the bottom lances had much more important influence on the mixing time at proper top lance parameters [8, 9]. In this work, the parameters of top and bottom lances were based on the real case in steel plant. The distance from the molten steel surface to the outlet of top lance was 0.2m, the gas flow rate was 76.6$Nm^3 \cdot h^{-1}$.The flow rates of bottom lance were designed as 1.40 $Nm^3 \cdot h^{-1}$, 1.80 $Nm^3 \cdot h^{-1}$ and 2.16 $Nm^3 \cdot h^{-1}$ respectively.

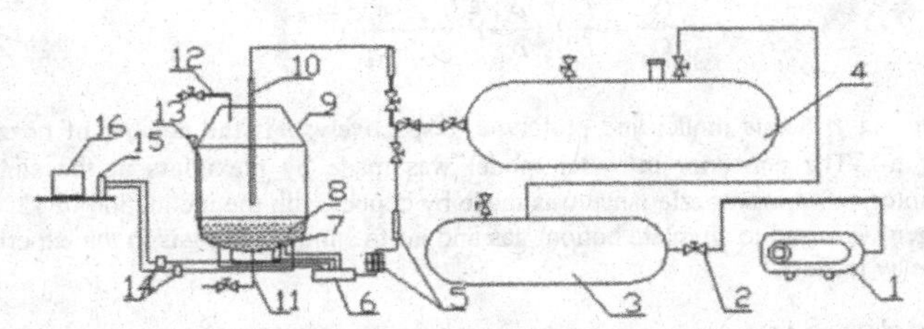

1.Air compressor 2.Valve 3.Gas tank 4. Buffer tank 5. Rotormeter 6.Bottom blowing tank 7 Electrode 8.Water 9.Converter model 10.Oxygen lance 11.Drain 12. Water inlet pipe 13.Funnel 14.Data collection card 15. Conductivity meter 16.Computer

Figure 4 Diagram of experimental apparatus

Experimental results

Figure 6 showed the comparison of the mixing time for four cases at three different gas flow rates. The mixing time was different among four cases which showed that the Coriolis force played an important role in the flow patterns. The mixing time in case a was shorter than case b in all three flow rates because of the effect of Coriolis force. The flow pattern in case a by the layout of the bottom lances created counter-clockwise trend which superimposed with the effect of the Coriolis force, thus strengthened the flow in the converter. In contrast, the layout of bottom lance in case b made the flow clockwise trend which was opposite to the influence of Coriolils force, thus the mixing time was prolonged.

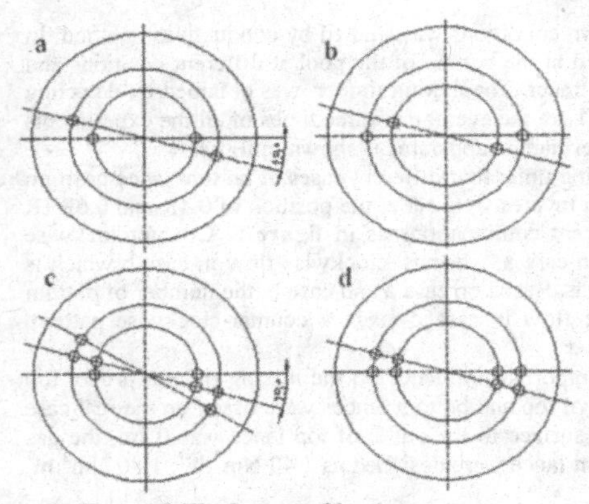

Figure 5 The layout of bottom tuyeres

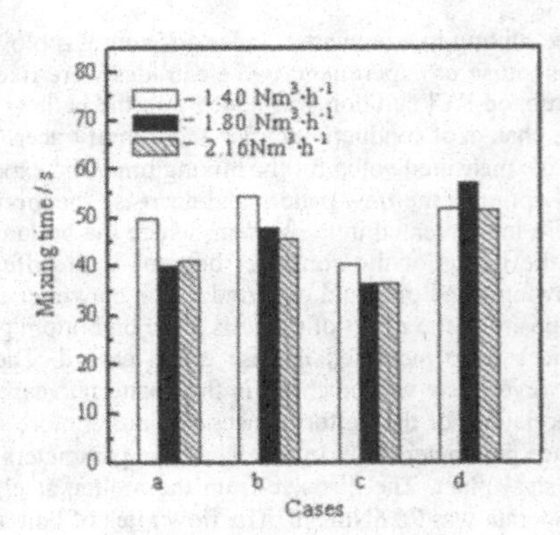

Figure 6 Mixing time of four cases at three flow rates

Comparing case c and case d, the mixing time of the former was far less than the latter. That was because the layout of bottom lances in case d didn't cause any initial deflection, which unable to be superimposed with the trend caused by Coriolis force.
The experimental results suggest that although Coriolis force is minimal, its action can't be ignored. The optimized layout of bottom lance can strengthen the stirring and mixing of bath and then decrease the mixing time combined with the effect of Coriolis force.

Conclusions

(1) The theoretical analysis and calculations show that Coriolis force can affect the flow field of a converter by the action on the gas flow of top and bottom lances.
(2) The influence of Coriolis force on the gas flow of bottom lance was much more than on top lance. The deflection angle of gas caused by Coriolis force from the outlet of lance to liquid surface is $6.02\times10^{-8}(^\circ)$ for top lance; however this value increased to $9.4\times10^{-5}(^\circ)$ for bottom lance.
(3) The counterclockwise trend caused by the layout of bottom tuyeres can be superimposed with the effect of Coriolis force and furthermore reduce the mixing time of bath.

Acknowledgements

The authors are very grateful for the financial support of the National Natural Science Foundation of China(No.51074021) and Educational Ministry Fund (No.20100006120008)

References

[1] Zhao S Q, Liu Z Q. Discussion on Coriolis Force. **JOURNAL OF JIANGXI NORMAL UNIVERSITY.** 1990, 14(3): 87.
[2] Guo L S, Cao J L. Analysizing Coriolis Force from mechanical view. **TIANJIN SHIDA XUEBAO,** 1992(2): 78.
[3] Zhuang C F. The Horizontal Coriolis Force. **THE JOURNAL OF FUJIAN TEACHERS UNIVERSITY.** 1987, 3(1): 91.
[4] S. Paul, D. N. Ghosh. Model study of mixing and mass transfer rates of slag-metal in top and bottom blown converters, **Metallurgical Transactions B**, 1986,17, (3) .461
[5] WU W, YANG L B, ZHENG C J, LIU L. Cold Simulation of Oxygen Transfer Rate in BOF, **JOURNAL OF IRON AND STEEL RESEARCH, INTERNATIONAL,** 2010, 17(9) : 7
[6] Ni H W, Yu S R, Qiu L H, Yan J X, Xie J Q, Luo M. Cold Model Study on 90t Top-Bottom Combined Blown Converter. **Steelmaking,** 2002, 18(3): 39.
[7] Tang P, Yu Y Y, Wen G H, Zhu M M, Zhou L, Long Y J, Liang Q. Study on the optimization of the combined blown converter process in Chongqing Iron and Steel Company. **Journal of University of Science and Technology Beijing,** 2008, 15 (1): 5.
[8] Lei Z L. Cold Model Study on Blowing Process Optimum for 120t Top-bottom Combined Blown Converter of Shao Steel. Master Dissertation. Shenyang: Northeast University, 2004.
[9] Sun L N, Wu G X, Tan M X. Research into Modelling Air Supply from the Bottom of 150ton Plural Blown Converter, **Journal of Liaoning Instiute of Science and Technology,** 2006, 8(2): 1.
[10] Fan Z H, Wang J G. University Physics (1st Volume). Xian: Northwestern University Press.
[11] Tang H Y, Li J S,Guo H J. Dynamic study of deoxidation and desulfurization with calcium treatment for molten steel. **J. Univ. Sci. Technol. Beijing,** 2009, 31(6): 690

[12] Shi C W, Wang J A, Jiang X Z. Mechanics effect study of a rising micro-bubble in still water. **Journal of Naval University of Engineering**, 2008，20 (3)：83
[13] Tang H Y, Li J S, Xie C H，Yang S F，Sun K M，Wen D S. Rational Argon Stirring for 150t Ladle Furnace. **International Journal of Minerals, Metallurgy and Materials,** 2009, 16(4): 383.

Materials Processing Fundamentals
Edited by: Lifeng Zhang, Antoine Allanore, Cong Wang, James A. Yurko, and Justin Crapps
TMS (The Minerals, Metals & Materials Society), 2013

MOTION CHARATERISTICS OF A POWDER PARTICLE THROUGH THE INJECTION DEVICE WITH SLATS AT FINITE REYNOLDS NUMBER

Zhongfu Cheng, Miaoyong Zhu*

School of Materials and Metallurgy, Northeastern University, Shenyang 110819, Liaoning, China

Keywords：particle, motion characteristic, powder injection, bottom-ladle

Abstract

The forces acting on a powder particle moving through the slat in the injection device were analyzed by taking account of the interaction between current and particle, and the drag force was found to play the most important role in flow direction and lift force acting in perpendicular to flow direction was most vital. A model was established describing the particle motion characteristics in the vertical flow, and the relationships between the particle velocity, translation displacement and time were deduced. Motion characteristics of particle along flow direction show that particle velocity increases slowly after acceleration in short time or displacement. Larger particle, as well as particle with greater density, passes farther horizontal displacements and the particle with horizontal velocity and in boundary layer moves instability, which collides with the slat wall easily, while the particle with higher inlet velocity is conveyed more smoothly and steadily.

Introduction

The motion characteristics of a powder particle are very important to many engineering applications. One of the most important examples is the application of metallurgical technology of powder injection which uses carrier gas to spray refined powder particles into molten bath [1]. As it increases the rate of reaction and yield rate of powders extremely, powder injection is widely adopted and takes an important role in secondary refining. Traditional powder injection method commonly employs spray lance made of refractory to immerse into molten iron or steel to inject [2-5].However, it brings a series of problems, for examples, seriously splashing of molten iron, high iron losses, and low utilization rate of desulfurizing agent, and so on. Moreover, if the refractory of lance drops, the molten steel will be contaminate. In order to overcome the lack of immersion lance, Zhu [6,7] developed a new technique that achieved powder injection from the bottom of blowing ladle. This new refining technology combines the bottom argon blowing and powder injection, and operation is simple and does not change the original layout. Powder particle motion characteristics can play a significant role in the process of powder transportation using pneumatic conveying system, and it determines whether powder injection process can run smoothly. Pan and Zhu [8] studied the motion characteristics of injection powder though porous brick mounted in the bottom refining ladle, and revealed the movement law of the particle in vertical flow. In their work, the forces, except drag, gravity, buoyant and virtual mass force, are all ignored. However, a series of theoretical studies [9-12] and experimental studies [13-15] indicate that some forces, especially lift force, acting on the direction perpendicular to the flow are significant to the motion characteristics. Therefore, the goal of this study is to reveal the motion characteristics of a powder particle through the injection device with slats at finite Reynolds number.

Analysis on Forces Acting on a Powder Particle

Drag Force

Drag force is due to the slip velocity between the particle and the surrounding fluid (where the fluid is gas), and it acts proportional to the slip velocity. The drag force can be expressed in the most general form as

$$F_{\mathrm{D}} = C_{\mathrm{D}} A \frac{\rho_{\mathrm{g}} \left| u_{\mathrm{g}} - u_{\mathrm{p}} \right| \left(u_{\mathrm{g}} - u_{\mathrm{p}} \right)}{2} \tag{1}$$

Where ρ_g is the density of gas, A is the projected area perpendicular to the flow direction, u_g is the velocity of gas, u_p is the velocity of particle, C_D is the drag coefficient which is a function of particle Reynolds number. In the range of finite Reynolds number ($0.1<Re_p<1000$), there is no analytical solution available for C_D. It is common to write the drag coefficient as expression (2) based on the study of Clif et al. [16] and Schiller-Neumann [17].

$$C_{\mathrm{D}} = \frac{24}{\mathrm{Re}_p}\left(1 + 0.15\,\mathrm{Re}_p^{0.687}\right) + \frac{0.42}{1 + (42500/\mathrm{Re}_p^{1.16})} \tag{2}$$

Where, Re_p denotes the particle Reynolds number. The expression is accurate to within 6% of experiments when C_D below the critical Reynolds number [18]. However, the drag coefficient C_D is influenced by fluid motion characteristics and particle rotation to some extent. For the case of linear shear, Loth yields a fit expression (3) based on resolved-surface simulations data of Kurose and Komori [9] for $Re_p \leq 500$ and $\omega^*_{shear} \leq 0.8$,

$$\frac{C_D}{C_{D,\omega=0}} = 1 + 0.00018\,\mathrm{Re}_p \left(\omega^*_{shear}\right)^{0.7} \tag{3}$$

Where ω^*_{shear} is termed the dimensionless continuous-phase vorticity of in form of linear shear. In order to examine the accuracy , datas of Bagchi and Balachandar [19] is exhibit ,which is consist with the Loth fit expression very well.
The drag coefficients C_D is found not to be significantly influenced for Re_p<100 and $\Omega^*_p \leq 0.8$ by the rotation based on the research of Rubinow and Keller [20], Kim and Choi [21]. Where Ω^*_p is termed the dimensionless particle angular velocity. Based on this, particle rotation can be elided in this study.

Lift Force

Lift force causes particle transference in the horizontal direction, and it's the most important force that prevents the particle from coming close to the wall. The lift force can be expressed in a general form as

$$F_L = C_L A \frac{\rho_g |u_g - u_p|(u_g - u_p)}{2} \tag{4}$$

Where C_L is the lift coefficient, which is a function of particle Reynolds number. Based on the previous work, Loth [18] proposed the two primary mechanisms leading to the generation of lift: vorticity in the continuous-phase and rotation of the particle. Since Saffman (1965) [22] gave an expansion for lift force which is restricted to creeping flow conditions, a series of useful work, investigating the value of $C_{L,shear}$ at finite Reynolds number, has been carried out. The empirical fit of $C_{L,shear}$ from Loth's work [18] is employed to calculate the shear-induced lift in our study and it's expressed as

$$C_{L,shear} \approx J^* C_{L,Saff} \qquad \text{for} \quad \mathrm{Re}_p \leq 50 \tag{5}$$

$$C_{L,shear} \approx -(\omega^*_{shear})^{1/3} \left\{ 0.0525 + 0.0575 \tanh\left[5 \log_{10}\left(\frac{\mathrm{Re}_p}{120} \right) \right] \right\} \qquad \text{for} \quad \mathrm{Re}_p > 50 \tag{6}$$

Where $C_{L,Saff}$ is Saffman coefficient [22], J^* is McLaughlin [23] lift ratio and the approximate result of Mei [24] is employed in our work.
For finite rotation rate and finite Reynolds number effects, Loth and Dorgan [12] gave an empirical correction for Re_p<2000 and Ω^*_p<20 based on theory, experiments, and resolved surface simulations as

$$C^*_{L\Omega} = 1 - \{0.675 + 0.15(1 + \tanh[0.28(\Omega^* - 2)])\} \times \tanh\left[0.18 \mathrm{Re}_p^{1/2}\right] \tag{7}$$

This model is more exact than those assuming $C^*_{L\Omega}$ as a constant [25,26]. Therefore, the empirical correction is employed to calculate the spin-induced lift in our study.

To simulate a particle motion, the contribution of fluid vorticity and particle rotation to the lift force should be taken into count simultaneously. Saffman [22] firstly proposed that the theoretical particle spin lift of Rubinow and Keller can be linearly combined with the first order shear-induced lift in the condition of creeping flow and Ω^*_p<<1 to compute the lift for combined fluid shear and particle spin. Based on RRS results of Bagchi and Balachandar [26], the simple linear combination was reasonable for ω^*shear and Ω^*_p values as high as 0.4 and Re_p as high as 100. Consistent with this, Loth [18] summarized a series of previous work and proposed that the lift for combined fluid shear and particle spin could be linear combination. In our work, Loth's results [18] were adopted to compute the influence of lift force.

Thermal Force

Slot-bottom powder injection device is equipped at the bottom of ladle; the temperature gradient in the vertical direction is significant. As powder particle size ranges in micron order, thermal force should not be ignored. Epstein proposed a theoretical calculation formula of thermal force in1929, later Derjaguin and Bakanov [27], Derjaguin [28], Brock [29, 30] proposed correction formulas fit for several conditions. To consider the influence of thermal force, the Brock's correction formula, suitable for most conditions, was chosen in our study. The thermal force is expressed by Brock as

$$F_T = \frac{9}{2}\pi \frac{\mu_g^2}{\rho_g T_g} d_p \cdot \frac{1}{1+3c_m \dfrac{2l}{dP}} \cdot \frac{\dfrac{k_g}{k_p} + c_t \dfrac{2l}{d_p}}{1+2\dfrac{k_g}{k_p} + 2c_t \dfrac{2l}{d_p}} \frac{dT_g}{dy} \tag{8}$$

Where T is the temperature along the vertical direction of slot-bottom powder injection device, and dT/dy is the temperature gradient.

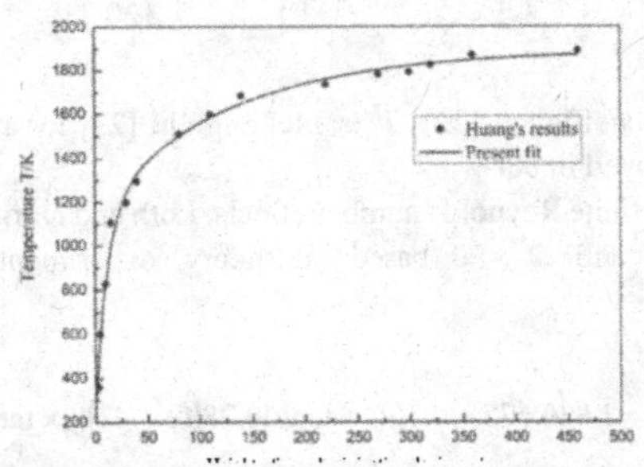

Fig.1 The temperature distribution along the vertical direction of injection device

Huang [31] did a great deal of work about the temperature distribution of bottom argon blowing device of refining ladle; and we used the results fitting a temperature curve and gave an approximate expression for temperature and temperature gradient distribution along the vertical direction. The results are showed as follows:

$$T = 267.3 + 727.5 \cdot \left(1 - e^{-\frac{y}{0.1287}}\right) + 897.4 \cdot \left(1 - e^{-\frac{y}{0.0118}}\right) \tag{9}$$

$$\frac{dT}{dy} = 5652.68 \cdot e^{-\frac{y}{0.1287}} + 76050.85 \cdot e^{-\frac{y}{0.0118}} \tag{10}$$

Other Forces

The external forces acting on the powder particle also include gravity force F_g, buoyancy force F_b, fluid stress gradient force F_{sg}, virtual force F_{Vm}, history force F_h and so on. You [32], proceeding from reality, came to the conclusion that the influence of history force is very insignificant for the motion of particles/droplets in gas. Similarly, compared with other forces, the fluid stress gradient force can also be ignored. The gravity force, buoyancy force, and virtual force can be expressed in the most general form as

$$F_g = \frac{1}{6} \pi \rho_p d_p^3 g \tag{11}$$

$$F_b = \frac{\pi}{6} \rho_g d_p^3 g \tag{12}$$

$$F_{Vm} = \frac{1}{12} \pi d_p^3 \rho_g \frac{d}{dt}(V_g - V_p) \tag{13}$$

Magnitude of Different Forces

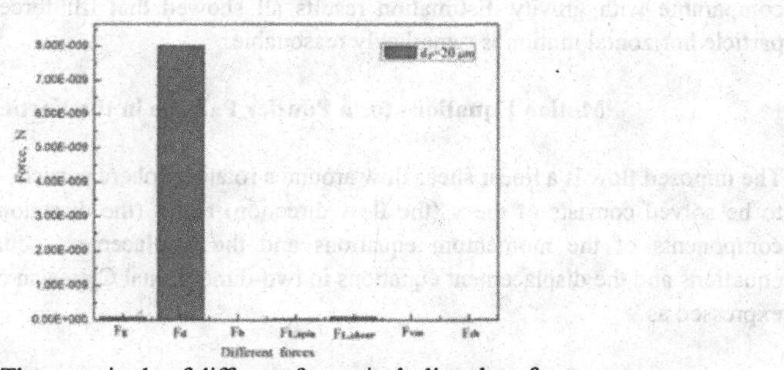

Fig.2 The magnitude of different forces including drag force

Based on above analyses, we can estimate the force magnitude. Supposing that the velocity values of particle are 85% those of fluid, estimating values of different forces are showed in the fig.2. As might be expected, compared with other forces, drag forces is more than three orders of magnitude higher than others. Therefore, drag force plays the leading role in the vertical direction.
In order to describe the relative magnitude of different forces more accurately, we eliminated the drag force and put the others in the same figure. Meanwhile, we examined the influences of particle size on forces.

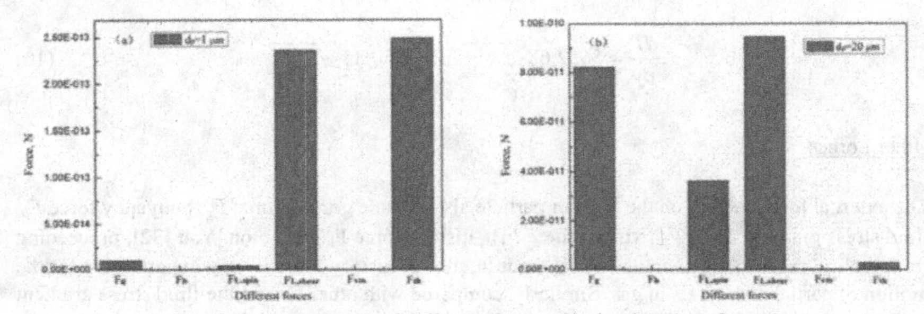

Fig.3 The magnitude of different forces eliminating drag force

The force magnitude comparison including the influences of particle size is put forth in fig.3.Obviously, the particle diameter in the range of 1 to 20 microns, buoyancy force F_b, fluid stress gradient force F_{sg} and virtual force F_{Vm} are so tiny that can be omitted compared with gravity F_g. Thermal force F_{th} takes effect only in the range of small particle size. That's because gravity value with particle diameter change is more sensitive than that of thermal force ($F_g \propto d_p^3$, $F_{th} \propto d_p$). While gravity grows rapidly from a lower value to a higher value with particle diameter rising, thermal force still keeps at the lower value. At this moment, thermal force becomes very insignificant referring gravity. Note that thermal force has been taken into account in the range of small particle size. Lift force, the most important force in the vertical flow direction, is constantly comparable with gravity. Estimation results all showed that lift force considered calculating particle horizontal motion is remarkably reasonable.

Motion Equations for a Powder Particle in the Vertical Flow

The imposed flow is a linear shear flow around a rotating sphere particle. The system of equations to be solved consists of the y (the flow direction) and x (the direction perpendicular to flow) components of the momentum equations and the displacement equations. The momentum equations and the displacement equations in two-dimensional Cartesian coordinate system can be expressed as

$$m_p \frac{dv_p}{dt} = F_{Dx} + F_L + F_{Vmx} + F_{hx} \tag{14}$$

$$\frac{dx}{dt} = v_p \tag{15}$$

$$m_p \frac{du_p}{dt} = F_{Dy} + F_g + F_b + F_{Vmy} + F_{hy} + F_{th} \tag{16}$$

$$\frac{dy}{dt} = u_p \tag{17}$$

Where, m_p is the mass of a powder particle, v_p is the velocity component and x is the displacement component in the direction perpendicular to the flow; similarly, u_p is the velocity component and y is the displacement component in the direction along with flow.

Numerical Results and Discussions

Model Verification

In order to verify the accuracy and dependability of the model, firstly, we computed the motion characteristics of lime particles in the vertical flow. The numerical calculation process is achieved through the science computation software Matlab. And then the vertical velocity and displacement results were compared to the previous work from Pan and Zhu [8]. Pan and Zhu's model only considered the motion of a particle along the fluid flow direction, and the drag coefficient was treated as a function about Reynolds number which was not affected by fluid shear and particle rotation. Because of the temperature in slot-bottom powder injection device regarding as constants, they didn't consider the influence of thermal force.

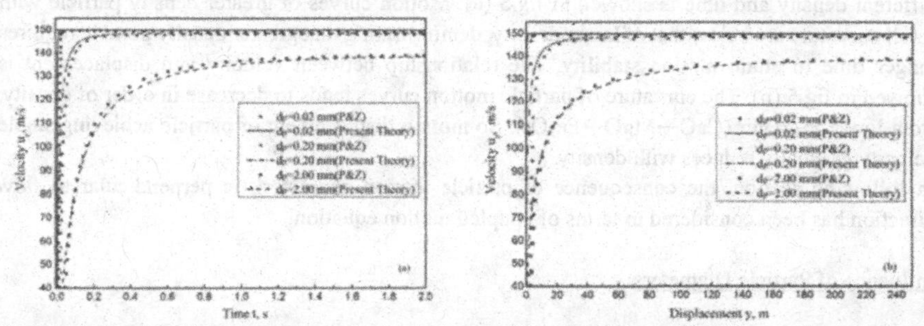

Fig.4 Relationships between the vertical velocities of lime particles with different diameters and time (a) and displacement (b) compared with results of Pan and Zhu [8].

In the nitrogen flow at the speed of 150m/s, the velocity of lime particles in range of different diameters with time was showed in fig.4 (a). The calculation results of particle velocity and displacement along the flow direction fits very well with the previous work. At the beginning of particle movement, particle acceleration is very large, and velocity grows rapidly. Later, velocity increases slowly and finally becomes stable. During the process, particle diameter has a significant impact on the initial acceleration: the particle diameter smaller, acceleration is larger, time required to achieve the stable velocity is shorter.The velocity of lime particles in range of different diameters with displacement was showed in fig.4 (b). Obviously, particle, in range of small diameter, achieving the stable velocity, passes a short displacement can, while the lager diameter particle achieving the stable velocity needs longer displacement.

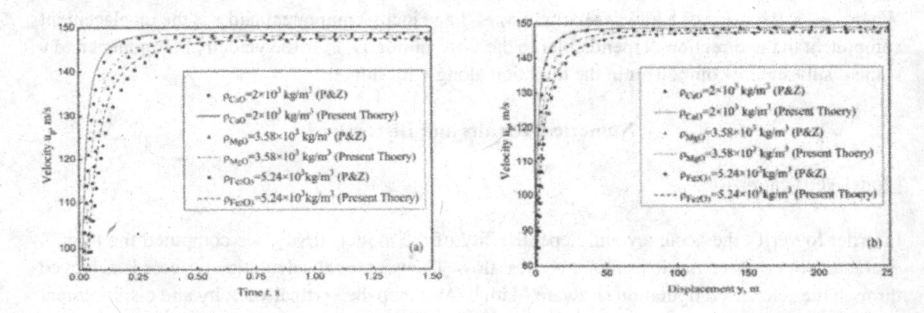

Fig.5 Relationships between the vertical velocities of particles with different density and time (a) and displacement (b) compared with results of Pan and Zhu [8].

In the nitrogen current at the speed of 150m/s, the relationship between velocity of particle with different density and time is showed in fig.5 (a). Motion curves of greater density particle with small curvature are below that of lower density, demonstrating that greater density particle requires longer time to attain motion stability. The relationship between velocity and displacement is showed in fig.5 (b). The curvature of particle motion curves tends to decrease in order of density, from lower to grater ($CaO \rightarrow MgO \rightarrow Fe_2O_3$). So motion displacement of particle achieving stable velocity gradually reduces with density.

In following section, the consequence of particle motion characteristic perpendicular to flow direction has been considered in terms of coupled motion equation.

Influence of Particle Diameters

During the motion of a particle, its translation velocity along the flow current approaches the local fluid velocity. Thus, the instantaneous Reynolds number decreases with time. Fig.6 shows the time dependence of particle translation velocity of x direction for the case of different diameters. From fig.6(a) it can be noted that particle translation velocity of x direction increases rapidly at the beginning, and then reaches the maximum value. However, maximum can not be held on, and soon, the value keeps dropping until reaching terminal. That is because the lift force pushes the rotating particle moving in the direction perpendicular to flow, and with the translation velocity rising, drag in opposite direction of lift force gradually goes up. Yet the lift force goes down by degrees due to rotation rate decreasing. Translation velocity reaches the maximum value when lift and drag attain balance. After the balance point, lift is less than the drag and keeps going down. Finally, the translation velocity gets to terminal.

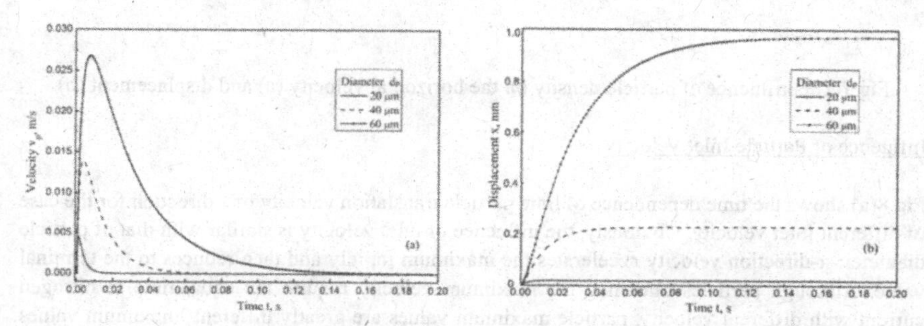

Fig.6 The influence of particle diameters on the horizontal velocity (a) and displacement (b)

Motion curves of different diameters change in the same way, which means that different particles have the same motion characteristics. For the diameter 20μm, the x translation velocity changes very rapidly. The particle attains the terminal velocity within t=0.013s. The particle with d_p=60μm responds much more slowly and attains the terminal velocity over t=0.13s. Particle displacements of x direction for this case are plotted in fig.6(b). Results are shown that particle x translation displacement are limited, and the displacement of big diameter particle is larger than that of small particle.

Influence of Particle Density

Fig.7(a) shows the relationship between x translation velocity of different density particle and time in nitrogen current (d_p=40μm). It is clear that particles with different density show a similar trend: their x velocity accelerates to the maximum value rapidly, and then slowly reduces to the terminal value (v_{xt}≈0); in addition they nearly have the same maximum velocity. But Fe_3O_4 particle needs more time to attain the maximum velocity, and moves longer achieving the terminal. It implies that great density particles go through longer displacement to reach dynamic balance than lower density ones. The results of relationship between displacement and time are plotted in Fig.7(b).

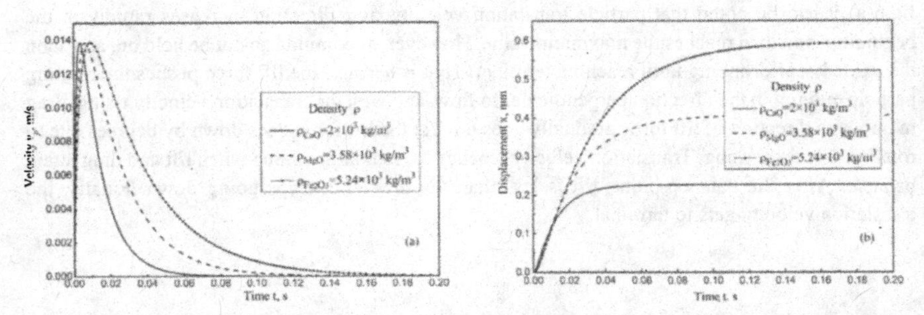

Fig.7 The influence of particle density on the horizontal velocity (a) and displacement (b)

Influence of Particle Inlet Velocity

Fig.8(a) shows the time dependence of lime particle translation velocity of *x* direction for the case of different inlet velocity. Obviously, the influence of inlet velocity is similar with that of particle diameter: x-direction velocity accelerates the maximum rapidly and then reduces to the terminal value. Although particles obtaining the maximum velocity require the equal time in nitrogen current with different velocity, particle maximum values are greatly different: maximum values decrease with increasing nitrogen velocity. It's interesting that all particles take equal time from the beginning to the motion terminal. As a result, x-direction displacement decreases with velocity of nitrogen current increasing (fig.8(b)), and that means particles in flow current with higher velocity move more stability than those in lower velocity current.

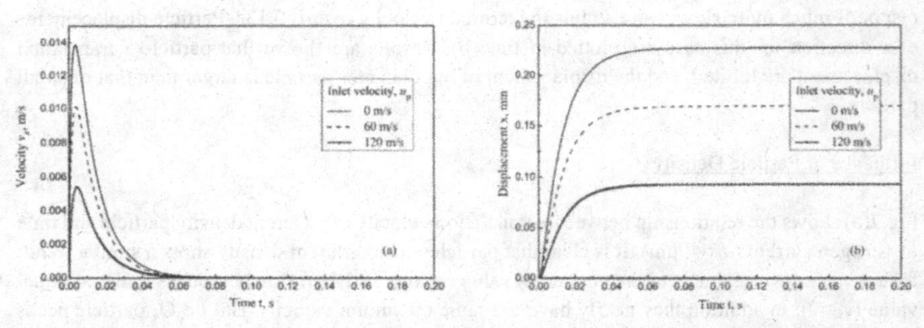

Fig.8 The influence of initial velocity of particle into slat on the horizontal velocity (a) and displacement (b)

Influence of Flow Boundary Layer

If particle is located in the flow boundary layer, the influence of continuous-fluid vorticity, in form of linear shear, and due to velocity gradient in flow direction, is significant. So the shear-induced lift plays a major role in particle motion characteristic of x-direction. Fig.9 (a) shows time dependence of translation velocity of x direction for the case of lime particle (d_p=40μm) in fluid boundary layer. At the same time, relationship between displacement and time is plotted in fig.9(b). Apparently, the motion characteristic of particles in boundary layer is of a unique style. Movement law of particles with inlet velocity below 100m/s is totally different with that of particles above 100m/s. The velocity direction of low-speed particle is reversed during the process of particle movement, while high-speed particle keeps moving towards positive direction. At the beginning of motion, the velocity of low-speed particle shows negative values in contrast with the high-speed one.

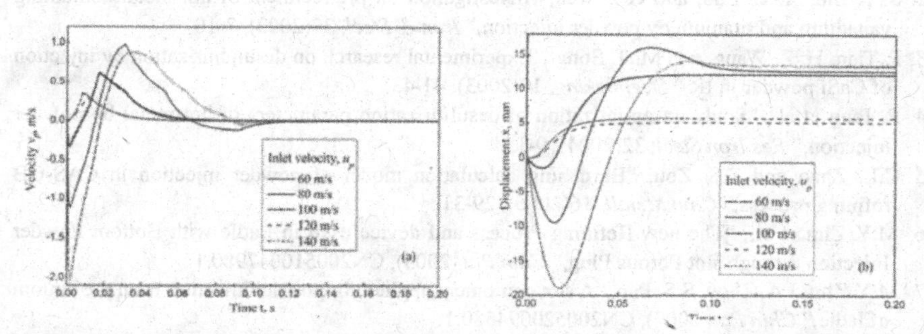

Fig.9 Relationships between time and the horizontal velocities (a) and displacement (b) of particles in flow boundary layer

Conclusions

Drag, gravity, buoyant, lift, Basset force, virtual mass force and thermal force etc, have been taken into considerations, and a model describing the motion characteristics of a powder particle in the slat of powder injection device at finite Reynolds number has also been established. According to the model, some parameters for slot-bottom powder injection device design and powder injection techniques can be established.

Motion characteristic of particle along flow direction shows that particle velocity increases slowly after acceleration in short time or displacement, which is consistent with previous work of Pan and Zhu. Furthermore, the influences on motion characteristic of particle in perpendicular to flow direction from particle diameter, particle density, particle inlet velocity, particle horizontal velocity and fluid boundary layer, were investigated. The results show that larger particle, as well as particle with greater density, passes farther horizontal displacements; particle with horizontal velocity and particle in boundary layer move instability, and particle-wall collision easily occurs; particle with higher inlet velocity is conveyed more smoothly and steadily.

Acknowledgement

The authors sincerely acknowledge the financial support from National Natural Science Foundation of China No. 51134009, Fundamental Research Funds for Central University of China No. N100102001 and financial support from Doctoral Fund No. 20110042110010.

References

1. Xinzhao Zhang, *Basic Principles of Powder Injection Metallurgy*, Beijing: Metallurgy Industry Press, 1988), 1-2.
2. S. P. Liu, X. G. Luo, and G.Y. Wen, "Investigation on pretreatment of hot metal containing vanadium and titanium by powder injection," *Iron & Steel*, 37(2002), 7-10.
3. P. Tian, H. Z. Wang, and M. T. Song. "Experimental research on desulphurization by injection of CaSi powder in IR," *Steelmaking*, 19(2003), 41-43.
4. P. Tang et al., "Study on optimization of desulfurization parameters of hot metal by powder injection," *Res Iron Steel*, 32(2004), 9-15.
5. C.L. Zhao and Z.S. Zou, "Burdening calculation model of powder injection in CAS-OB refining process," *Chin Metall*, 16(2006), 29-31.
6. M.Y. Zhu et al., "The new Refining Process and device used in Ladle with Bottom Powder Injection through slot Porous Plug," *Chin Pat,* 2005), CN200510047980.1.
7. M.Y. Zhu, J.A. Zhou, S.S. Pan, "A slot equipment applied for powder injection from the bottom of ladle," *Chin Pat*, 2007), CN200520094580.1.
8. S.S. Pan and M.Y. Zhu, "Motion Characteristics of injection powder through porous brick module in the bottom of refining ladle," *Acta Metallurgy Sinica*, 43(2007), 553-556.
9. R. Kurose and S. Komori, "Drag and lift forces on rotating sphere in a linear shear flow," *Journal of Fluid Mechanics*, 384(1999), 183-206.
10. S. Mukherjee, M.A.R. Sharif and D.M. Stefanescu, "Liquid convection effects on the pushing-engulfment transition of insoluble particles by solidifying interface: Part II. Numerical calculation of drag and lift forces on a particle in parabolic shear flow," *Metallurgical and Materials Transactions A*, 34A (2004), 623-629.
11. P. Bagchi and S. Balachandar, "Effect of free rotation on the motion of solid sphere in linear shear flow at moderate Re," *Physics of Fluids*, 14(2002), 2719-2737.
12. E. Loth, A.J. Dorgan, An equation of motion for particles of finite Reynolds number and size, Environ Fluid Mech. 9(2009): 187-206.
13. Y. Tsuji, Y. Morikawa, O. Mizuno, Experimental measurement of the Magnus force on a rotating sphere at low Reynolds numbers, Journal of Fluids Engineering. 107(1985) 484-488.
14. J.M. Foucaut, M. Stanislas, Experimental study of saltating particle trajectories, Experiments in Fluids, 22(1997) 321-326.

15. B. Oesterle, T.B. Dinh, Experiments on the lift of a spinning sphere in a range of intermediate Reynolds numbers, Experiments in Fluids. 25(1998) 16-22.
16. R. Clift, J. R. Grace, M. E. Weber, Bubbles, Drops and Particles, Academic Press, New York, 1978.
17. L. Schiller, A. Z. Naumann, "Über die Grundlegenden Berechungen bei der Schwerkraftaufbereitung," VDI-Zeitschrift, 77(1993) 318–320.
18. E. Loth, Lift of a Solid Spherical Particle Subject to Vorticity and/or Spin, AIAA, 46(2008) 801-809.
19. P. Bagchi, S. Balachandar, Effect of Free Rotation on Motion of a Solid Sphere, Physics of Fluids, 14(2002) 2719–2737.
20. S.I. Rubinow, J.B. Keller, The Transverse Force on Spinning Spheres Moving in a Viscous Liquid, Journal of Fluid Mechanics, 11(1961) 447.
21. J. Kim, H. Choi, Laminar Flow Past a Sphere Rotating in the Streamwise Direction, Journal of Fluid Mechanics, 465(2002) 354–386.
22. P.G. Saffman, The Lift on a Small Sphere in a Slow Shear Flow, Journal of Fluid Mechanics, 22(1965) 385–400.
23. J.B. McLaughlin, Inertial migration of a small sphere in linear shear flows, Journal of Fluid Mechanics, 224(1991) 261-274.
24. R. Mei, J.F. Klausner, C.J. Lawrence, A Note on the History Force on a Spherical Bubble at Finite Reynolds Number, Physics of Fluids, 6(1994) 418–420.
25. B.D. Tri, B. Oesterle, F.Deneu, Premiers Resultants Sur la Portance D'Une Sphere en Rotation aux Nombres de Reynolds Intermediaries, Comptes Rendus de l'Académie des Sciences, Série II: Mécanique, Physique, Chimie, Sciences de l'Univers, Sciences de la Terre, 311(1990) 27.
26. P. Bagchi, S. Balachandar, Effect of Free Rotation on Motion of a Solid Sphere, Physics of Fluids, 14(2002) 2719–2737.
27. B.V. Derjaguin, S.P. Bakanov, Kolloid Zh., Vol. 21, 1959.
28. B.V. Derjaguin, Y.J. Yalamov, Theory of Thermophoresis of Large Aerosol Particles, Journal of Colloid Science, 20(1965) 555-570.
29. J.R. Brock, On the theory of thermal forces acting on aerosol particles, Journal of Colloid Science, 17(1962) 768–780.
30. J.R. Brock, The thermal force in the transition region, Journal of Colloid and Interface Science, 23(1967) 448–452.
31. A. Huang, Mathematical and Physical Simulation on Application Process and Optimization of Structure and Performance of Purging Plug in Refining Ladle. Wuhan University of Science and Technology. 2010.4: 80-81.
32. C.F. You, H.Y. Qi, X.C. Xu, Progresses and Applications of Basset Force, Chinese Journal of Applied Mechanics, 19(2002) 31-33.

Materials Processing Fundamentals
Edited by: Lifeng Zhang, Antoine Allanore, Cong Wang, James A. Yurko, and Justin Crapps
TMS (The Minerals, Metals & Materials Society), 2013

Study on Internal Cracks in Continuous Casting Slab of AH36 Steels

Shufeng Yang[1], Yubin Li[2], Lifeng Zhang[1]

[1]School of Ecological and Metallurgical Engineering
University of Science and Technology Beijing, Beijing 100083, China
Correspondence author: Lifeng Zhang, zhanglifeng@ustb.edu.cn
[2]Steelmaking Department, SINOSTEEL MECC, Beijing 100080, China
Email: liyubin@mecc.sinosteel.com

Keywords: Continuous casting, slab, internal cracks.

Abstract

The internal cracks often occurred on the continuous casting slab of AH36 steels, which caused serious quality problems. In this work, a great number of production data were collected to analyze the primary factors that caused the internal crack. Various techniques such as acid etching, metallographic microscope and SEM-EDS detection were used to analyze the macro- and micro- characteristics of the cracks. High temperature mechanical properties of the steel were studied using Gleeble1500. The results showed that the impurity element segregation, such as S, P and Cu, overheat and inclusion accumulation are the main reasons for the generation of internal cracks in the AH36 steel slabs.

Introduction

It is an increasing requirement of high strength and impact toughness for ship plate steels. Various kinds of defects of slab should be controlled and avoided. Internal crack is one of the main defects in continuously cast slabs and is of prime importance for the control of the properties and quality of the cast product. The formation of the internal crack is very complex and is the result of heat transfer, mass transfer, and stresses interaction. The formation and control of internal crack has been extensively studied. It has been reported that the internal cracks are resulted from the excessive tensile strains produced at the solidifying front of the slab [1–2]. The formation of the cracks also depends on the mechanical properties of the solidifying front, especially strength and ductility, which are closely related to the microsegregation near the solidifying front [3]. The composition of steel is an important factor affecting the formation of internal cracks. With the increase of carbon and sulphur content in steel, the internal crack appearing probability increased [4]. Fujii studied the effect of continuous casting process parameters on formation of internal crack and found that internal crack was much easier to form in the area of columnar grains than in the area of equiaxed grains [5]. Cai et al studied the effect of rolling reduction of rolls on the formation of internal crack and found that improper rolling reduction would cause internal cracks [6].

In the current paper, a great number of production data were collected to analyze the primary factors causing the internal crack. The external and internal causes of internal crack was analyzed using various techniques such as acid etching, metallographic microscope and SEM-EDS to provide evidence for stabilizing the continuous casting and improving product quality.

Surveys on Affecting Factors of Internal Cracks

Production Process

In this paper, samples and production data of shipbuilding steels of AH36 from a steel plant was collected and studied. The production process was 100t EAF→LF→VD→Slab continuous casting. The main operation parameters of continuous casting were listed in the Table I. Table II shows the composition of AH36 steel.

Table I. Process parameter of continuous casting

Type of casting machine	Machine×strand	Radius arc (m)	Metallurgical length (m)	Length of mold (mm)
Vertical-Arc	1×1	R=8	L=24.9	900
Size of slab (mm)	Casting speed (m/min)	Second cooling system	Method of Straightening	
200×1500	1.0-1.4	Automatic water distribution model	Continuous Bending	

Table II. Compositions of AH36 steel (%)

Steel grade	C	Si	Mn	P	S	Cu	Nb	Als
AH36	0.14-0.18	0.15-0.5	1.2-1.45	≤0.025	≤0.015	<0.2	0.015-0.025	≥0.02

In order to investigate the effect of different factors on internal cracks formation, plenty of data was collected for statistical analysis, and to find the main causes of internal cracks. Total 1440 heats production data were analyzed. The slabs of 292 heats had internal cracks, and 82 heats had surface cracks and a few of heat slabs had edge cracks or more than one kind of cracks. Figure 1 shows the statistical results and the occurring probability of internal cracks was 14.2%.

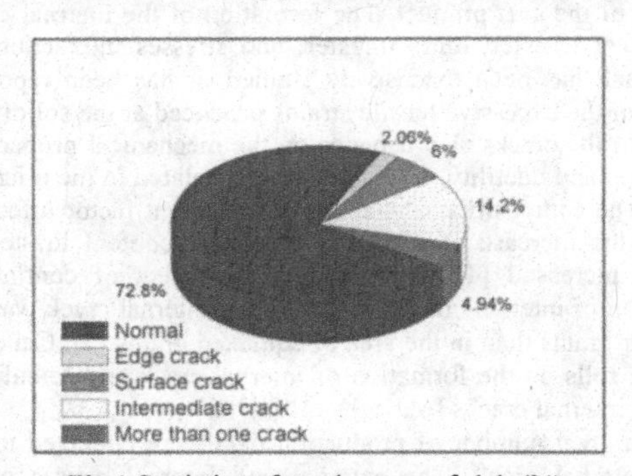

Fig.1 Statistics of crack occur of slab (%)

Effect of Compositions on Internal Cracks

Figure 2 shows internal crack probability at different carbon content and it indicates when carbon is 0.13%~0.15%, cracks happen easily. When carbon is 0.10%~0.14% in the steel, peritectic reaction happens during solidification and δ phase transforms into γ phase. During the transformation, the volume of steel shrinks, gas gap forms when slab shell and mold wall copper separates, heat flow conduced decreases and slab shell becomes thin, so the depression would form on the surface.[1] The shear concentrates on the depression area easily and causes internal cracks.

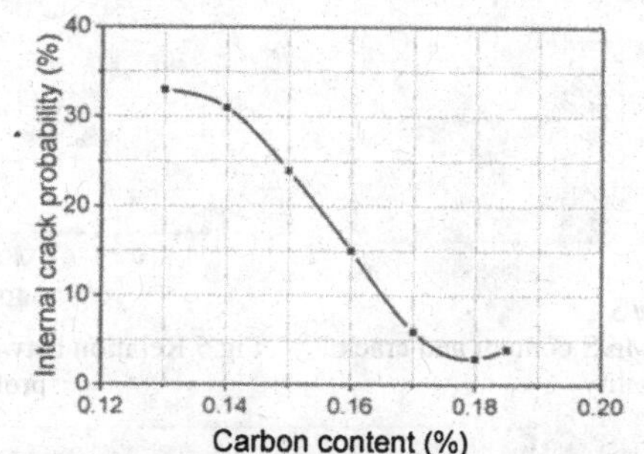

Fig.2 Relation between C content and crack probability

Figure 3 shows the relation between sulfur content and crack probability. There are two areas divided by [S]=0.010% in the Figure 2, when [S]>0.010%, the internal crack is very serious and when [S]<0.010%, the internal crack probability is low, especially when [S]< 0.005% ,the internal cracks nearly happen. Sulfur is an element segregating easily in steel, when sulfur segregates and (Mn, Fe)S precipitates on the grain boundary, zero deformation temperature decreases much. So, when carbon content is certain, internal crack probability increases with sulfur content increasing.

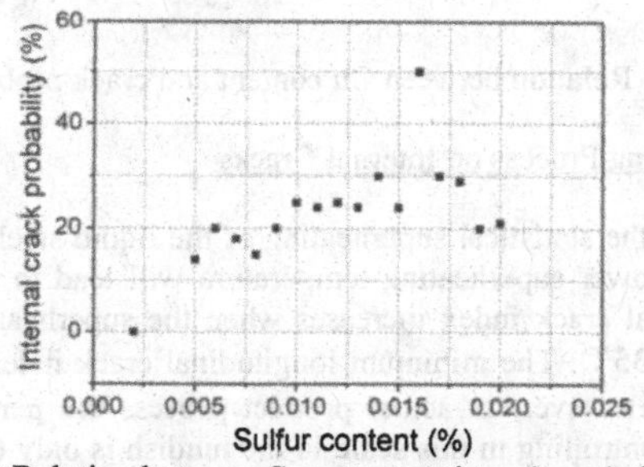

Fig.3 Relation between S content and crack probability

As shown in the Figure 4, the internal crack probability decreases obviously with increasing Mn/S in the steel. Increasing Mn/S in the steel is favor of MnS distributing on the grain boundary or matrix granularly and improves the high temperature property of the steel. Figure 5 shows the relation between sulfur content and internal crack probability. When P content is 0.020%, internal crack probability increases. This is because phosphorus also is segregated element as sulfur element. Copper element in the also influence internal crack probability, as shown in the Figure 6, when Cu content is0.06～0.09%, internal crack probability is higher.

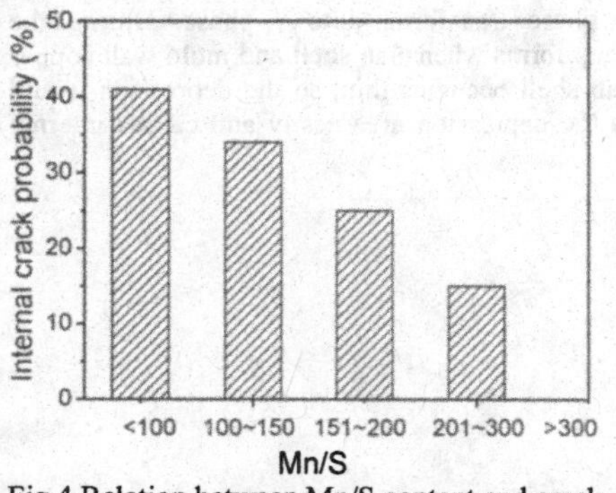

Fig.4 Relation between Mn/S content and crack probability

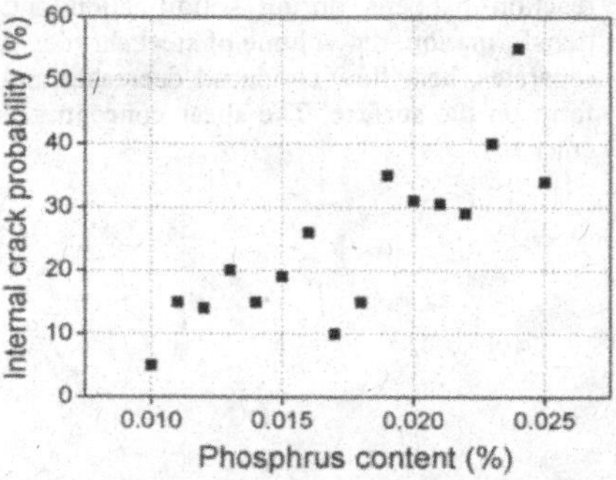

Fig.5 Relation between P content and crack probability

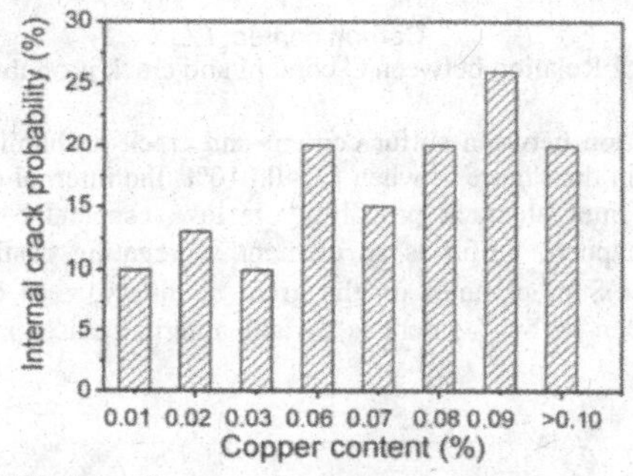

Fig.6 Relation between Cu content and crack probability

Effect of Continuous Casting Process on Internal Cracks

The relationship between the statistical superheating of the liquid steel and slab crack shows as Figure 7. The higher or lower superheating temperature will lead to slab crack. The Figure 7 shows that the longitudinal crack index increases when the superheating temperature is lower than 15℃ or higher than 35℃. The minimum longitudinal crack index got as the superheating temperature is 15-30℃. However, in actual product process the percentage of the statistical liquid steel temperature controlling in this scale in the tundish is only 60% in the steel plant. So the management should be strengthened and the operant level should be improved to keep the temperature stable in the tundish.

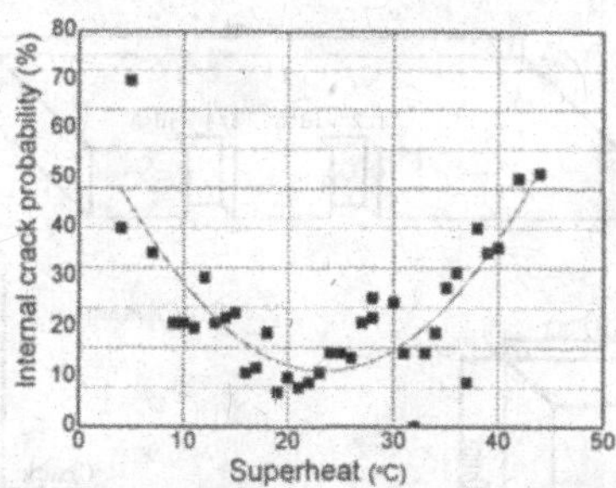

Fig.7 Relation between superheat and internal crack probability

The casing speed on the slab crack influence shows as Figure 8. The too high or too low casing speed will lead to slab crack increase, the proper casing speed is 1.2-1.3m/min. Figure 9 show the effect of cooling intensity on internal crack probability, when specific water flow from 0.35L/Kg increases to0.738L/Kg, the internal crack probability from 15.5% increases to 32.5%.So in order to decrease internal crack probability, the specific water flow should be decreased in current production process.

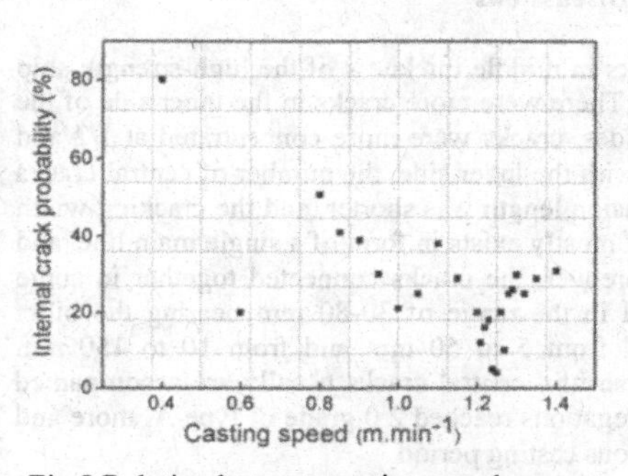

Fig.8 Relation between casting speed content and internal crack probability

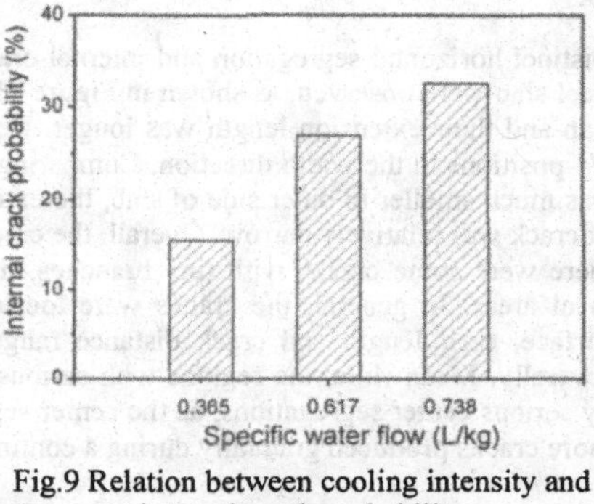

Fig.9 Relation between cooling intensity and internal crack probability

Analysis of Internal Cracks in AH13 Steel Slabs

Method of Sampling and Analysis

Figure 10 shows scheme of sampling, two big samples(130mm×50mm×200mm) were taken at 1/2 width and at 1/4width on slabs for one heat, then internal areas were secreted and the samples were cut into 15cm×5cm×5cm for etch test and some for SEM observation.
A solution 4% alcohol of hydrogen nitrate was used to erode cracked specimens with the generation of macrostructure in cracks. The distribution of crack and the length data can be got from observing the morphology of the crack section. SEM was used to observe micro-

morphology of internal crack and inclusions and XRD was identified the distribution of elements and inclusions compositions.

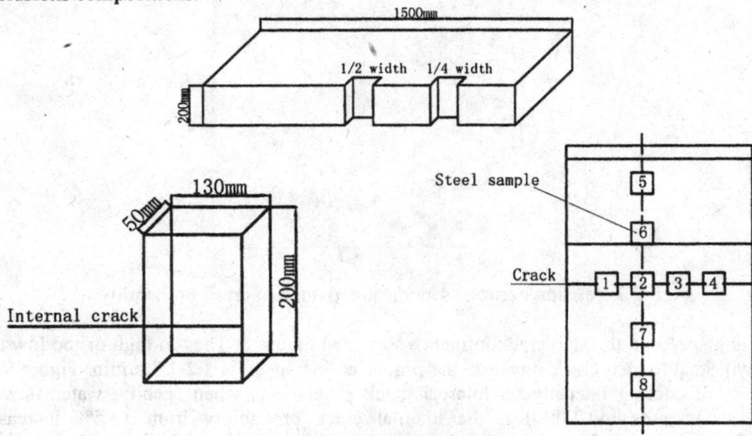

Fig.10 Scheme of sampling

Results and Discussions

Distinct horizontal segregation and internal cracks in middle thickness of the high-strength ship steel slab were observed, as shown in Figure 11. There were more cracks in the inner side of the slab and their extension length was longer. Besides, cracks were more concentrated at 1/2 and 1/4 positions in the width direction. Comparing with the inner side, the number of central cracks was much smaller in outer side of slab, the extension length was shorter, and the cracking width of crack was relatively narrow. Overall, the crack mostly exists in form of a single main line, and there were some cracks with tiny branches, moreover, the cracks connected together in some local areas. In general, the cracks were located in the range of 30-80 mm nearing the billet surface, their length and crack distance ranged from 5 to 50 mm and from 10 to 150 mm severally. Meanwhile, the regions with serious harmful central cracks usually are accompanied by serious center segregations, as the center segregations reached 2.0 grade of type·A. more and more cracks produced gradually during a continuous casting period.

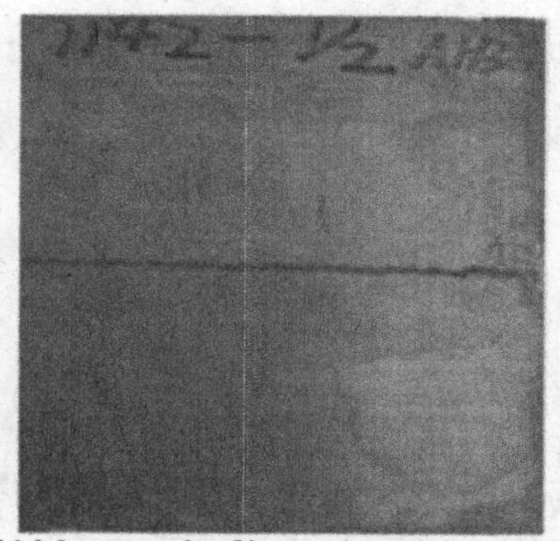

Fig.11 Macrograph of internal crack in AH36 slab

Figure 12 shows micrograph of internal crack in AH36 slab, indicating an obvious accumulation zone. The main reasons are elements of the segregation and selected crystallization in solidification process. Figure 13 shows the result of line scan near the internal crack and it indicates sulfur, manganese and carbon segregate seriously and cooper and phosphorus segregate a little.

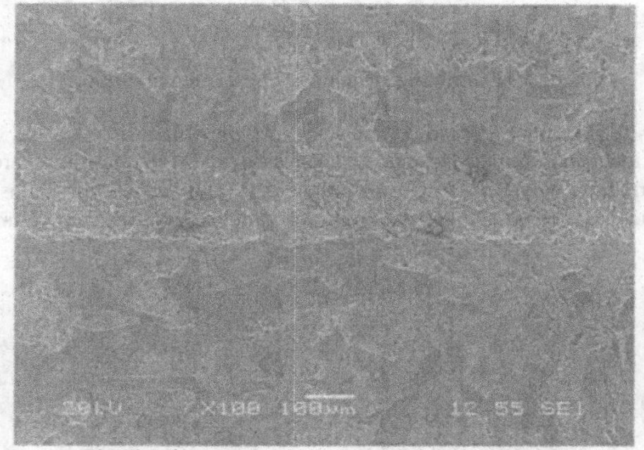

Fig.12 Micrograph of internal crack in AH36 slab

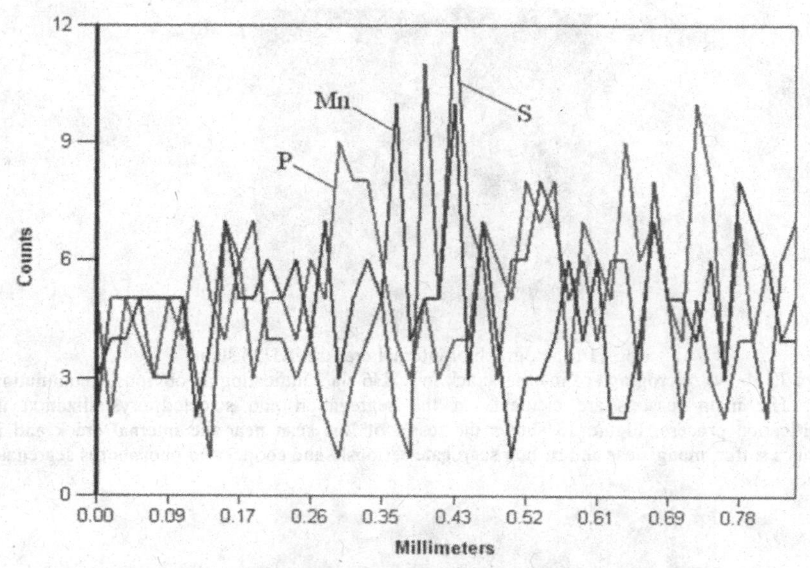

Fig.13 Distribution of element near internal crack

Figure 14 shows the morphology and composition of typical inclusions on the internal cracks. MnS are main inclusions on the internal cracks, sizes of MnS inclusions are 5-10μm and morphology is irregular polygon（as shown in the Fig.14(a). Another type inclusion is complex inclusions of oxide and sulfide and the sizes of these complex inclusions are 10μm（as shown in the Fig.14(b).

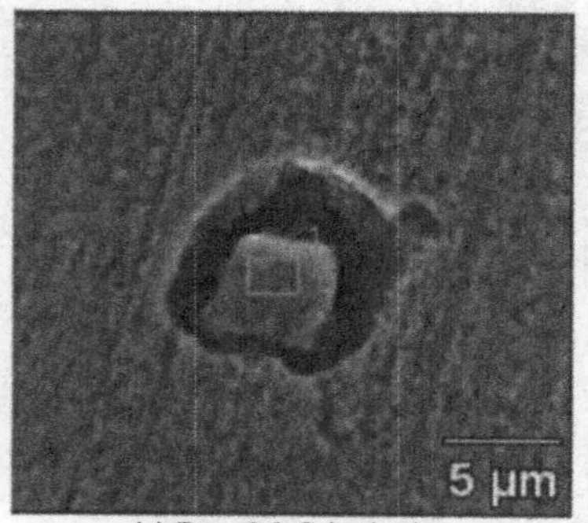

(a) Pure MnS inclusion

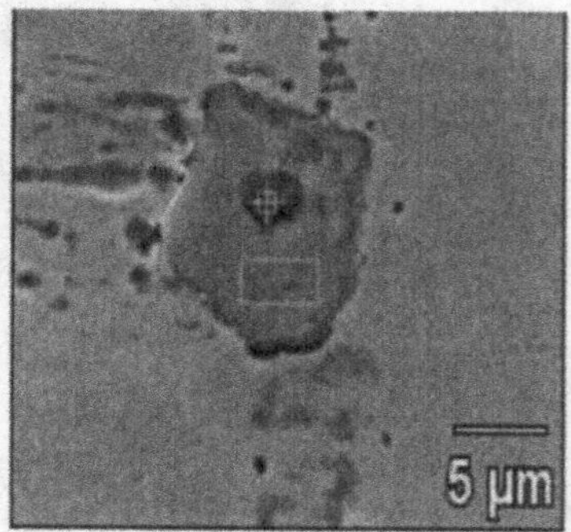

(b) MnS inclusion with Al_2O_3-MgO core

Fig.14 Morphology and composition of inclusions

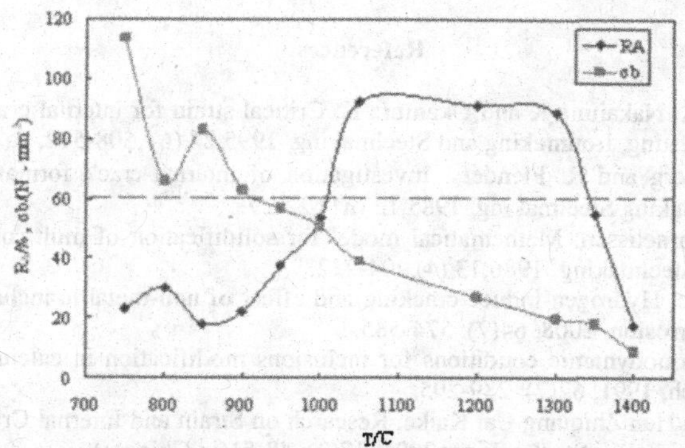

Fig.15. High temperature mechanical properties of AH36 slab

High Temperature Mechanical Properties

High temperature mechanical performance tests were performed using a Gleeble-1500 thermal stress / strain simulation machine. For AH36 samples, tensile experiments were performed from 750℃ to 1400℃ at a constant strain rate. Figure 15 shows the temperature mechanical properties test results of AH36 slab. σ_b is the maximum stress in various test temperatures and R_A stands for the reduction of area. Test results show that the low temperature brittleness zone of AH36 is a wide region ranging from 750 to 1000℃. So the straightening temperature in the continuous casting process had better avoid this temperature range. The slab is in the plastic zone between 1000 and 1300℃. When the temperature is over 1300℃, the tensile strength decreases, and the reduction of area sharply declines, at the same time, the slab is in high temperature brittle region. According to the test results of the high temperature performance, various target temperature of the billet in the secondary cooling zone was determined, and a reference for the rational design of the cooling curve also was provided.

Conclusions

1) The main factors causing cracks are S, P and Cu which havn't been controlled strictly, the molten steel overheated, the unstable casting speed and the second cold systems and so on;
2) There exists serious element segregation such as C, P, Mn and S in the internal crake band. The inclusions at crack position are mainly MnS, spherical or approximately spherical compound deoxidization products Al_2O_3-MgO and Al_2O_3-MgO-CaO, the polygon or approximately spherical compounds with oxide wrapped by sulfide, and certain amount of Si;
3) The high temperature mechanical properties test shows that the cold brittleness area is in the range of 750-1000℃, the temperature range is very wide. High temperature brittle area is in 1300-1400℃.

Acknowledgements

The authors are grateful for the support from the National Science Foundation China (Grant No. 51274034), the Laboratory of Green Process Metallurgy and Modeling (GPM^2), the High Quality Steel Consortium at University of Science and Technology Beijing (China).

References

[1] Yamanaka. A, Nakajima K and Okamura K. Critical strain for internal crack formation in continuous casting, Ironmaking and Steelmaking, 1995,22 (6) ,508-512.

[2] K. Wunnenberg and R. Flender, Investigation of internal-crack formation,using a hot model, Ironmaking Steelmaking, 1985,12 (n1),22-29.

[3] M. C. M. Cornelissen: Mathematical model for solidification of multicomponent alloys. Ironmaking Steelmaking, 1986,13 (4),204-212.

[4] Koh S.U., etc. Hydrogen-induce cracking and effect of non-metallic inclusion in linepipe steels [J]. Corrosion, 2008, 64(7): 574-585.

[5] Fujii H. Thermodynamic conditions for inclusions modification in calcium treated steel. Steel Research, 1991, 62(7): 289-295.

[6] Yuan Weixia Han Zhiqiang Cai Kaike. Research on Strain and Internal Crack in Cast Slab during Solidification. Steelmaking, 2001, 17(2), 48-51(in Chinese).

AUTHOR INDEX
Materials Processing Fundamentals

SUBJECT INDEX
Materials Processing Fundamentals

S

T

U

V

W

X

Z

Printed in the United States
By Bookmasters